Design and Operation of Container Ships 2008

London, United Kingdom
3 - 4 July 2008

ISBN: 978-1-5108-8557-8

Printed from e-media with permission by:

Curran Associates, Inc.
57 Morehouse Lane
Red Hook, NY 12571

Some format issues inherent in the e-media version may also appear in this print version.

The Royal Institution of Naval Architects
8-9 Northumberland Street
London, WC2N 5DA
United Kingdom

Phone: 020 7235 4622
Fax: 020 7259 5912

publications@rina.org.uk

Additional copies of this publication are available from:

Curran Associates, Inc.
57 Morehouse Lane
Red Hook, NY 12571 USA
Phone: 845-758-0400
Fax: 845-758-2633
Email: curran@proceedings.com
Web: www.proceedings.com

CONTENTS

* Presentation Only.

RINA

DESIGN AND OPERATION OF CONTAINER SHIPS

3 - 4 July 2008

AUTHORS' CONTACT DETAILS

O. el Moctar
Germanischer Lloyd, Germany
ould.el-moctar@gl-group.com

S. Malenica
Bureau Veritas Research Dept., France
sime.malenica@bureauveritas.com

G. Storhaug
Det Norske Veritas, Norway
gaute.storhaug@dnv.com

G. de-Jong
Bureau Veritas, France
gijsbert.de-jong@nl.bureauveritas.com

N. Vladimir
University of Zagreb, Croatia
nikola.vladimir@fsb.hr

M. Toyoda
IHI Marine United Inc, Japan
masanobu_toyoda@ihimu.ihi.co.jp

G. Storhaug
Det Norske veritas, Norway
eivind.steen@dnv.com

A. Kahl
Germanischer lloyd, Germany
adrian.kahl@gl-group.com

J. Carlton
Lloyd's Register, UK
john.carlton@lr.org

R. P.Dallinga
MARIN, The Netherlands
r.p.dallinga@marin.nl

Y. Yamaguchi
Nippon Kaiji Kyokai (ClassNK), Japan
yamaguchi@classnk.or.jp

M. Landamore
Newcastle University, UK
m.j.landamore@newcastle.ac.uk

C. B. McKesson
CCDoTT, USA
chris@mckesson.us

S. Whitworth
Lloyd's Register, UK
stewart.whitworth@lr.org

N. Rattenbury
Lloyd's Register, UK
Norman.rattenbury@lr.org

H. Schmid
Wartsila Switzerland Ltd, Switzerland
heinrich.schmid@wartsila.com

M. Eijssen
DSM Dyneema B.V., The Netherlands
marc.eijssen@dsm.com

T. Flauger
KRAL AG, Austria
t.flauger@kral.at

ASSESSMENT OF WHIPPING EFFECTS ON GLOBAL LOADS OF ULTRA LARGE CONTAINERSHIPS

O. el Moctar and **J. Oberhagemann,** Germanischer Lloyd, Germany
H-H. Lee, T-H. Park and **H.-R. Ryu**, Hyundai Heavy Industries Co., Ltd, Korea

SUMMARY

We present a computational procedure that considers large-scale hydroelastic effects to assess slamming-induced hull whipping on sectional loads of a large containership in waves. The purpose was to compare computed results with model test measurements.

In the computations, rigid body motions are superimposed on elastic deformations, where the former can be of large amplitude while the latter have to be of small amplitude relative to rigid body motions. The slamming impact causes the ship structure to vibrate with a wide range of frequencies. Assuming that low eigenmodes in bending contribute most to total vibration energy, we apply a finite element Timoshenko beam, oriented in the longitudinal direction of the ship, to account for bending and shear deformation. The structural model is coupled to a RANS solver. This code solves the Reynolds-averaged Navier-Stokes equations in their integral form for a finite number of control volumes constituting the solution domain. An additional conservation equation is solved to consider the transport of the two fluids water and air. The integrated pressure and friction forces on the hull surface excite rigid body hull motions as well as elastic hull deformations. In return, these motions and deformations influence the kinematics of the surrounding fluid. To account for this influence, the fluid grid is transformed not only according to the translations and rotations of the ship, but also adapted to the vibratory displacements of the hull surface.

Comparative measurements of slamming pressures, vertical bending moment and shear forces were obtained on the segmented model of a 13,000 TEU containership designed by Hyundai Heavy Industries. Representative time series of computed normalized vertical bending moments and vertical shear forces for the rigid hull and the elastic hull are shown.

NOMENCLATURE

(B)	=	Damping matrix
(C)	=	Stiffness matrix
$\{F(t)\}$	=	Load vector
$\vec{F}$	=	Force vector
G	=	Center of gravity
G,x,y,z	=	Newtonian reference frame
G,ξ,η,ζ	=	Ship-fixed reference frame
I_L	=	Inertia tensor
m	=	Ship mass
(M)	=	Mass matrix
$\vec{M}$	=	Moment vector
S	=	Transformation matrix
t	=	Time
w	=	Nodal displacement
γ	=	Nodal rotation
$\{U\}$	=	Nodal displacement vector
$\{\dot{U}\}$	=	Nodal velocity vector
$\{\ddot{U}\}$	=	Nodal acceleration vector
VBM	=	Vertical bending moment
VSF	=	Vertical shear force
v	=	Design speed
x	=	Distance from aft perpendicular
$\vec{x}$	=	Translational position vector
$\ddot{\vec{x}}$	=	Translational acceleration vector
$\vec{\xi}$	=	Global position vector
$\vec{\omega}$	=	Rotational velocity vector
$\dot{\vec{\omega}}$	=	Rotational acceleration vector

1. INTRODUCTION

The life expectancy of a ship depends on the experienced stresses, which in turn depend on the external loads and the structural design. In this regard, dynamic loads and fluid-structure interaction have grown in importance over the last years. Dynamic loads influence both fatigue strength and experienced maximum loads for ultimate strength. The associated accelerations (rigid-body and elastic vibrations) may influence cargo safety and reduce comfort of passengers in passenger ships and yachts.

Containerships have grown rapidly in size. Now "mega carriers" with more than 12,000 TEU capacity are widely established. These ships have ship lengths around 380 m and speeds around 25 knots. Ship seakeeping usually considers the ship hull as rigid. Global strength analyses are usually based on quasi-static approach. Safety factors in regulations account then for the effect of dynamic loads on the elastic ship hull, i.e. the hydro-elastic interaction between fluid and structure. Improved and validated simulations could serve to quantify dynamic loads directly. Such simulations could support a better evaluation of unconventional designs. They could also serve to guide adjustment of safety factors in regulations.

To assess whipping of a ship's hull structure induced by impact-related slamming, we present a numerical procedure that accounts for large-scale elastic effects during transient fluid dynamic computations.

We selected a modern twin screw containership designed by Hyundai Heavy Industries to investigate large-scale hydroelastic effects caused by slamming-induced hull whipping. Table 1 lists principal particulars of the ship.

Table 1: Principal particulars of containership

Capacity	13 000 TEU
Length between perpendiculars	366.0 m
Molded breadth	54.2 m
Depth	27.7 m
Scantling draft	15.0 m
Service speed	25.0 knots

2. COMPUTATIONAL PROCEDURE

The reliable prediction of loads is crucial for an accurate global structural analysis of a ship. The problem is to define these loads in a manner useful for the assessment of the ship's structural strength, for example, via the analysis of a finite element (FE) model of the hull. It is impractical to apply a straight-forward nonlinear time domain simulation for a long-term assessment. Instead, based on the so-called equivalent design wave approach, we present here a numerical procedure that combined the use of a boundary element method (BEM), an extended Reynolds-averaged Navier-Stokes equation (RANSE) solver, and a finite element Timoshenko-beam . First, the BEM GLPANEL [9] obtained the hydrodynamic data base needed for a statistic/probabilistic long-term analysis of the ship's operation, yielding so-called equivalent regular design waves that represented design load conditions relevant for the dimensioning of the hull structure. This analysis yielded long-term loads that did not account for impact-related wave effects. For those wave conditions and ship speeds that led to the highest design loads, we subsequently performed RANSE-FE computation to demonstrate the effects of impact-related loads on global hull girder loads. To accomplish this task, the RANSE solver COMET [11] was extended to obtain ship motions, the corresponding pressures distribution acting on the ship under these design wave conditions and the hull elastic deformations by solving the nonlinear rigid body six degrees of freedom motion equations and the Timoshenko beam equations in the time domain. In this way, wave-induced loads included whipping effects caused by slamming and green water on deck.

Our procedure represents the continued development of a technique developed by El Moctar et al. [2][3][4].

3. NUMERICAL METHODS

3.1 RANSE SOLVER

The RANSE solver COMET, a code that implements interface-capturing techniques of the volume-of-fluid (VoF) type, proved to be suitable for handling strong nonlinearities. Today, this kind of code is the obvious choice for computing complex free-surface shapes with breaking waves, sprays and air trapping, hydrodynamic phenomena that should be considered to predict slamming pressures. The conservation equations for mass and momentum in their integral form serve as the starting point. The fluid is assumed to be viscous. The solution domain is subdivided into a finite number of control volumes that may be of arbitrary shape. The integrals are numerically approximated using the midpoint rule. The mass flux through the cell face is taken from the previous iteration, following a simple Picard iteration approach.

The remaining unknown variables at the center of the cell face are determined by combining a central differencing scheme (CDS) with an upwind differencing scheme (UDS). A second-order central difference scheme (CDS) can lead to unrealistic oscillations if the Peclet number exceeds two and large gradients are involved. On the other hand, an upstream difference scheme (UDS) is unconditionally stable, but leads to higher numerical diffusion. To obtain a good compromise between accuracy and stability, the schemes are blended. Near the ship hull, the blending factor is chosen between 0.8 and 0.9. The three time level methods are used to integrate in time. Pressure and velocity are coupled by a variant of the SIMPLE algorithm [11]. All equations except the pressure correction equations are under-relaxed using a relaxation factor of 0.8. The pressure correction equation is under-relaxed using a relaxation factor between 0.2 and 0.4 for unsteady simulations, finding in each case a suitable compromise between stability and convergence speed. The two-fluid system is modelled by a two-phase formulation of the governing equations. No explicit free surface is defined during the computations, and overturning (breaking) waves as well as buoyancy effects of trapped air are accounted for.

The spatial distribution of each of the two fluids is obtained by solving an additional transport equation for the volume fraction of one of the fluids. The free surface is smeared over two to three control volumes [7].

3.2 FLUID-STRUCTURE INTERACTION

We define two interdependent solution domains, namely, the fluid solution domain $\mathbf{W}$ and the structure solution domain $\mathbf{U}$. We use two coordinate systems, namely, a Newtonian coordinate system $\mathbf{O}'(\xi,\eta,\zeta)$ moving with the ship's mean velocity and a local coordinate system $\mathbf{O}(x,y,z)$ moving with the ship's center of gravity. The undeformed hull with mass m and mass moment of inertia $\mathbf{I}$ about its center of gravity is the reference configuration; the deformed hull with its shape and position at time instant t_n, the actual configuration. **I** is expressed in the local coordinate system **O**. To obtain the position of any material point of the ship, its motions are split into a translatory and a rotary component derived from rigid body motions and another component derived from the elastic displacement relative to the local coordinate frame, see Figure 1. Position vector $\vec{r}$ of a point on the undeformed ship is expressed in local coordinates. Thus, $\vec{r}$ is time independent. Vector $\vec{u}$ defines the displacement of $\vec{r}$ caused by deformations of the hull, and vector $\vec{u}$ $\vec{u}_{(\vec{r},t)}$ defines the distinct relation between the undeformed hull and the actual configuration.

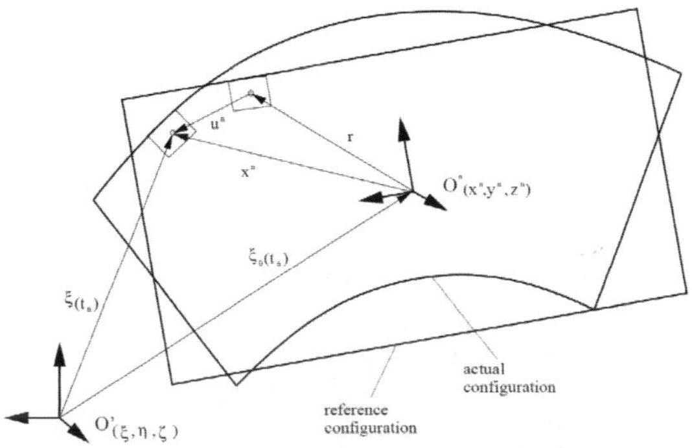

Figure 1: Global and local reference frame of undeformed and actual configuration at time t_n

Rotational matrix **S** performs the transformation from local to global coordinates by means of the three consecutive Euler rotations Ψ, Θ, and Φ:

$$\mathbf{S}_{(\Psi,\Theta,\Phi)} = \mathbf{S}_\Psi \mathbf{S}_\Theta \mathbf{S}_\Phi \quad (1)$$

Vector $\vec{x}$ defines the location of a point in the actual configuration. This vector is the sum of location vector $\vec{r}$ in the reference configuration and the actual displacement vector $\vec{u}$. Vectors $\vec{x}$ and $\vec{\xi}$ are then related as follows:

$$\vec{\xi} = \vec{\xi}_O + \mathbf{S}\,\vec{x} = \vec{\xi}_O + \mathbf{S}\,(\vec{r} + \vec{u}) \quad (2)$$

where $\vec{\xi}_O$ is the vector extending from origin **O'** to the ship's center of gravity, which is located at **O**. Translatory and angular accelerations of the center of gravity, $\ddot{\vec{\xi}}_O$ and $\dot{\vec{\omega}}$, respectively, are obtained by solving the following equations:

$$\vec{F} = m\,\ddot{\vec{\xi}}_O \quad (3)$$

$$\vec{M} = \mathbf{SIS}^T \dot{\vec{\omega}} + \vec{\omega} \times \mathbf{SIS}^T \vec{\omega} \quad (4)$$

where $\vec{F}$ and $\vec{M}$ are, respectively, the vector of external forces at and the vector of external moments about the ship's center of gravity. Using a second order implicit scheme, accelerations are integrated twice to obtain the position of the center of gravity [2].

The ship's elastic deformations were computed by means of an FE beam extending in the longitudinal direction of the ship. To expedite the formulation, the ship's hull was separated lengthwise into several sections. Considering the equilibrium of external fluid and inertia forces on one section, we write

$$\vec{f} - \mu\ddot{\vec{\xi}}_g = \vec{f}_c \quad (5)$$

$$\vec{m} - \vec{\omega} \times \mathbf{SI}_s\mathbf{S}^T \vec{\omega} - \mathbf{SI}_s\mathbf{S}^T \dot{\vec{\omega}} = \vec{m}_c \quad (6)$$

where $\vec{f}$ is fluid force per unit length, μ is mass per unit length, $\vec{\xi}_g$ is the center of gravity of the section, $\vec{m}$ is the external moment per unit length acting on the section, and $\mathbf{I}_s$ is the mass moment of inertia of the section about its center of gravity $\vec{\xi}_g$. Vectors $\vec{f}_c$ and $\vec{m}_c$ represent, respectively, forces and moments that cause the elastic deformation of the section.

A Timoshenko beam modeled the ship's structure. Vertical displacements and rotations about the ship's horizontal axis were considered. A section's rotation accounted for shear by assuming a constant transverse distribution of shear stresses over the section's length, which is equivalent to a planar rotation of all sections.

For the FE formulation, motion equations of the Timoshenko beam sections were derived in the local coordinate system. We first defined f_c^* and m_c^* as the transverse and tangential components, respectively, of the exciting terms $\vec{f}_c$ and $\vec{m}_c$, transformed into the local coordinate system. Transverse force F_c and bending moment M_c are related to f_c^*, m_c^*, vertical displacement w, and rotation about horizontal axis γ as follows:

$$w' - \gamma = \frac{F_c}{G\ A_s}\ ;\ \gamma' = \frac{M_c}{E\ I_y} \quad (7)$$

$$\frac{dF_c}{dx} = f_c^*\ ;\ \frac{dM_c}{dx} = m_c^* \quad (8)$$

where As is effective shear area, E is Young's modulus, G is shear modulus, and Iy is sectional area moment of inertia. Derivatives w' and γ' are spatial derivatives in the x-direction. Formulating force and momentum equilibrium led to the following differential equations for the displacement and rotation of the beam sections:

$$(GA_s(w' - \gamma))' - \mu\,\ddot{w} = 0 \quad (9)$$

$$(EI_y\,\gamma')' + GA_s(w' - \gamma) - \bar{\mu}\,\ddot{\gamma} = 0 \quad (10)$$

where $\bar{\mu}$ is the moment of inertia per unit length. The finite elements of length l_E of the Timoshenko beam were assumed to have constant properties E, G, I_y, A_s, μ, and $\bar{\mu}$. Every element had four degrees of freedom, namely, vertical displacement and rotation about horizontal axis at both ends of the element. Third order shape functions described the distribution of vertical displacement and rotation about horizontal axis along an element's length. Terms of the stiffness matrix **K** and the mass matrix **M** as well as the force vector $\hat{\vec{f}}$ were obtained on the bases of these shape functions [10], resulting in the following system of linear equations:

$$\mathbf{M}\vec{a} + \mathbf{K}\vec{d} = \hat{\vec{f}} \quad (11)$$

where $\vec{d}$ is the vector of nodal displacements and rotations, and $\vec{a}$ is the vector of their second time derivatives. The solution was obtained by solving for $\vec{a}$ and time integrating using the second order Newmark scheme [8].

Solutions of $\mathbf{U}^n$ and $\mathbf{W}^n$ were solved for at every time step, allowing for implicit as well as explicit coupling. (Index n refers to time step number.) Because of their interdependence, implicit coupling required iterative solutions of $\mathbf{U}^n$ and $\mathbf{W}^n$. Hence, at each time step, the numerical grid had to be updated several times, which caused a significant increase in computing time.

We considered only small deformations in the vertical x,z-plane of the Timoshenko beam, thereby allowing for a simpler grid deformation algorithm without significant loss of mesh quality. The updated vector $\vec{d}^n$ of the beam's nodal displacements w_j and rotations γ_j determined the actual deflection line of the ship's hull.

3.3 3D PANEL CODE

GL PANEL is a linear frequency-domain panel code that uses zero-speed Green functions and a forward speed correction based on the so-called encounter frequency approach. A velocity potential is found by distributing singularities (sources and sinks) of constant strength over the mean wetted surface of the hull. The velocity potential is separated into a time-independent steady contribution caused by the ship's forward speed and a time-dependent part associated with the incident wave system and the oscillating ship motions. For a ship advancing at constant mean forward speed, the incident waves undergo scattering (diffraction), leading to a diffraction wave potential that oscillates harmonically and induces a wave field of the same frequency radiating away from the ship. The ship moving in the wave field creates body motion potentials corresponding to three translational and three angular ship motions. These body potentials and the diffraction potential are superimposed on the incident wave potential.

The source strengths are found by satisfying the body boundary conditions, leading to integral equations of the second Fredholm type for each of the source strength contributions. These equations are solved numerically by replacing them with a system of linear equations. The wetted hull is discretized into a finite number of small triangular or rectangular surface patches (panels) in a way that represents the hull surface without creating `leakage' gaps. The integral equations are replaced by sets of linear equations from which the desired source strengths are determined. Dynamic pressure forces resulting from ship motions follow from the pressure on the hull surface according to the linearized Bernoulli equation. Integration of pressures over the hull surface yields hydrodynamic force and moment amplitudes.

3.4 MODEL TESTS

Measurements with a model of the investigated containership were carried out by Hyundai Heavy Industries for comparison with computed results. The model, shown in Figure 2, was constructed at a scale of 1:73.2 and segmented at stations x/L of 0.25 (station 5), 0.50 (station 10), and 0.75 (station 15). It was equipped with load cells at the segmented joints, accelerometers at the fore and aft perpendiculars, and pressure sensors at 15 locations. Analyzing accelerations from a hammering test of the floating model at its aft perpendicular, its midship section, and its forward perpendicular yielded 2-, 3-, and 4-node natural frequencies listed in Table 2.

Figure 2: The segmented model

Table 2: Natural frequencies of scale model obtained from hammering test

Mode shape	2-Node	3-Node	4-Node
Aft perp.	0.32 Hz	0.78 Hz	1.21 Hz
Midship	0.34 Hz	0.84 Hz	1.23 Hz
Forward perp.	0.33 Hz	0.81 Hz	1.21 Hz

4. NUMERICAL SIMULATION

4.1. FLOW NUMERICAL GRID

The numerical volume grid surrounding the ship comprised about two millions hexahedral control volumes. To avoid flow disturbances at outer grid boundaries, these boundaries were located at a distance of one ship length to the side, 1.5 L ahead of the bow, 2.5 L aft, 1 L above the deck and 1.5 L beneath the keel. Near the ship hull and ahead of the ship grid density was high to resolve the wave, whereas aft of the ship the grid became coarse to dampen the waves. Figure 3 shows the numerical hull-surface grid of the investigated ultra-large containership.

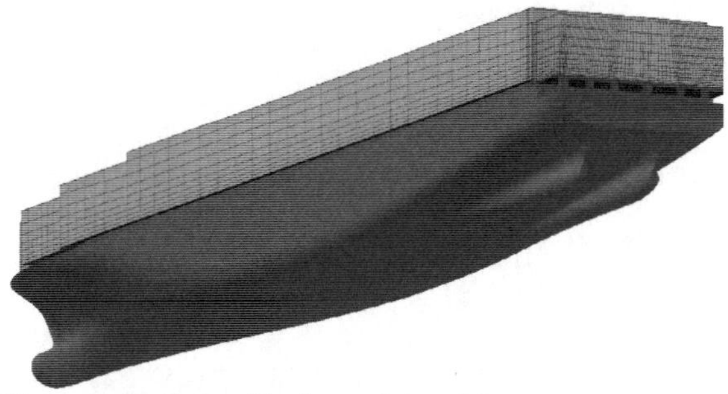

Figure 3: Surface grid of a containership

4.2. FLOW BOUNDARY AND INITIAL CONDITIONS

Front, side, bottom and top flow boundaries were specified as inlets of known velocities and known void fraction distributions defining water and air regions. On the hull surface a no-slip condition was enforced on fluid velocities and on the turbulent kinetic energy. The wake flow boundary was specified as a zero-gradient pressure boundary according to hydrostatic pressure. The time step size was chosen such that the Courant number was smaller than unity on average. The momentum equations were discretized using 80 percent central differences and 20 percent upwind differences. Ship motions were realized by moving the entire grid at each time step. Thus, each outer iteration all boundary conditions were newly computed.

Volume fractions and velocities that initialized the flow field [1] arose from superposition of ship speed and orbital particle velocities of the nonlinear second order Stokes waves. The influence of numerical damping on the wave height was taken into account. Numerical diffusion caused by the course grid aft of the ships dampened the incident wave to such an extent that no significant wave reflection occurred at the outlet boundary.

4.3 TIMOSHENKO BEAM

The beam model of the ship structure was derived from the global FE model of the ship by condensing the hull sectional properties to a six element finite beam model. Attention was paid to preserve the fundamental bending modes and their respective natural frequencies. Figure 4 shows a comparison of mode shapes and dry natural frequencies of the original model and the six element beam model used in the subsequent computations. Here, heavier lines denote mode shapes obtained from the six element model.

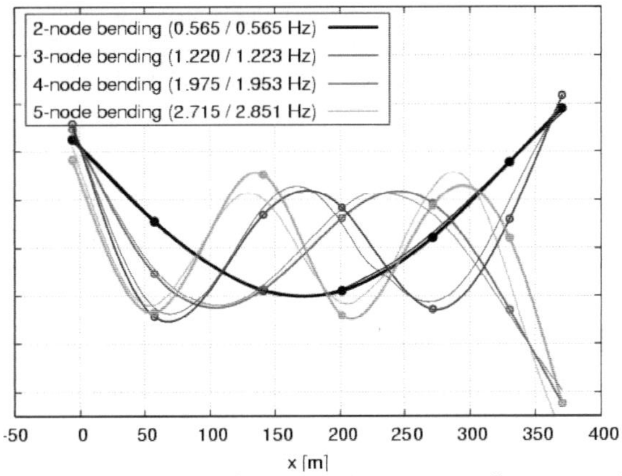

Figure 4: Mode shapes and corresponding natural frequencies of beam model

4.3. CONVERGENCE HISTORY

Simulation of the flow field and the ship responses continued until a periodic solution was reached. After a simulation time of several encounter periods, depending on ship motions, ship speed, wave height and position of the investigated plate fields, periodically converging solutions were obtained. For each time step up to ten outer iterations were needed.

5. COMPUTATIONAL RESULTS

Results for four different design wave conditions will be presented. Table 3 summarizes conditions of the four regular design waves we analyzed. For each run are listed wave heading β, wave length λ, wave height H, and ship speed v. A wave heading of β = 0 deg. denotes following waves; of β = 180 deg. head waves.

Table 3: Regular waves analyzed

Run	β [deg.]	λ [m]	H [m]	v [knots]
1	0	400	19.6	0
2	180	292	28.2	0
3	180	306	18.0	16.7
4	180	400	19.0	17.3

5.1 RUN 1

This run was intended to validate computed pressures against measured data. As seen in Figure 5, stern waves subjected the hull to two consecutive wave-induced slamming impacts in its stern region. The first impact occurred at the transition area between flat bottom and propeller hoses; 0.4 s later, the second impact occurred directly underneath the flat stern area.

Heave and pitch motions from model tests as well as slamming-induced pressure peaks compared favorably with RANS computed results as shown in Figure 6 and Figure 7, respectively.

5.2 RUN 2

During run 2, with the ship in head waves at zero speed, slamming underneath the stern area as well as green water impact loads on the forecastle occurred, as shown by the coupled RANS computed pressure distributions in Figure 8, vertical accelerations, vertical shear forces, and vertical bending moments were obtained for the rigid as well as the elastic hull girder. Resulting time series of accelerations are presented in Figure 9 and Figure 10, while global loads are presented in Figure 11 and Figure 12, respectively, together with comparable model test measurements.

5.3 RUN 3

This run for the ship with positive forward speed in head waves was performed for two different configurations of the hull to investigate the effect of green water on deck. In the model tests no deck containers were considered but a wave breaker at approx. $x/L = 0.75$, see Figure 2.

Again, vertical accelerations, vertical shear forces, and vertical bending moments were obtained for the rigid as well as the elastic hull girder. Resulting time series of accelerations and global loads are presented in Figure 13 to Figure 16, respectively, together with comparable model test measurements. Figure 17 shows both configurations during a typical slamming event at $t = 85.2\ s$.

5.4 RUN 4

To assess the influence of different loading conditions on the dynamic amplification of sectional loads, run 4 was simulated using two loading conditions for the same draft and trim, namely *MinHog* and *MaxHog* loading condition according to sagging and hogging condition. No measurement data was available for this run.

Resulting time series of vertical bending moment and vertical shear force are presented in Figure 18 and Figure 19, respectively.

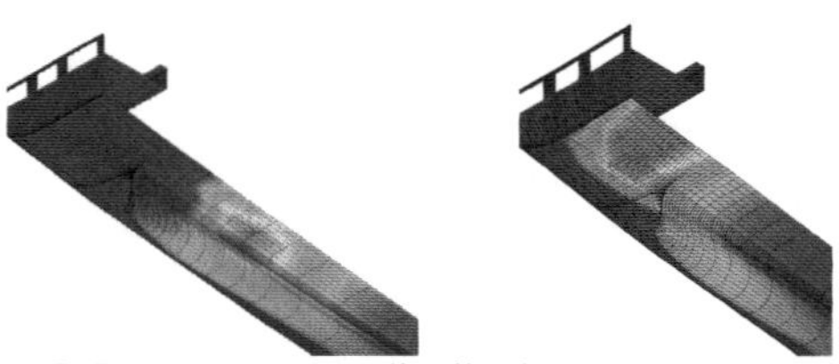

Figure 5: Impact pressure distributions, run 1

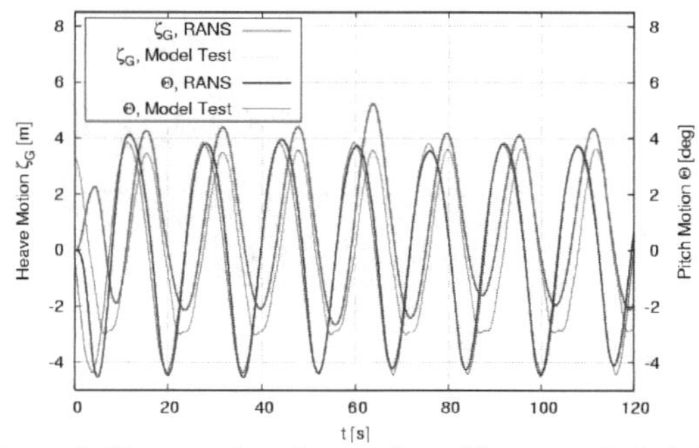

Figure 6: Comparative time series of heave and pitch for run 1

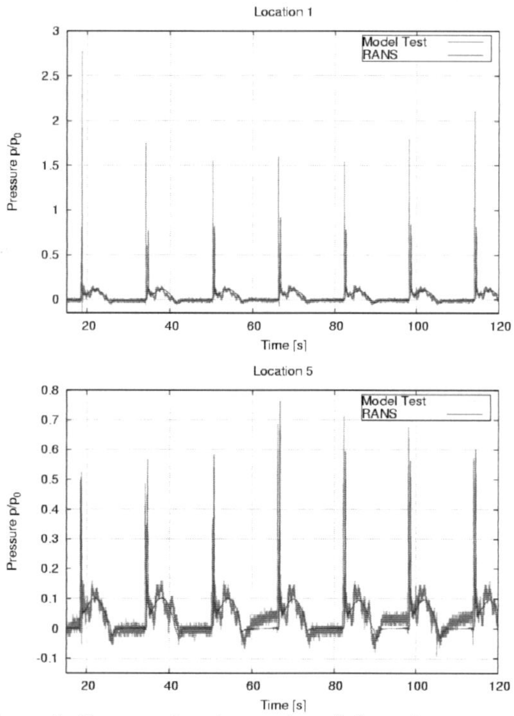

Figure 7: Comparative time series of slamming pressures, run 1

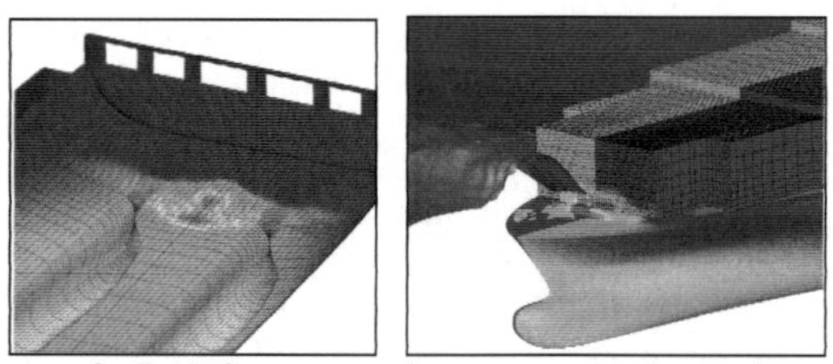

Figure 8: RANS computed pressures underneath stern (left) and on forecastle (right) for run 2

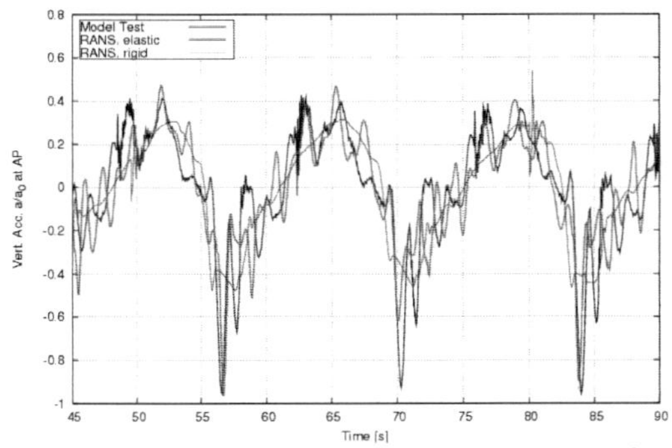

Figure 9: Vertical acceleration at aft perpendicular for run 2

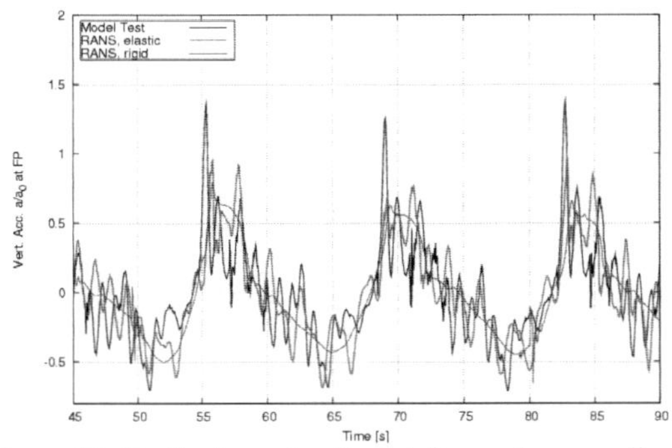

Figure 10: Vertical acceleration at forward perpendicular for run 2

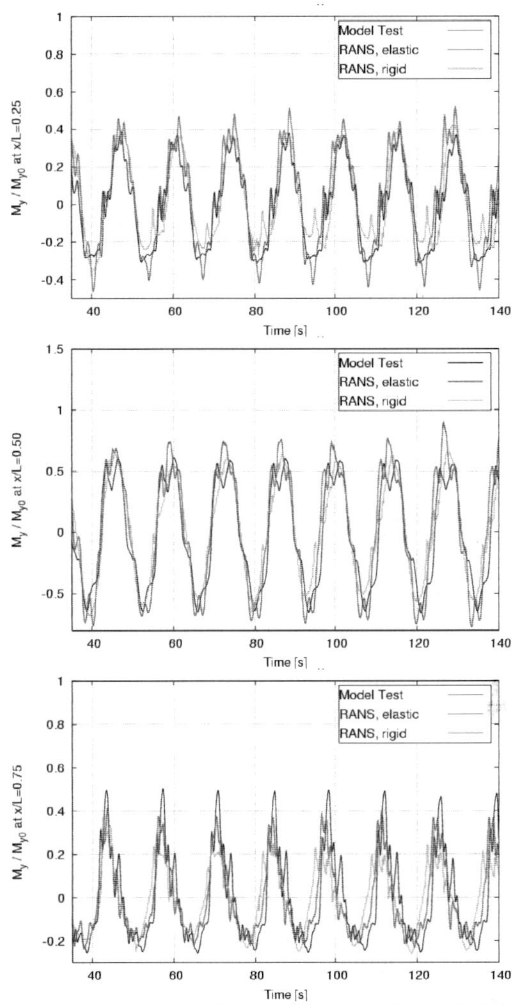

Figure 11: Vertical bending moments at $x/L = 0.25$, $x/L = 0.50$ and $x/L = 0.75$ (top to bottom) for run 2

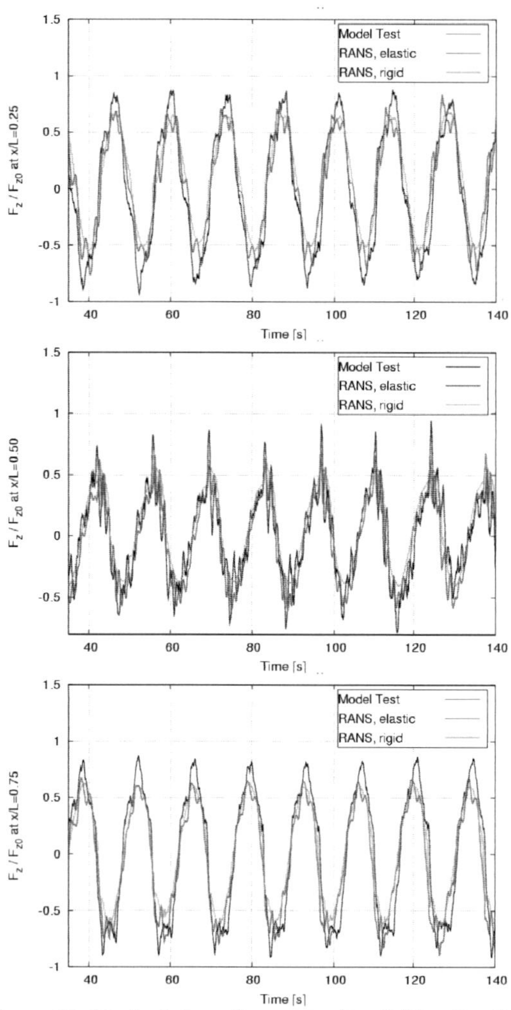

Figure 12: Vertical shear forces at *x/L = 0.25*, *x/L = 0.50* and *x/L = 0.75* (top to bottom) for run 2

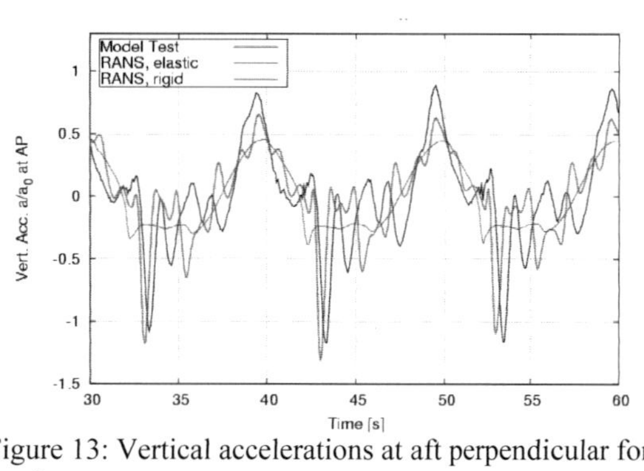

Figure 13: Vertical accelerations at aft perpendicular for run 3

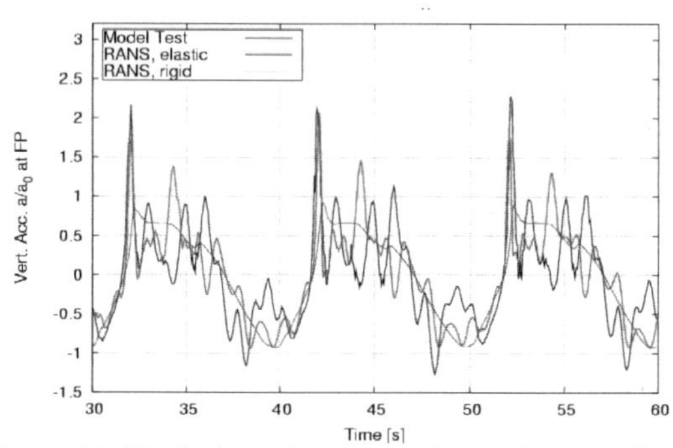

Figure 14: Vertical acceleration at forward perpendicular for run 3

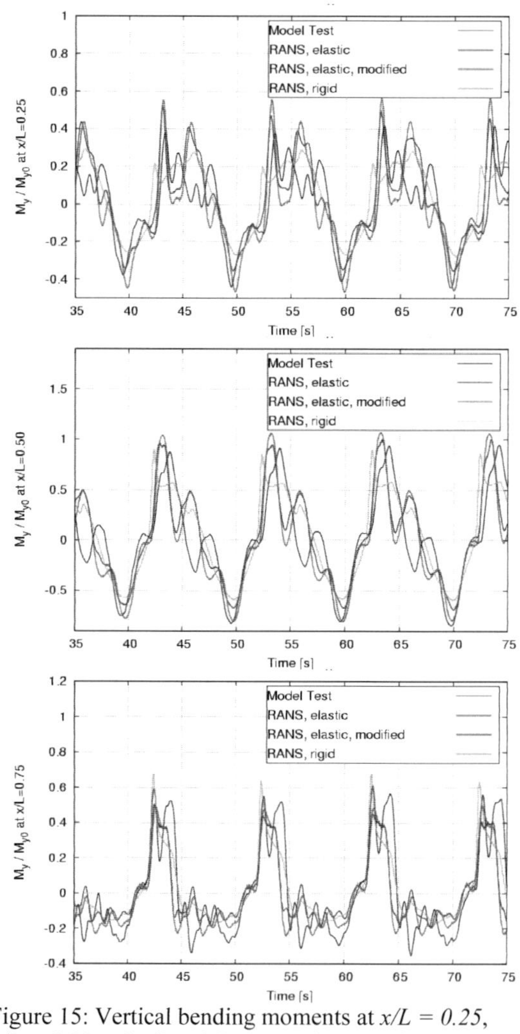

Figure 15: Vertical bending moments at *x/L = 0.25*, *x/L = 0.50* and *x/L = 0.75* (top to bottom) for run 3

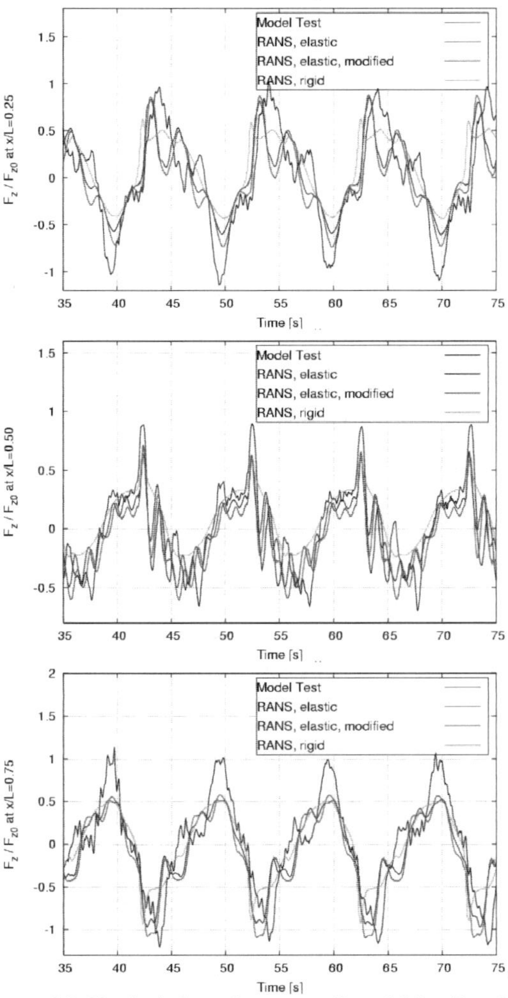

Figure 16: Vertical shear forces at $x/L = 0.25$, $x/L = 0.50$ and $x/L = 0.75$ (top to bottom) for run 3

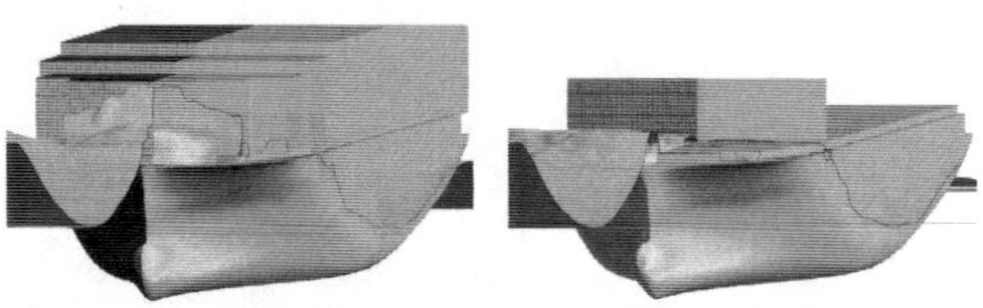

Figure 17: Slamming event at $t = 85.2\,s$ during run 3; comparison of hull with containers (left) vs. hull with model test wave breaker (right)

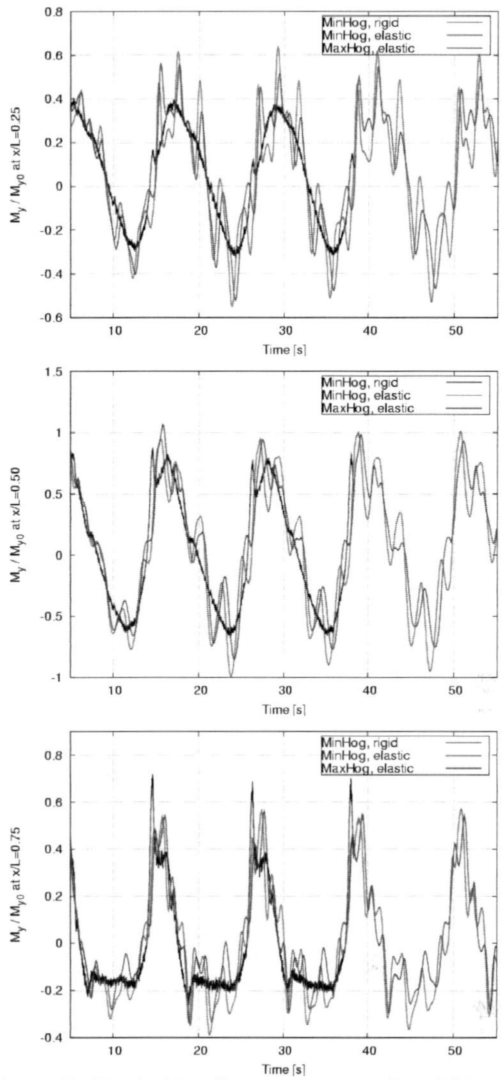

Figure 18: Vertical bending moments at $x/L = 0.25$, $x/L = 0.50$ and $x/L = 0.75$ (top to bottom) for run 4

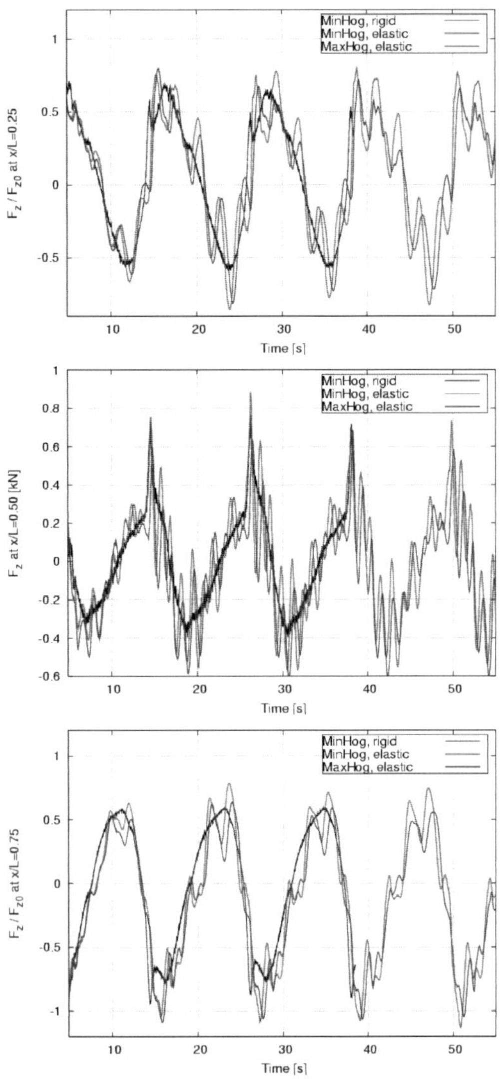

Figure 19: Vertical shear forces at *x/L = 0.25*, *x/L = 0.50* and *x/L = 0.75* (top to bottom) for run 4

6. CONCLUSIONS

Accounting for strong nonlinearities associated with impact load-induced whipping impelled us to solve the fluid-structure interaction problem in the time domain and to extend the RANSE code by solving the nonlinear 6DoF equations of motion and the Timoshenko beam equations. This extension of the RANSE solver made it possible to accurately predict whipping effects.

The computational procedure and numerical methods used here were suitable for handling the strong nonlinearities associated with slamming-induced whipping. Today, this kind of methods is the obvious choice for computing complex free-surface flows with breaking waves, sprays and air trapping, phenomena that should be considered to predict whipping. The favorable comparison of computations with model test measurements validated the methods used.

The definition of wave loads was based on the advanced design wave concept. With this concept, equivalent regular design waves were defined to model the seaway. Although these design waves only approximated the real wave conditions, there is at present no alternative available for practical application. A more accurate way, such as modeling the natural seaway by superimposing a large number of regular waves, would have made it necessary to deal with a prohibitively large number of variables.

We systematically investigated the large containership and selected worst case scenarios with respect to maximum stern slamming pressure and sectional loads. Acceleration, vertical shear force and vertical bending moment were found to be suitable as dominant parameters to globally assess the whipping effects on the ships' hull structure.

To investigate hull girder whipping, it was important to account for effects of fluid-structure interaction between the wave-induced loads and the elastic hull structure. This was accomplished by two-way coupling. The ship responses obtained for the elastic hull girder were much higher compared to those obtained for the rigid hull girder. This demonstrates the importance of dynamic amplification. Comparisons of different ship configurations (mass distribution and amount of water on deck) for the computations outlined the sensitivity of vibration amplitudes to variations of loading conditions.

7. REFERENCES

[1] Albina, F.O.: A procedure to set volume fraction for calculations with Comet. Interner Bericht, Technische Universität Hamburg-Harburg, Hamburg, 2000

[2] Brunswig, J. and El Moctar, O., "Prediction of Ship Motions in Waves Using RANSE," Proc. 7th Numerical Towing Tank Symp., Bertram, V. and Söding, H., Edts., Hamburg, 2004.

[3] El Moctar, O., Brehm, A., and Schellin, T.E., "Prediction of Slamming Loads for Ship Structural Design Using Potential Flow and RANSE Codes," Proc. 25th Symp. on Naval Hydrodynamics, 2004, St. John's.

[4] El Moctar O., Schellin T.E., Priebe T.: CFD and FE Methods to Predict Wave Loads and Ship Structural Response. In: Proceedings of the 26th Symposium on Naval Hydrodynamics, Rome, 2006.

[5] Guedes Soares, C. and Schellin, T.E., "Long-Term Distribution of Nonlinear Wave-Induced Vertical Bending Moments on a Containership," J. Marine Structures, Vol. 9, 1995, pp. 333-352.

[6] Hachmann, D., "The Calculation of Pressures on a Ship's Hull in Waves," J. Ship Technology Research, Vol. 38, 1991.

[7] Muzaferija, S. and Peric, M., "Computation of Free-Surface Flows Using Interface-Tracking and Interface-Capturing-Methods," Nonlinear Water Wave Interaction, Computational Mechanics Publ., Southampton, 1998, pp. 59-100.

[8] Oberhagemann J., el Moctar O.: A Simplified Approach to Investigate Fluid-Structure Coupling Effects on Slamming Loads of Ships. In: Proceedings of the 10th Numerical Towing Tank Symposium, Hamburg, 2007.

[9] Papanikolaou, A. D. and Schellin, T. E., "A Three-Dimensional Panel Method for Motions and Loads of Ships with Forward Speed," J. Ship Technology Research, Vol. 39, 1991, pp. 147-156.

[10] Söding, H.: Vibrationen von Schiffen I, Vorlesungsmanuskript Nr. 23, Institut für Schiffbau, Hamburg, 1987.

[11] Star-CD: COMET User Manual 2.0. CD-Adapco, Nürnberg, 2002.

8. AUTHORS' BIOGRAPHIES

Ould el Moctar started studying naval architecture at the University of Hamburg in 1990. During his studies he gathered research experience at the Fluid Dynamics and Ship Theory Department of the University of Technology Hamburg-Harburg (TUHH) and at the University of Kiel. After graduation in 1997, he became a research assistant at the TUHH, specializing in ship hydrodynamics. In 2000 he joined the Hamburg Model Basin as a research engineer. He earned his PhD in 2001. In 2002 he joined Germanischer Lloyd, where he currently is the head of the Department Fluid Dynamics.

Jan Oberhagemann studied naval architecture at the University of Technology Hamburg-Harburg (TUHH). Since his graduation in 2006, he is working in the fluid dynamics department at Germanischer Lloyd. Focus of his activities is whipping and slamming-related phenomena and the development of fluid-structure coupling techniques.

Hyun-ho Lee earned his Bachelor's degree in Naval Architecture and Ocean Engineering (NAOE) at Seoul National University (SNU) in 1995 and earned his Master's degree in NAOE at SNU in 1997. Currently he is a senior research engineer in the Ocean Industry Research Department of Hyundai Heavy Industries Co., Ltd.

Tae-hyun Park earned his Bachelor's degree in Forest Products Engineering at Seoul National University in 1999 and earned his Master's degree in Mechanical Aerospace Engineering at Seoul National University in 2002. At present he is employed as a chief researcher in the Ocean Industry Research Department of Hyundai Heavy Industries Co., Ltd.

Hong Ryeul Ryu earned his Bachelor's degree in Naval Architecture and Ocean Engineering at Seoul National University in 1994 and earned his Master's degree in Naval Architecture and Ocean Engineering at Ulsan University in 2006. At present he is working as a senior engineer of Hull Initial Design Department of Hyundai Heavy Industries Co., Ltd. He has an experience in the area of ship structural analysis and hull initial design of more than 14 years.

3DFEM-3DBEM MODEL FOR SPRINGING AND WHIPPING ANALYSES OF SHIPS

Š. Malenica, Bureau Veritas, France
J. T. Tuitman, Delft University of Technology, The Netherlands

SUMMARY

The paper deals with the modeling of the wave induced ship structural vibrations. The presented method is fully integrated in both the linear frequency domain and non-linear time domain seakeeping calculations. Due to the transient non-linear nature of slamming loading, the corresponding whipping response is calculated in the time domain only, while the non impulsive springing response is evaluated both in frequency and time domain. A method based on the full coupling of 3DFEM structural model and 3DBEM hydrodynamic model is presented. The choice of the full coupling with the 3DFEM structural model is necessary due to the particular structural characteristics of the container ships, in particular their open midship section, which is hard to model using non-uniform beam model. In addition, the 3DFEM model gives the direct access to the structural response (stresses and strains) at any required location.

The stresses at the structural details are investigated using a so called top-down analysis, and the total stress response is separated into quasi static and dynamic parts in order to improve the convergence.

1. INTRODUCTION

Increased size of the recent Ultra Large Container Ships (ULCS) and LNG ships, reactualizes the hydroelastic wave induced type of ship structural responses. Usually the hydroelastic ship response is separate into springing and whipping types, springing being defined as the steady ship vibrations induced by non-impulsive wave loading and whipping as a transient vibratory ship response induced by impulsive loading such as slamming. In this paper only slamming induced whipping is considered.

2. FREQUENCY DOMAIN HYDROELASTIC SEAKEEPING MODEL

The general methodology for hydroelastic seakeeping model is rather well known and the first developments can be attributed to Bishop & Price [2]. In their work they used the simplified, so called Timoshenko beam, model for structural modeling and strip theory for seakeeping part. Since then several more or less sophisticated models were proposed (e.g. [4, 12, 13, 14]).

Below we briefly recall the basic principles of the model used in this study. The 3DBEM model for the seakeeping is coupled to a 3DFEM model of the ship structure. A more detailed description of the applied 3DBEM model can be found in [6] and [9].

2.1 HYDROELASTIC MODEL

In contrast to the well known rigid body seakeeping model, the hydroelastic model basically extends the motion representation with the additional modes of motion/deformation chosen as a series of the dry structural natural modes.

We write:

$$H(x,y,z,t)=\sum_{i=1}^{N}\xi_i(t)\ \ h^i(x,y,z)$$
$$=\sum_{i=1}^{N}\xi_i(t)[h_x^i(x,y,z)\ \ i+h_y^i(x,y,z)\ \ j+h_z^i(x,y,z)\ \ k] \quad (1)$$

Where $h^i(x,y,z)$ denotes the general motion/deformation mode which can be either rigid or elastic.

The above decomposition leads to the additional radiation boundary value problems (BVP) for elastic modes, with the following change in the body boundary condition:

$$\frac{\partial\varphi_{Rj}}{\partial n}=h^j\ \ n \quad (2)$$

After solving the different BVP-s the resulting pressure is calculated using Bernoulli's equation and integrated over the wetted surface in order to obtain the corresponding forces, so that the following coupled dynamic equation can be written:

$$\left\{-\omega_e^2([\ m\ ]+[\ A\ ])-i\omega_e[\ B\ ]+[\ k\ ]+[\ C\ ]\right\}\{\ \xi\}=\{\ F^{DI}\} \quad (3)$$

Where:

$[\ m\ ]$ - modal genuine mass
$[\ k\ ]$ - modal structural stiffness
$[\ A\ ]$ - hydrodynamic added mass
$[\ B\ ]$ - hydrodynamic damping

$[\,C\,]$	- hydrostatic stiffness
$\{\xi\}$	- modal amplitudes
$\{\;F^{DI}\}$	- modal excitation

The solution of the above equation gives the motion amplitudes and phase angles ξ_i and the problem is formally solved. Note that the motion equation includes both 6 rigid body modes and a certain number of elastic modes.

Several technical difficulties need to be solved before arriving to the above motion equation (3). Certainly the most difficult one is the solution of the corresponding hydrodynamic BVP. In this paper we do not enter into the detailed description of the methods used to solve the seakeeping problem at forward speed and we just mention that these difficulties remain the same, both for the rigid and elastic body. It is fair to say that the numerical methods which are used to solve the seakeeping problem nowadays, are not fully ready yet for a general combination of speed, heading and frequency. However, most of the methods have approximate solutions to account for the forward velocity. The method used in this paper is the so called encounter frequency approximation which was reasonably well validated for rigid body case.

The second technical difficulty concerns the application of the body boundary condition (2) for the general mode of motion. In the next section we briefly explain how this was done in the present work.

2.2 BODY BOUNDARY CONDITION

In the case of the simplified beam structural model, the transfer of the modal displacements of the beam onto a hydrodynamic mesh can be done relatively easy, but in the case of 3DFEM structural model this transfer requires very careful attention. Indeed, the modal displacements are known on the structural 3DFEM mesh which is usually completely different from the hydrodynamic mesh, so that the special interpolation procedure is necessary. In the present work, the following procedure is adopted.

For each hydrodynamic point (panel center) the following steps are performed, these steps are illustrated in figure 1:

1. Looking for the 3 closest structural points. In the case that one of the 3 points is within small enough distance (ε) from the considered hydrodynamic point, we retain that structural point for interpolation.
2. The list of the structural finite elements containing at least one of the above defined points is created.
3. The hydro point is projected onto the surfaces created by the retained structural elements. If the projection falls inside the element, the corresponding distance is calculated. The element with smallest distance from the hydro point is retained for interpolation.
4. In the case when the hydro point do not project on any element from the list, the projection on the sides of the considered elements is performed. If the projection falls onto the side, the corresponding distance is calculated. The side with the smallest distance from the hydro point is retained for interpolation.
5. In the case when there is neither point at ε distance, nor element nor the element side on which point project, the structural point closest to the hydro point is retained for interpolation.
6. The interpolation using the shape functions of the retained finite element, of the structural displacements is performed on the projection of the hydro point and the calculated displacements are associated to the hydro point.

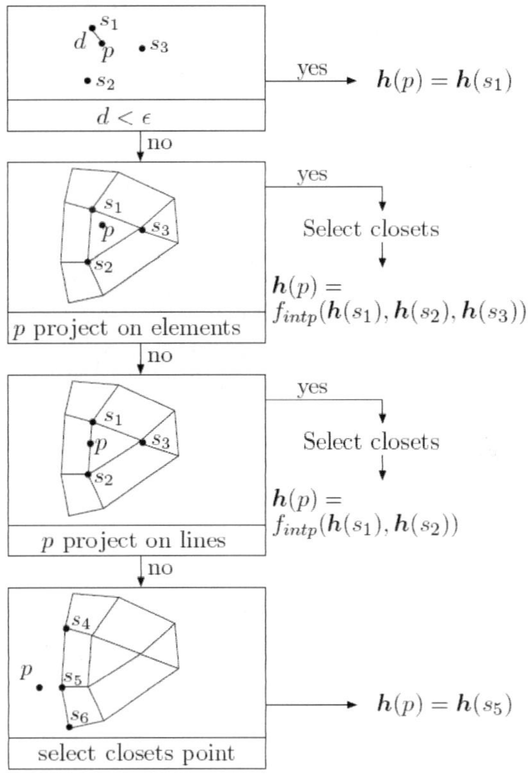

Figure 1: *Mapping procedure*

This procedure was verified on several ship types and showed to be very efficient. In Figure 2 we present one example for the first torsional mode of typical ULCS.

2.3 SPRINGING

The solution of the hydroelastic motion equation (3) includes the linear springing response automatically. However, there is no clear justification for considering springing as a linear phenomena, at least as far as the wave excitation part is considered. Indeed, experience shows that many type of ship suffer from more or less important structural response around their first structural natural frequencies even if there is no important wave energy around these frequencies. The explication for existence of the structural response can be explained only by introducing the nonlinearity into the hydrodynamic model. The well known example is the springing of the tendons of the TLP platforms where the hydrodynamic excitation was successfully explained using the weakly non-linear second order theory. In principle similar kind of methods should also be applied in the case of the ships, but unfortunately the state of the art in ship hydrodynamics do not allow for that so that only approximate and empirical methods can be used.

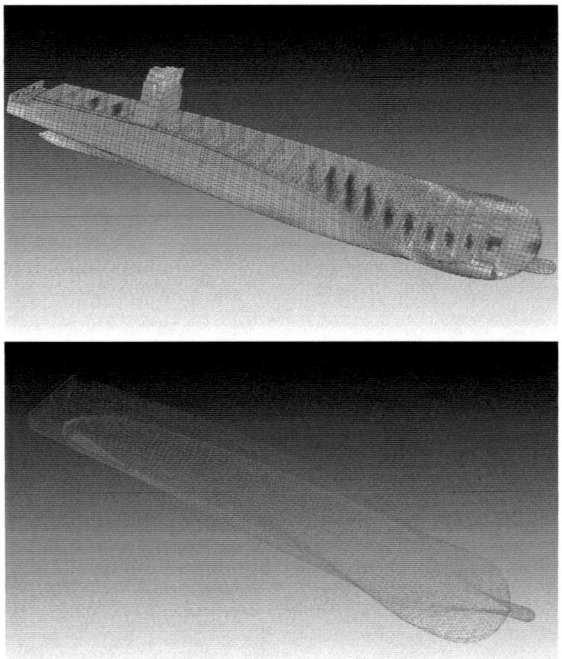

Figure 2: *First torsional mode of ULCS and the transfer of modal displacements from structural to hydrodynamic model.*

However, due to their huge dimensions and particular structural properties, the case of ULCS is quite particular because their first structural natural frequencies become quite low so that it can be excited in a linear sense too! This means that the linearity excited springing might become dominant which justify the linear model for its assessment.

The non-linear hydroelastic model in time domain which will be presented in section 4 does also include the linear springing response. The Froude Krylov loading is the only non-linear wave excitation in this model. Some non-linear springing due to this Froude Krylov is observed in the results of the time domain calculations, but to the authors opinion a more sophisticated non linear model is necessary to correctly calculate the non-linear springing response.

3. CALCULATION OF THE STRESSES IN FREQUENCY DOMAIN

Once the motion equation (3) solved, all necessary quantities (motions, accelerations, stresses, ...) can now be calculated. Indeed, the modal decomposition remains valid for any quantity and, in particular, we can write for the stress distribution $\Sigma(x, y, z, \omega)$:

$$\Sigma(x, y, z, \omega) = \sum_{i=1}^{N} \xi_i(\omega)\sigma^i(x, y, z) \tag{4}$$

Where $\sigma^i(x, y, z)$ represents the spatial distribution of the stresses corresponding to each mode of motion/deformation. Equation similar to (4) remains valid in the time domain model too, (10). The only change concerns the replacement of ω by t.

3.1 SEPARATION OF THE QUASI STATIC AND DYNAMIC CONTRIBUTIONS

In the present context, it is important to note that the rigid body modes do not contribute to any stress, and that the above expression (4) is slowly convergent in general. This is especially true for the structural details which are affected by the local structural effects. A good example of such detail is a hatch corner of the container ship. The longitudinal response is well described by the first few longitudinal flexible modes. However, in order to obtain any contribution of the transversal (side shell) loading, one need to include natural mode shapes which describes the deformation in transversal direction. The 3DFEM methods calculates the structural natural frequencies of the ship from the lowest natural frequency. By using a proper 3DFEM model of ship it is relatively easy to obtain the first few torsional and longitudinal modes. However, after these lowest global structural modes, the numerous local modes with the similar natural frequencies will quickly "pollute" the solution and it will be almost impossible to obtain a subset of modes necessary for accurate representation of the transversal stresses of the hatch corner. Fortunately, the global structural dynamic response is well described by the first few lowest modes. In order to obtain the converged stress distribution, the total structural response should be properly separated into the quasi static and dynamic parts. The stresses due to the quasi static ship response will be calculated, after separation, by the so called direct method which calculates the structural response after the "rigid body" hydrodynamic pressure is transferred on the wetted finite elements [8]. It is also important to note that the decomposition into the quasi static and dynamic response is made with respect to the structural response, which means that the quasi static structural response does also includes the dynamic rigid body response and the dynamic response includes the dynamic structural response only.

In the present approach the decomposition of the different parts of the response is done by first schematically rewriting the motion equation (3) in the following form:

$$\left(\begin{bmatrix}[\,RR\,] & [\,RE\,]\\ [\,ER\,] & [\,EE\,]\end{bmatrix} + \begin{bmatrix}[\,0\,] & [\,0\,]\\ [\,0\,] & [\,\mathrm{k}\,]\end{bmatrix}\right)\begin{Bmatrix}\xi^R\\ \xi^E\end{Bmatrix} = \begin{Bmatrix}F^R\\ F^E\end{Bmatrix} \qquad (5)$$

Where R stands for the rigid body parts, E for the elastic ones and k is the structural modal stiffness matrix.

At the same time we separate the total response amplitudes into the quasi static and dynamic parts:

$$\xi^R = \xi_0^R + \xi_d^R \quad , \quad \xi^E = \xi_0^E + \xi_d^E \qquad (6)$$

The quasi static part of the responses is defined by the following equations:

$$[\,RR\,]\{\xi_0^R\} = \{F^R\} \qquad (7)$$

$$[\,\mathrm{k}\,]\{\xi_0^E\} = \{F^E\} - [\,ER\,]\{\xi_0^R\} \qquad (8)$$

After inserting (6), (7) and (8) into (5), the following linear system of equations for dynamic parts is obtained:

$$\left(\begin{bmatrix}[\,RR\,] & [\,RE\,]\\ [\,ER\,] & [\,EE\,]\end{bmatrix} + \begin{bmatrix}[\,0\,] & [\,0\,]\\ [\,0\,] & [\,\mathrm{k}\,]\end{bmatrix}\right)\begin{Bmatrix}\xi_d^R\\ \xi_d^E\end{Bmatrix} = -\begin{Bmatrix}[\,RE\,]\xi_0^R\\ [\,EE\,]\xi_0^E\end{Bmatrix} \qquad (9)$$

As mentioned before, the above procedure was adopted in order to be able to keep the classical direct approach for the quasi static part [8] and to clearly identify the dynamic part as a correction of the quasi static one. Anyhow, the proposed decomposition completely removes the convergence problems mentioned before.

3.2 TOP DOWN ANALYSIS OF STRUCTURAL DETAILS

Springing hydroelastic analysis is usually performed using the relatively coarse structural mesh and some additional manipulations are necessary in order to obtain the local stresses in the critical structural details. Usually the so called top down analysis is applied. The top down analysis is well developed for the classical quasi static structural responses. The principles are relatively simple and the calculations are performed in two steps:

1. Calculation of the global structural response on the coarse mesh
2. Application of the structural deformations of the coarse mesh on the fine mesh and calculation of the local stresses

One typical situation is shown in Figure 3.

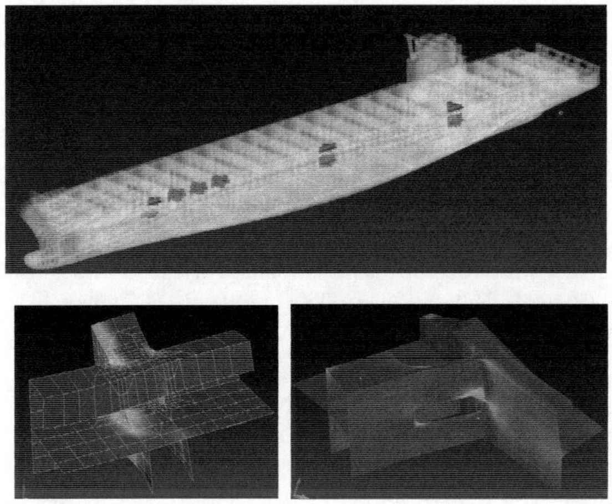

Figure 3: *Typical structural details and their position on the ship structure.*

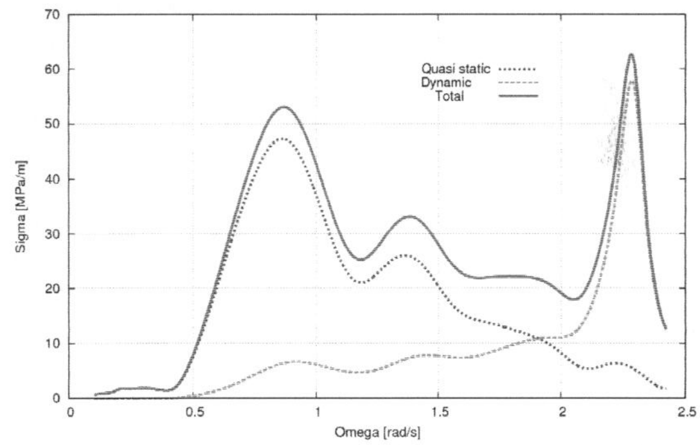

Figure 4: *Typical stress RAOs.*

As already mentioned before, in the case of the quasi static structural analysis, the loading case for the structural FEM model is built by considering the ship as a rigid body. After solving the rigid body seakeeping boundary value problems, the resulting pressure is applied on the FEM mesh together with the inertia loads resulting from the rigid body accelerations. Since, in the rigid body case, the hydrodynamic and structural parts of the problem are independent they can be performed separately and the transfer, of the resulting deformations of the coarse mesh, onto the fine mesh is relatively straightforward. In the case of dynamic structural analysis, which is of main concern here, the additional loading cases for fine meshes need to be created. These additional loading cases corresponds to each structural natural mode deformations. The resulting local stresses should be added to the quasi static ones, after being previously multiplied by the associated dynamic modal amplitudes ξ_d^E. One typical example of the local stresses in the hatch corner of the classical ULCS are shown in Figure 4.

3.3 HYDROELASTICITY

It is interesting to observe that the dynamic part of the response contributes to the stresses even for the frequencies well away from the natural frequency. This might seems a bit strange, but one should remind that these are very flexible ships. The static deformation of the structure, due to the still water bending, between midship and the fore perpendicular is of the order of $1m$ for the ULCS. The wave bending moment is of the same order as the still water bending moment and will cause the same order of deflections. This deflection will influence the heave response and the total bending moment.

All this suggests the importance to utilize a hydroelastic analysis for ULCS's even when springing and whipping are not explicitly considered.

4. NON LINEAR TIME DOMAIN SIMULATIONS

In contrast to the linear springing calculations which can be performed in frequency domain, the nonlinear springing and whipping simulations require the time domain simulations. The procedure used in this paper is based on the method proposed in Cummins [3] and elaborated in Ogilvie [10]. This method uses the frequency domain hydrodynamic solution and transfers it to the time domain using the inverse Fourier transform. In this way the following time domain motion equation is obtained:

$$([\mathbf{m}]+[\mathbf{A}^{\infty}])\{\ddot{\xi}(t)\}+([\mathbf{k}]+[\mathbf{C}])\{\xi(t)\} + \int_0^t [\mathbf{K}(t-\tau)]\{\dot{\xi}(\tau)\}d\tau = \{\boldsymbol{F}^{DI}(t)\}+\{\boldsymbol{Q}(t)\} \qquad (10)$$

Where overdots denote the time derivatives and:

$[A^{\infty}]$	-infinite frequency added mass matrix
$[K(t)]$	-matrix of impulse response functions

It was shown by Ogilvie [10], that the impulse response functions can be calculated from the frequency dependent damping coefficients:

$$K_{ij}(t)=\frac{2}{\pi}\int_0^{\infty}B_{ij}(\omega)\cos\omega t\, d\omega \qquad (11)$$

After the impulse response functions K_{ij} have been calculated, the motion equation (10) is integrated in time using the Runge Kutta 4th order scheme.

The main advantage of the time domain method lies in the fact that we can introduce non-linear components in the excitation forces $Q(t)$. There are many non linear effects which are missing in the linear model and the state of the art in the numerical seakeeping, do not allows for the consistent inclusion of all of them. In this paper we include the so called Froude Krylov correction for the wave excitation and the non linear slamming forces. Here below we briefly explain the way in which the Froude Krylov and slamming loads are calculated.

4.1 FROUDE KRYLOV APPROXIMATION

It is well known that the linear hydrodynamic model evaluate the pressure up to $z=0$ only so that we can end up with the negative pressure close to the waterline, as well as with zero pressure above the waterline and below the wave crest which is unphysical. The so called Froude Krylov approximation is usually employed to correct this pressure distribution close to the waterline. The "simplest" Froude Krylov variant is rather intuitive and consists in adding the hydrostatic part of the pressure ($-\rho gz$) below the wave crest (in the linear sense) and by putting zero total pressure above the wave trough. Even if the methodology might appear quite simple, the correct practical implementation in the numerical codes might become very tricky. To the authors knowledge, the details of the numerical implementation of the Froude Krylov method are usually not discussed in the literature and that is why we are briefly elaborating on it here below. It is also important to note that some other extrapolation procedures, like Wheeler stretching, for the pressure close to the waterline are sometimes employed, but we do not discuss them here.

According to the linear theory the linear hydrodynamic pressure at any point below $z=0$, is composed of two parts:

- pure hydrodynamic part associated with the velocity potential and its derivatives
- hydrostatic pressure variation due to ship motions

For the wave of unit amplitude we can write:

$$p(x,y,z,t)=\Re\{[i\omega_e\rho\varphi-\rho U\nabla(\bar{\phi}-x)\nabla\varphi]e^{-i\omega_e t}\} + \Re\{-\rho g\zeta_v e^{-i\omega_e t}\} \qquad (12)$$

Where $\varphi(x,y,z)$ is the linear unsteady potential, $\bar{\phi}(x,y,z)$ is the steady potential due to the forward speed in calm water, and ζ_v is the vertical displacement of the considered point on the ship surface.

In order to simplify the notations, it is useful to express the pressure in meters of water height and write:

$$\tilde{p}(x,y,z,t)=\frac{Ap(x,y,z,t)}{\rho g}=\tilde{p}^z \qquad (13)$$

In this way the relative wave elevation (usually denoted by η^R) is equal to $\tilde{p}$ at $z=0$

$$\eta^R(x,y,t) = \tilde{p}^0 \qquad (14)$$

Now we consider separately the case $\tilde{p}^0 > 0$ and $\tilde{p}^0 < 0$.

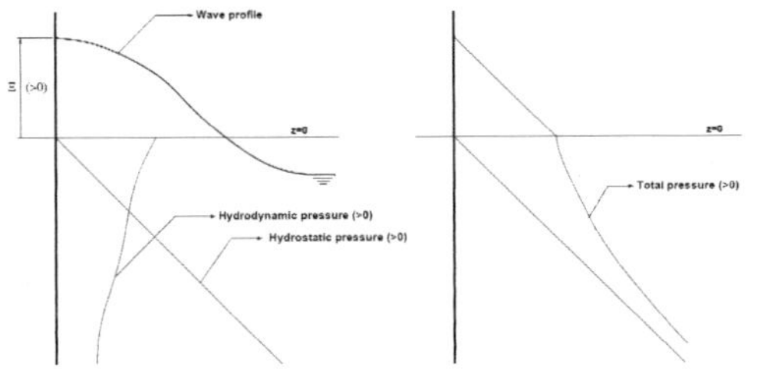

Figure 5: *Modified pressure distribution for* $\tilde{p} > 0$.

In the case of $\tilde{p} > 0$ (Fig. 5) we can identify two different wetted regions with the following total pressure distribution:

$$\tilde{P}(z,t) = \begin{cases} \tilde{p}^0 - z & \text{for} \quad 0 < z < \Xi \\ \tilde{p}^z - z & \text{for} \quad z < 0 \end{cases} \qquad (15)$$

The choice of the maximal wetted point Ξ is quite natural:

$$\Xi = \tilde{p}^0 \qquad (16)$$

In the case of $\tilde{p} < 0$ (Fig. 6) the situation is more complicated and we need to be careful.

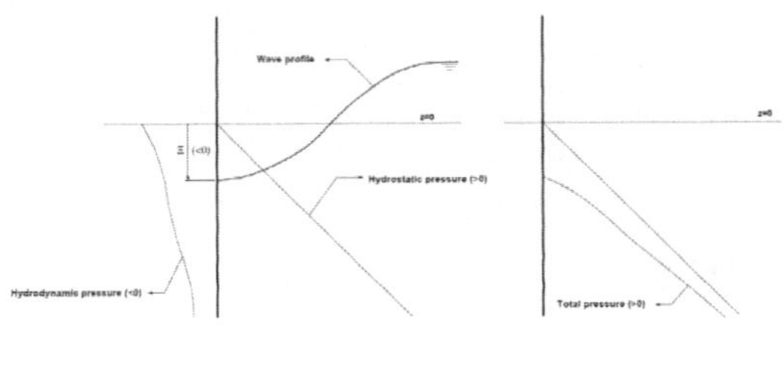

Figure 6: *Modified pressure distribution for* $\tilde{p} < 0$.

As far as the pressure distribution is concerned, the situation is similar to the previous case, and the choice is quite natural (even if, strictly speaking, not fully consistent):

$$\tilde{P}(z,t) = \begin{cases} 0 & \text{for} \quad 0 > z > \Xi \\ \tilde{p}^z - z & \text{for} \quad z < \Xi \end{cases} \qquad (17)$$

The main problem appears to be the choice of the minimal wetted point Ξ. The usual choice for Ξ is the same as in the previous case:

$$\Xi = \tilde{p}^0 \qquad (18)$$

The problem is that this choice introduce the jump in the pressure distribution because:

$$\tilde{p}^0 \neq \tilde{p}^z \quad , \quad for \quad z = \Xi \qquad (19)$$

Which means that the pressure at the point imediatly below the wave profile will not be zero but:

$$\tilde{P}(\Xi,t) = \tilde{p}^\Xi - \Xi = \tilde{p}^\Xi - \tilde{p}^0 \qquad (20)$$

The reason for this is the inconsistency of the adopted approach which do not include all non-linear terms, so that all necessary conditions can not be satisfied at the same time and we need to "sacrify" something.

Let us now choose the value of Ξ as a solution of the following equation:

$$\Xi = \tilde{p}^\Xi \qquad (21)$$

It is important to note that an iterative procedure should be used to obtain the solution of this equation. It is easy to see that this choice eliminates the jump in the pressure distribution. The disadvantage is that we have to solve the equation (21). The "physical" consequence of the above choice is that the wave profile is changed from the assumed one (η^R). However, this slight move of the free surface is within the order of the approximations we adopted at the begining. Strictly speaking both choices are not fully consistent but the second one has the advantage to allow for more physical pressure distribution.

In summary, we can say that the Froude Krylov approximation is not so clear and we need to be very careful in the implementation and in the interpretation of the results. Indeed, as we have seen, several approximations are involved in the above described procedure and, in spite of the quite complicated considerations, it is impossible to remain fully consistent. It is also important to note that several variants of the Froude Krylov method are in use and the simplest one consists in taking only the incident wave part of elevation in the definition of the relative wave elevation η^R. This choice highly simplify the numerical implementation and that is the reason why it is most often employed. In that case we need to be extremely careful when interpreting the results. This, lets call it incident wave Froude Krylov approximation, is likely to be valid only in the case of rather long waves where the diffraction and radiation parts play less important role in the overall contribution to relative wave elevation η^R . However, in the case of ULCS which have quite a low block coefficient i.e. rather slender form the incident wave Froude Krylov approximation can be reasonably justified for head or almost head wave conditions for which the springing and especially whipping are most likely to occur.

4.2 COURSE KEEPING

The non-linear part of the force in equation (10) will cause a mean loading at the ship. The ship will not hold course and speed due to this mean force. This mean force is not an error in the calculation. The mean force by the Froude Krylov calculation is a part of the added resistance in waves (RAW) and slamming forces cause also a loading in the horizontal direction.

The drift of the ship in the time domain calculation should be compensated in order to obtain a proper calculation. Springs and dampers are often added to keep the ship at the correct positions. An other solution is to model the compete resistance, propulsion and maneuvering system of the ship. Both method require quite some tuning to be able to make a realistic simulation. To the authors opinion the correct non-linear surge, sway and yaw response will not affect the structural loading much. Also one could question if using a spring, damper system would give the correct non-linear horizontal motions. In the present approach Lagrange multipliers are added to equation (10). These Lagrange multipliers impose the surge, sway and yaw motions as is calculated using the frequency domain solution (3). This gives a reasonable horizontal motions without any tuning.

4.3 SLAMMING INDUCED WHIPPING

The time domain calculation will automatically include the linear and non linear springing. In order to calculate the slamming induced whipping response, the slamming forces should be calculated during the seakeeping calculation. This forces are added to the force vector $\{ Q(t)\}$ of equation (10).

4.3 (a) Slamming Force

The hydrodynamic modeling of slamming is extremely complex and still no fully satisfactory slamming model exists. However, the 2D modeling of slamming is well mastered today and 2D models are usually employed to assess the slamming loads on ships. Within the potential flow approach, which is of concern here, several more or less complicated 2D slamming models exist, starting from simple von-Karman model and ending by the fully nonlinear model. In between these two models there are several intermediate ones such as Generalized Wagner Model (GWM) [11] and Modified Logvinovich Model (MLM) [5] which are implemented in the presented method. We will not go into too much details of these models and we refer to [7]. The advantage of the MLM method when compared to GWM method lies in the lower requirements of CPU time. Most often, at least when rigid body impact is concerned, two methods give comparable results as shown in Figure 7

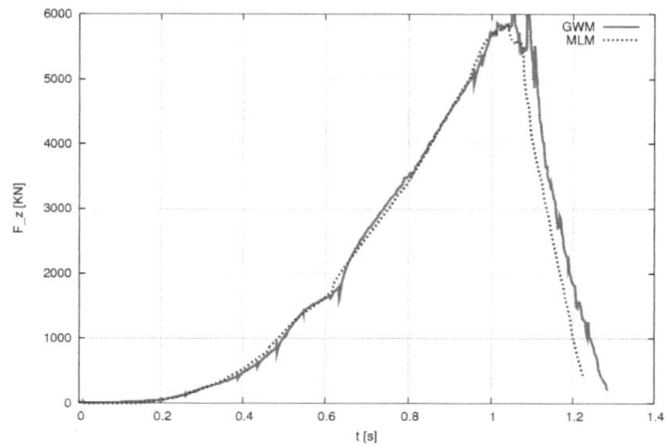

Figure 7: *Slamming forces on typical 2D section obtained by GWM and MLM method.*

The GWM method is used for the presented examples because this method has a wider range of validity and the number of slamming calculations remains limited. If many calculations, for example a complete scatter diagram, have to be preformed the MLM method might become more attractive. The validity of the MLM calculation can be checked using the GWM results for the some cases.

The use of 2D methods implicitly requires the employment of the so called strip approach for 3D simulations. The usual procedure consist in cutting the 3D mesh into a certain number of 2D sections as shown in Figure 8.

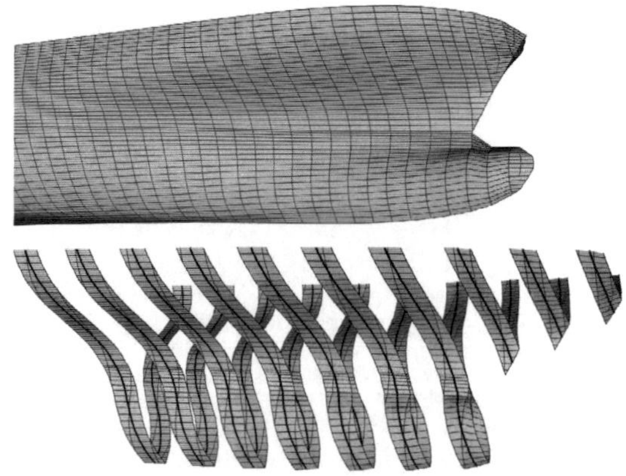

Figure 8: *3D hydrodynamic mesh and corresponding 2D sections.*

For each particular section, independent calculation of slamming loading are performed and the overall slamming is obtained by summing up different contributions. The implementation of this procedure is different for, the most often used, beam structural model and for 3DFEM model. Indeed, in the case of the nonuniform Timoshenko beam model, only the overall slamming force on the slamming sections is required, while in the case of the 3DFEM model the transfer is slightly more complicated. Here below we briefly describe the method which is likely to be very efficient for 3DFEM case.

The general expression for pressure excitation force projected on the motion/deformation mode h^i, can be written in the following form:

$$F^i = \iint_S p \; h^i \; ndS \tag{22}$$

Where S denotes the part of the ship surface on which the pressure is applied and that regardless if the pressure is calculated using the 2D or 3D methods. In the present case, and since the slamming pressure is calculated on the 2D sections, the above expression is rewritten in the following form:

$$F^i = \sum_{i=1}^{N_s} \iint_{S_s} p_s \; h_s^i \; n_s dS_s \tag{23}$$

Where S_s denotes the surface associated with the corresponding slamming section, n_s the associated normal vector, p_s the calculated 2D slamming pressure and h_s^i the projection of the 3D deformation mode h^i onto 2D slamming section. The differential element of the surface of the slamming section can be rewritten as:

$$dS_s = b_s(l) dl_s \tag{24}$$

Where $b(l)$ is the normalized width of the section:

$$b_s(l) = \frac{dS_s}{dl_s} \tag{25}$$

Which can be easily calculated from the 3D sectional mesh extracted from 3D mesh (Figure 8). The same is true for the associated normal vector, so that the following final expression, for the slamming induced modal excitation force, can be written:

$$F^i = \sum_{i=1}^{N_s} \int_{L_s} p_s \; h_s^i \; n_s b_s dl_s \tag{26}$$

This expression can be easily implemented into the existing rigid body slamming 2D module based on GWM or MLM method.

We note here that, due to the lack of time, for the examples presented in this paper a simplified method is used. The modal excitation is calculated by multiplying the 2D slam force with the area of the 3D section and the average mode shape over the sections. This methods works well for the first longitudinal modes.

4.3 (b) Whipping

One typical whipping event from full scale measurements is shown in Figure 9 (taken from [1]). The curve represents the time history of the vertical bending moment after slamming event. Two important points should be observed: whipping can increase the wave bending moment significantly and the transient whipping vibrations can last for quite a long time because of low damping. This is very important because it implies that whipping will be relevant both for extreme structural response as well as for the fatigue loading.

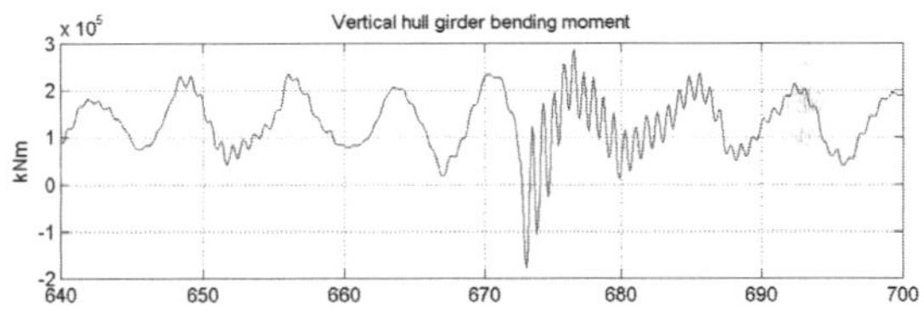

Figure 9: *Whipping from full scale measurements.*

Anyway, once the above procedure implemented, it is possible to calculate the time history of the stresses at any particular finite element. The decomposition into quasi static and dynamic contribution do not seems to be straightforward. The most practical way is to perform two separate simulations: one purely quasi static and another with the structural dynamics included. The above described procedure is applied on 11400 TEU ship sailing with speed $U = 10m/s$ in the irregular head waves (Jonswap spectrum $H_s = 5m$ and $T_p = 16s$). The typical situation at one particular time instant is shown in Figure 10 and the typical stress time history around a slamming event in Figure 11. Typical transitory whipping behavior can be observed. It should be noted that only modal superposition is used for this figure and no separation into quasi static and dynamic parts as only the longitudinal loading will contribute to the stress for this element. In the present case the local structural stress is presented but the exactly same procedure can be applied for the global quantities as well. In particular the vertical bending moment at a particular ship section can be evaluated. Calculation of the vertical bending moment is usually required in order to have quick idea of the influence of whipping on the overall ship design. In Figure 12 one typical simulation result for vertical bending moment is presented. This example is very illustrative and shows that, even if the whipping usually appears in sagging conditions, the vertical bending moment in hogging conditions is also affected due to relatively low structural damping.

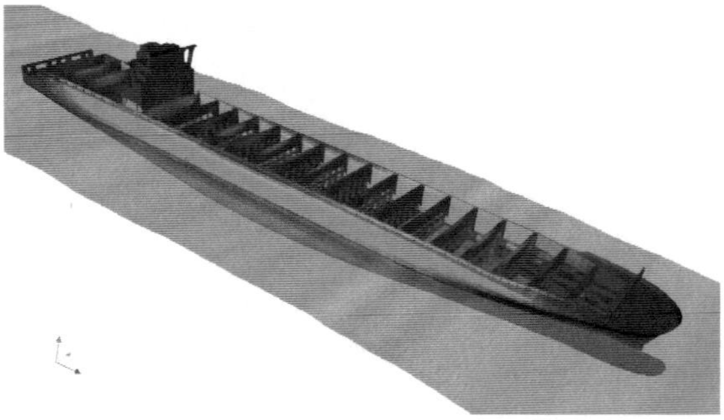

Figure 10: *Stress distribution in the ship structure at one particular time instant. The total stress includes the linear and non linear quasi static, springing and whipping contributions at the same time.*

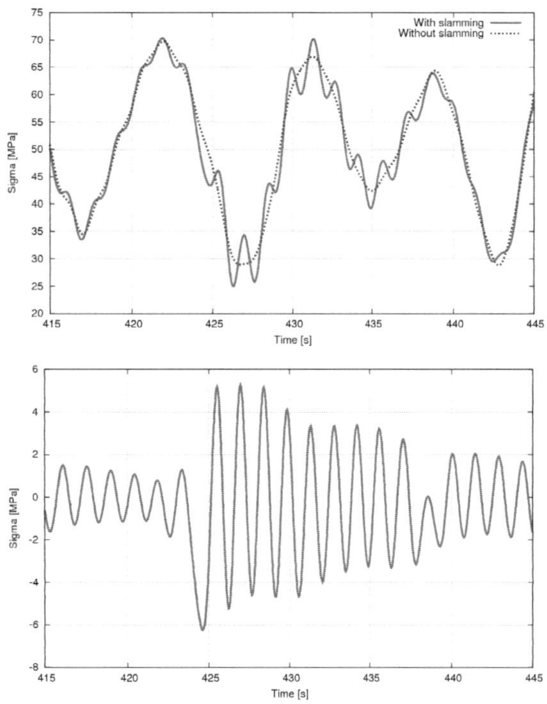

Figure 11: *Typical stress time history during the whipping event (top - total, bottom - filtered).*

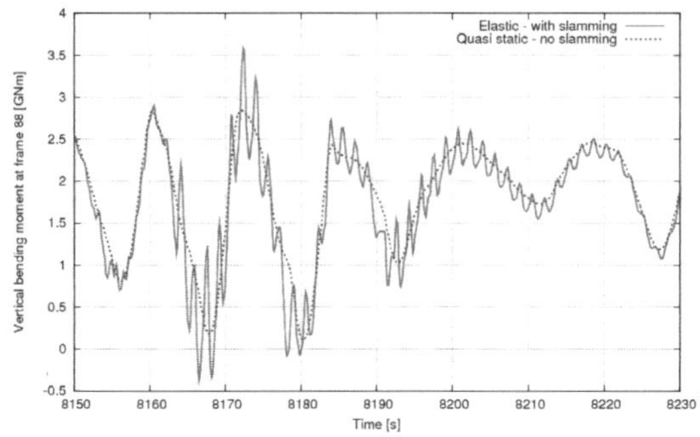

Figure 12: *Typical time history of the vertical bending moment during the whipping event.*

The stress in the structural detail is calculated for 2000 second in the same sea state to obtain a better quantification of the effect of whipping. The non-linear calculation is run with and without slamming using the same wave train. Figure 13 shows the probability distribution of the stress in the structure. The rainflow cycles and the fatigue damage calculated using a standard SN-curve is shown in figure 14. The slamming induced whipping clearly increase the amplitude of the stress and reduces the fatigue life of the structural detail.

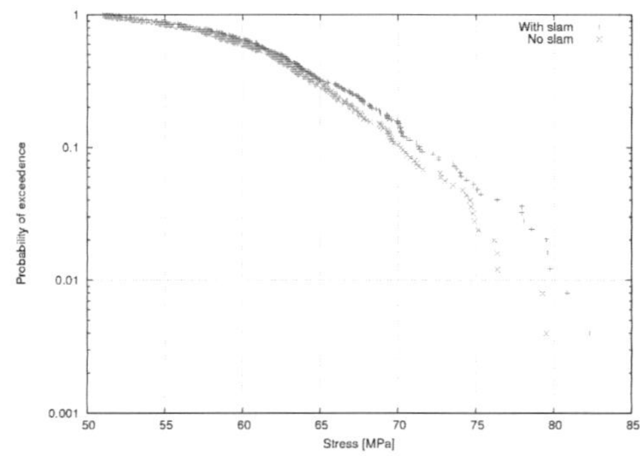

Figure 13: *Probability of exceedence of stress in the structural detail*

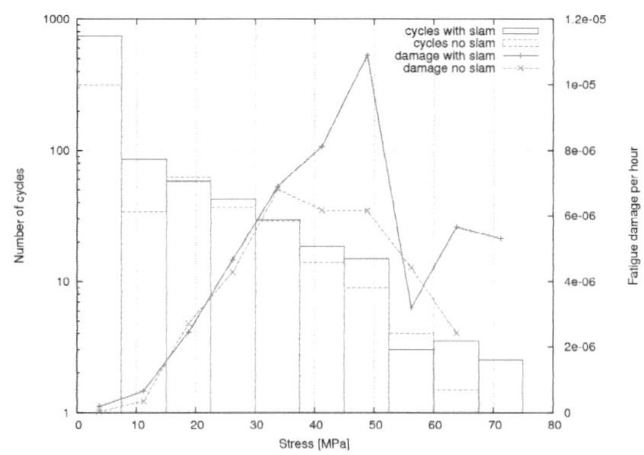

Figure 14: *Rainflow count and fatigue damage of the structural detail*

5. CONCLUSION

The method for hydroelastic analysis using the 3DFEM 3DBEM coupling procedure is presented. Both frequency and time domain approaches are discussed. The overall problem is extremely difficult and only approximate solution is possible up to now. In spite of numerous technical difficulties the present model seems to produce quite realistic results. More detailed validations and calibrations are necessary before employing the method in practice.

6. REFERENCES

[1] Aalberts P.J. & Nieuwenhuijs M., 2006. : "Full scale wave and whipping induced hull girder loads.", 4th. Int. Conf. on Hydroelasticity, Wuxi, China.

[2] Bishop, R.E.D. & Price, W.G., 1979. : "Hydroelasticity of ships", *Cambridge University Press.*

[3] Cummins W.E., 1962. : "The impulse response function and the ship motions.", *Schiffsctechnik*, Vol. 47, pp.101-109.

[4] Jensen J.J. & Wang Z., 1999. : "Wave induced hydroelastic response of high speed monohull displacement ship.", 2nd. Int. Conf. on Hydroelasticity, Fukuoka, Japan.

[5] Korobkin A.A., 2005. : "Analytical models of water impact.", *Euro Journal on Applied Mathematics.*

[6] Malenica Š., Molin B., Remy F. & Senjanović I., 2003. : "Hydroelastic response of a barge to impulsive and non impulsive wave loads.", 3rd. Int. Conf. on Hydroelasticity, Oxford, UK.

[7] Malenica Š. & Korobkin A.A., 2007. : "Some aspects of slamming simulations in seakeeping.", 9th. Int. Conf. on Numerical Ship Hydrodynamics, Ann Arbor, USA.

[8] Malenica Š., Stumpf E., Sireta F.X. & Chen X.B., 2008. : "`Consistent hydro-structure interface for evaluation of global structural responses in linear seakeeping.", 26th OMAE Conf.

[9] Newman, J.N., 1994. : "Wave effects on deformable bodies", Applied Ocean Research, Vol. 16, pp.47-59.

[10] Ogilvie T.F., 1964. : "Recent progress toward the understanding and prediction of ship motions.", 5th Symposium on Naval Hydrodynamics.

[11] Zhao R., Faltinsen O.M. & Aarsnes J.V., 1996. : "Water entry of arbitrary two dimensional section with and without flow separation.", 21st Symposium on Naval Hydrodynamics.

[12] Wu M.K. & Moan T., 1996. : "Linear and nonlinear hydroelastic analysis of high speed vessels.", J. of Ship Res., Vol.40/2, pp.149-163.

[13] Wu Y. & Price W.G., 1986. : "A general form of the interface boundary condition of fluid structure interaction and its applications.", CSSRC Rep. 86010.

[14] Xia J. & Wang Z., 1997. : "Time domain hydroelasticity theory of ships responding to waves.", J. of Ship Res., Vol.41/4, pp.286-300.

ULTIMATE STRENGTH ASSESSMENT OF ENGINE ROOM AREA OF THE CONTAINER VESSEL MSC NAPOLI – JANUARY 2007 INCIDENT

E. Steen, G. Storhaug, J. Kippenes, O. Sadó, H. Austefjord and **G. Dugstad,** Det Norske Veritas AS, DNV Maritime, Høvik, Norway

SUMMARY

The strength of the container vessel MSC Napoli is reassessed with focus on the hull girder wave loads and hull capacities as relevant for the sea states in the English Channel at the 18th of January 2007. Results from a set of hydrodynamic wave loads analyses and from non-linear structural FE analyses of the hull girder sections in way of the engine room bulkhead are documented. It is shown that for a severe hogging state, the vessel may collapse into a failure mode very similar to the one observed on the vessel

1. INTRODUCTION

1.1 GENERAL

In stormy weather conditions, the container vessel MSC Napoli experienced severe problems in the English Channel on the 18th of January 2007, with large areas of the hull skin suddenly cracking with severe water ingress and significant flooding of the engine room following. The vessel had to be abandon, while all the crew members were airlifted to safety by Royal Navy helicopters.

After the incident the vessel started drifting with a severe structural damage and large cracks extending across almost the whole cross-section in way of the engine room. Two tug vessels started immediately manoeuvring the vessel toward a harbour on the English side, however, it become soon clear that the vessel was seriously damaged and on the limit of being teared into two parts. Thus, in order to avoid a major environmental disaster, the vessel was beached in the Lyme Bay area.

1.2 DAMAGE CLASSIFICATION

From the available pictures published on the internet shortly after the incident, a localized plate folding damage was evident, extending vertically at both starboard and port sides just aft of the engine room bulkhead, Figure 1. Moreover, a significant change of the deck line slope at the engine room bulkhead was observed which implies that the bottom structure in this area must have shortened significantly in the longitudinal direction in order for this hull girder rotation (hinge) to develop. Such a bottom shortening is likely to be the result of severe buckling and collapse of highly compressed elements in this region.

It is well known that structural hull elements compressed well into their post-collapse region will develop a very localized failure pattern. Moreover, such localized buckling and plate folding mechanism will imply large plastic strains exceeding the material fracture limit. Thus large cracks may develop almost instaneously with water ingress and engine room flooding as the result.

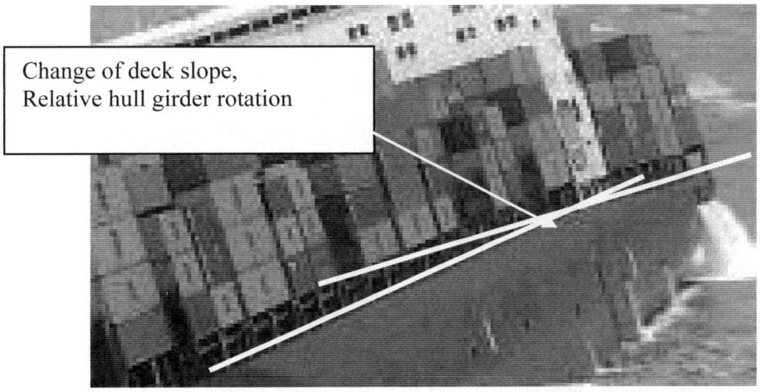

Figure 1: Structural damage in ship side. Pictures taken 18th Jan 2007 after ship has been abandoned. (© Cargo Law)

Pictures taken in dry dock after the ship was split in two parts, also clearly indicated that the collapse was very localized in the longitudinal direction while extending across the whole ship breadth in the transverse direction, see Figure 2 (dry dock in Belfast),. Only a narrow band of the order of a frame distance suffered from the severe deformations and cracking and the band extended from around (# 81) in the bottom area to the engine room bulkhead (#88) high up in the shipside.

1.3 SHIP PARTICULARS AND SHIP CONDITION DATA

The MSC Napoli was built to BV Class in 1991. At that time the vessel with its 4400 TEU container capacity and main dimensions as specified in Table 1, was among the largest container vessels setting new standards. It was transferred to DNV Class in 2002.

The hydro and structural models documented herein are based on drawings as provided by the builder. Moreover, ships in service data documents a well maintained ship in general and for the engine room in particular no corrosion and reduced plate thicknesses have been reported. This implies that the "as built" dimensions were used for the FE models.

With respect to the loading condition and mass distribution at the time of the incident, the present analyses are based on the available loading computer data as provided by the MAIB-UK, [1].

Table 1 Ship main particulars

Length over all, L_{OA}	275.6	[m]
Length between perpendiculars, L_{PP}	261.4	[m]
Scantling length	258.31	[m]
Breadth moulded, B	37.1	[m]
Depth moulded, D	21. 5	[m]
Block coefficient, C_B	0.609	
Top of hatch coaming above BL	23.3	[m]
Draught, T (scantlings)	13.5	[m]
Draught, T (design)	12.4	[m]

Figure 2: Forepart of vessel in drydock, Belfast.

2. METHOD OF ANALYSES

All the damage observations, and information on the vessel speed and heading in the severe waves, indicated that the ship had broken its back and collapsed due to a global overload in the forward part of the engine room. In order to determine whether this failure scenario was a likely one, the global hull girder strength margins of the engine room area had to be evaluated. This involved assessment of the extreme load and comparisons against the structural capacity of the hull sections in the collapsed area. It would also be of crucial importance to show that a global overload could lead to a localized collapse mode with excessive deformations, strains and cracks across the whole sections as the pictures showed.

It is difficult to assess the acting wave load and the corresponding hull capacity considering all the uncertainties and variations in input data. In particular this is true in relation to variations in environmental and response data like wave heights, wave periods, hull whipping effects etc. Therefore, the present approach is to assess ranges of the global loads and hull capacities within which the MSC Napoli most likely has been operated around the time of the incident.

A major uncertainty with respect to the hull girder load assessment is the possibility of a 2-noded hull vibration response (whipping) adding on top of the normal hull girder wave loads. However, due to the complexity and lack of validated methods for assessing such responses, no explicit assessment of the whipping effect for MSC Napoli is given herein. A discussion of whipping effects is given the summary Section 6.2

For the load assessment the hydrodynamic tool WASIM [2] was used and for the structural capacity assessment a non-linear FE analyses using ABAQUS [3] was used.

3. LOAD ANALYSES

The wave conditions around the time of the incident have been reported with quite a large variation from different sources such as the vessel, Ocean Systems Incorporated (OSI), Argoss and the UK Met office. A summary of the wave conditions at the time 10:00 – 12.00 are given in Table 2. In particular it is noted that the wave height observations can vary from 7 to 9 m. The up-crossing period Tz is, however, more uniquely identified to be around 10 to 11 seconds.

Table 2: Wave and wind data from the different references at 18th of January 2007, time 10:00- 12.00

	Vessel	OSI	Argoss	UK Met Office
H_S [m]	-	9.0	6.9	7.3
T_Z [s] Up-crossing		10	9.6	11
Wind speed [kn]	60 (B_n 10-11)	42		39
Wind/wave direction	SSW, 203°	254/259°	258°	WSW 247.5°
Vessel Speed [kn]	11			
Vessel course [°]	240			

3.1 WAVE LOAD SENSITIVITY STUDY

The wave load sensitivity study was made in order to quantify the main response characteristics of the MSC Napoli operating under environmental conditions specific for the English Channel at the time of the incident and assess the importance of parameters such as zero up-crossing period, water depth, wave spreading, wave energy concentration and steepness, non-linear effects etc.

The most important findings were:

- A wave period of approximately 10 seconds gives the highest vertical bending moment (VBM) at the engine room bulkhead at frame #88. This is very close to the actual wave periods reported for MSC Napoli around the incident.
- The water depth is a moderately important parameter. VBM is of the order 10% higher at 70 m water depth than at infinite depth (#88). A water depth of around 70 m is representative for the location of the MSC Napoli at the time of incident.
- Assuming an almost uni-directional (long-crested) wave train hitting the vessel gives of the order 10 % higher VBM (#88) than the more normal (short crested) sea. It is not unlikely that a rather uni-directional wave train hit the MSC Napoli.
- The energy concentration in the waves, i.e. the concentrated "narrow band" sea ($\gamma = 5$, under developed sea) versus more "broad band" wave ($\gamma = 1$, fully developed sea) (γ in JONSWAP wave spectra) is found to increase the wave moment of the order of 15%. From the reported conditions of opposite wave and current directions, and considering the shallow water effects, it is not unlikely that a rather steep "regular" wave train hit MSC Napoli.
- A ship heading angle against the main wave direction varying between 0 and 15 degree has negligible effect on the VBM (#88). The MSC Napoli had most likely a close to head sea direction around the time of incident.
- A variation of the average speed of 5 knots up-or-down from 11 knots implies approximately a wave load change of the order 10%. The higher speed the higher wave load. The MSC Napoli had most likely an average speed around 11 knots at the time of incident as informed by MAIB.
- In order for realistic wave loads to be assessed on slender container ship hulls, non-linear effects from the ship moving in and out of large waves have to be accounted for. The non-linear effect for a hogging wave condition, i.e. a moment in time where the vessel is positioned with a wave crest around mid-ship area (see Figure 4), reduce the wave loads relative to standard linear methods. For the MSC Napoli this reduction is assessed to be of the order 20-25% for the environmental conditions as specific for the English Channel at the time of the incident.
- The main effect of longitudinal container mass variations is the change in the still-water moment and not the wave bending moment. This implies that by varying the container mass distributions up or down slightly this may change the still-water moments significantly, but it will have a smaller effect on the wave loads.

3.2 HULL SECTIONS LOADS – ENGINE ROOM

The loading condition used herein is based on the on-board loading computer data at departure River Schelde Antwerp, [1]. The still-water loads for this loading condition are summarized in Table 3 below, giving both the vertical hogging bending moment (VBM_{SW}) and corresponding shear force (VSF_{SW}) at the engine room bulkhead position (#88).

Based on a thorough examination of the existing data, judgement of possible sources of uncertainties and using in-house experience, two sea states were defined in order to cover the total range of important parameters:

1) Case 1 "high""
2) Case 2 "low"

The complete set of environmental parameters used for each of the two sea states are summarized in Table 4 below.

Case 1 "high" is considered to be a maximum environmental condition giving severe global hull loads. In particular, assuming a very narrow wave spreading function ($\cos^8$) means that almost all wave energy is localized into an uni-directional wave train meeting the vessel in head sea. Moreover, a $\gamma = 5$ in the JONSWAP wave spectrum reflects an energy rich, "regular" and steep wave profile. A significant wave height of 9 m is high, but not unrealistic for a short duration at the time of the incident.

Case 2 "low" represents a less severe set of environmental conditions. By setting the JONSWAP wave spectrum parameter to $\gamma = 1$ (PM), the waves are fully developed and broad banded.. A significant wave height of 7 m is very likely and as reported by different met offices. Moreover, a vessel speed of 11 knots at a 15 degrees angle off head sea is also within the variations as reported.

The maximum loads at the engine room bulkhead are illustrated as probability distributions (Rayleigh based extreme values) in Figure 3 for the two sea states analysed, Case 1 "high" and Case 2 "low". It is to be noted that the probability distributions in Figure 3 have been scaled down by 27% and 20 % (Case1, Case 2) according to the non-linear hogging wave effect relative to linear methods. This is done in order to reflect the real life vessel response as closely as possible.

In Table 5 the total likely load range is summarized for overview. The total moment range defined includes the likely span of the extreme short term wave loads assuming for the Case 1 "high" Case 2 " low". A 10% probability of exceedence value has been used for the upper limit wave load for the Case 1 distribution. This limit corresponds to a total moment of 4300 MNm. A similar definition of 90% probability is used for the Case 2 "low" giving a lower bound value of 3400 MNm.

Global shear loads are also assessed corresponding to the VBM ranges. The shear loads are typical in the range 30 - 40 MN for a VBM range of 3400 - 4300 MNm. The shear force at the time of the maximum hogging bending moment is also very close to being the maximum shear force (# 88).

Table 3 Still-water loads (NAPA, [4]) at engine room bulkhead (# 88)

Loading condition	Still-water moment (MNm) Hogging VBM_{SW}	Still-water shear force (MN) VSF_{SW}
Departure River Schelde, [1]	**2243**	**20**

Table 4 Definition of sea states; Case 1 "high" and Case 2 "low"

Parameters	**Case 1 "high"**	**Case 2 "low"**
Significant wave height ($H_{1/3}$)	9	7
Wave up-crossing period (T_z) (sec)	11	10
Vessel heading (degrees from aft)	180 (head sea)	165
Vessel speed (knots)	11	11
Wave spreading ($\cos^n$)	8	6
Wave spectra (JONSWAP, γ)	5	1
Water depth (m)	54	73

Table 5 Total load range at #88; still-water + wave load (no whipping)

Load range	Lowest load Hogging VBM (MNm)	Highest load Hogging VBM(MNm)
Vertical bending moment at # 88	**3400**	**4300**

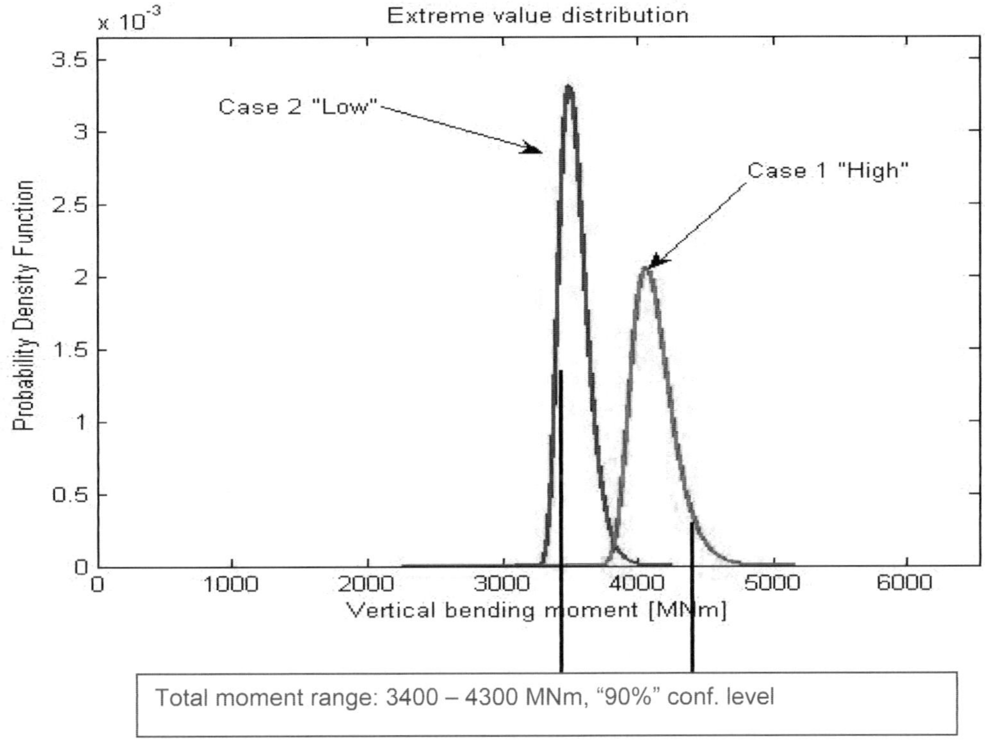

Figure 3: Total load ranges: still-water + wave load (VBM) at engine room bulkhead (#88). Non-linear wave effects are included in all curves while whipping is excluded.

4. LOAD TRANSFER TO FE MODEL

The wave loads for the vessel as presented herein are representative for the ship "frozen" in the waves as illustrated in Figure 4. In such a moment in time, typically with a wave crest approximately mid-ship, the hull girder will set up a significant hogging condition, giving high longitudinal compressive stresses in the bottom and bilge area.

For the structural FE analyses the hydrodynamic wave loads as representative for such a maximum hogging state, is automatically transferred to the structural FE model and it is further scaled until collapse is identified, see Section 5.

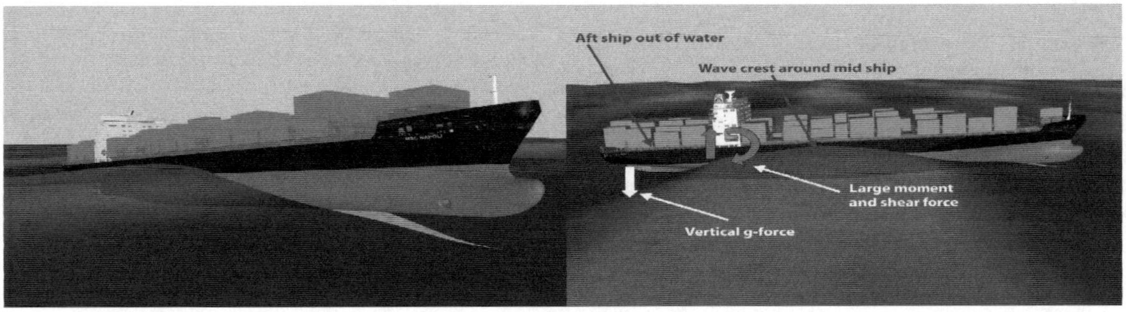

Figure 4: Ship "frozen" in waves in a moment time giving high hull girder hogging loads

5. NON-LINEAR FE MODEL

5.1 MODEL DESCRIPTION

Linear stress analysis methods are not capable of identifying buckling and material yielding in the structure, effects which are crucial for hull capacity and progressive failure mechanism assessments. Therefore a non-linear stress and capacity model was made and analysed using the FE code ABAQUS [3]. The total FE model is shown in Figure 5.

The global FE model used in this analysis consists of two super-elements:

- A local, very fine mesh non-linear super-element representing the area between frame 78 and frame 92. Here both non-linearity in geometry and material behaviour is included.
- A global super-element representing the rest of the vessel. This is constrained to behave linearly elastic and it is referred to as the linear super-element.

The non-linear super-element is modelled using 4-node plate element (ABAQUS, S4) for both plating, stiffeners, girder webs etc. The meshing is very fine with typically.

- 5 to 6 elements between stiffeners
- 3 elements across a stiffener height
- quadric element shape is chosen for most of the structure
- element size is of the order 150 mm
- openings, cut-outs and local secondary stiffeners are modelled in detail

This mesh density is capable of describing very localized buckling and collapse behaviour of plating, stiffeners and girder structures.

The model size is

- non-linear super element; 1.06 mill degrees of freedom (dof)
- total model; 1.22 mill degrees of freedom (dof), linear + non-linear super-element

Figure 6 and Figure 7 show some close up views of bilge and double bottom parts respectively.

The assumed pattern of geometrical imperfections is important for non-linear capacity analyses. In particular the imperfections have to be included such that the structure is not prevented from buckling and collapsing into the most critical and detrimental modes.

The imperfection pattern used in the present analyses is shown Figure 8. The pattern is strongly scaled up in this figure (200 times) for illustration purposes. In the non-linear analyses the imperfections are set to be small in order for the structure to reflect a strength and ultimate capacity limit consistent with normal fabrication and tolerance levels.

As seen the pattern is rather harmonic, covering the whole lower part of the model, i.e. the outer bottom, tank top plating and bilge area are all given an evenly distributed slight imperfection pattern.

In order to account for non-linear material behaviour, i.e. plastic flow and strain hardening effects, a bi-linear material model has been used. Moreover, in order to reflect uncertainties and possible yield stress variations in the steel, analyses are carried out for a range of three different sets of material properties, see Table 6.

- Lower 5% quantile yield stress (characteristic yield strength used in Rules)
- Expected, mean value yield stress
- Upper 5% quantile yield stress

The plastic tangent modulus is set to 1000 N/mm^2 and the material characteristics are all in line with properties as defined by standard Class rules. The material data and variations used in the analyses are shown in Table 6. This selected range of material characteristics gives the corresponding range of ultimate capacity as found using the non-linear FE model.

Most of the critical and important load bearing structures within the engine room has a material grade of high strength steel 32. Stress-strain curve for this material is shown in Figure 9, illustrating the range of possible variations according to the Table 6 data.

The loads applied to the ABAQUS super-elements are transferred from the hydrodynamic WASIM analyses and the loading is based on a split into two load categories

i) still water loading
ii) loads on the vessel in waves, Figure 4

Hull sea pressures and mass forces representative for both load categories are included.

The non-linear analyses are performed in steps as follows (load-history):

i) First the still-water load set is applied
ii) Next the wave load distribution is added and scaled further until the collapse limit is found.

This way of splitting the loads into two sets will take care of the differences in sea pressures and mass force distributions along the ship length for the still-water and wave load condition respectively.

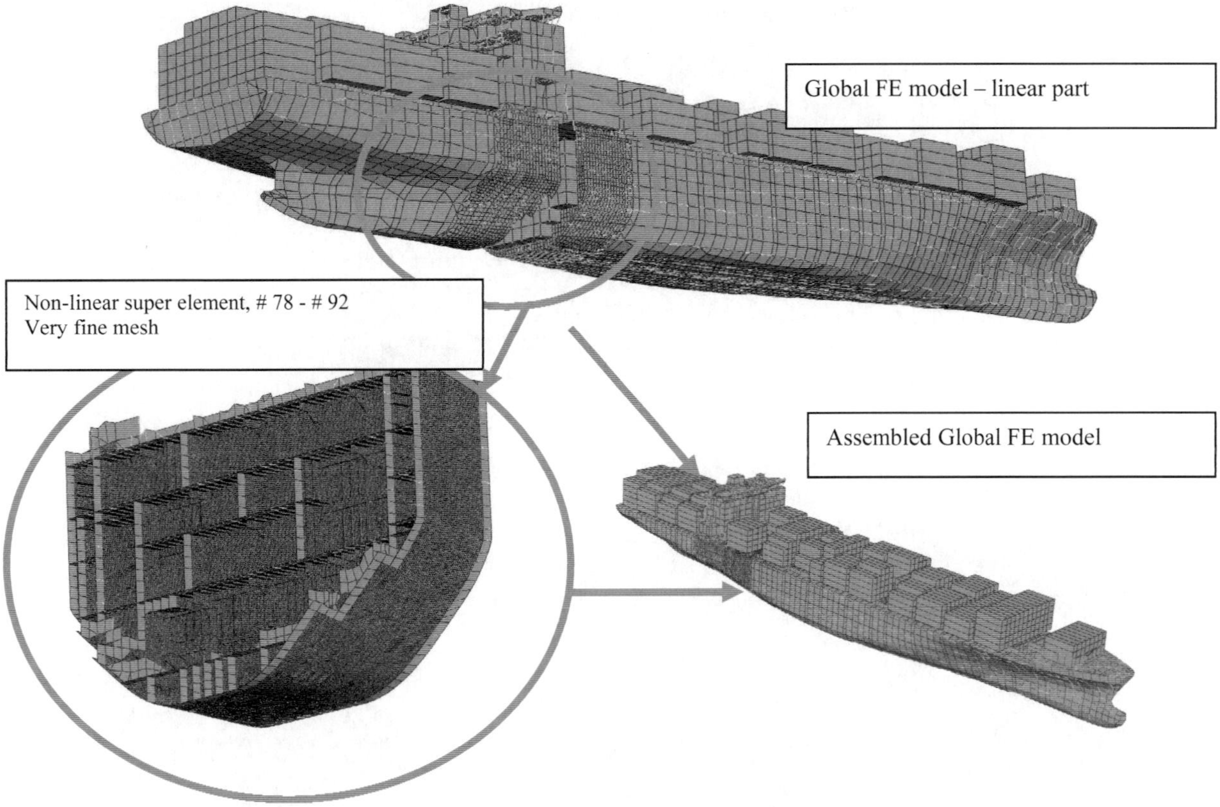

Figure 5 Non-linear super-element with very fine mesh inserted into Global FE model

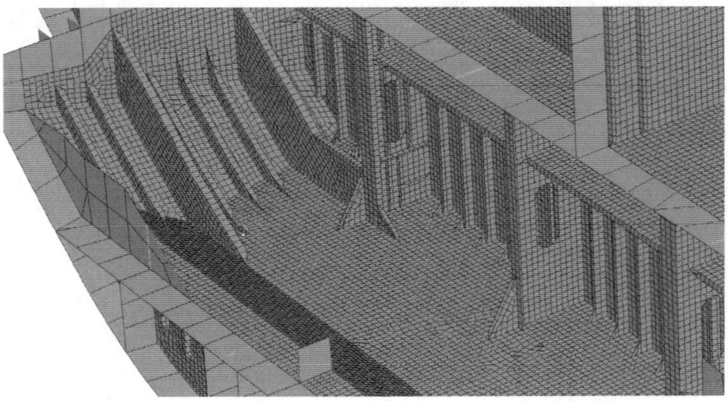

Figure 6: Close up – non-linear super-element, engine room bilge area

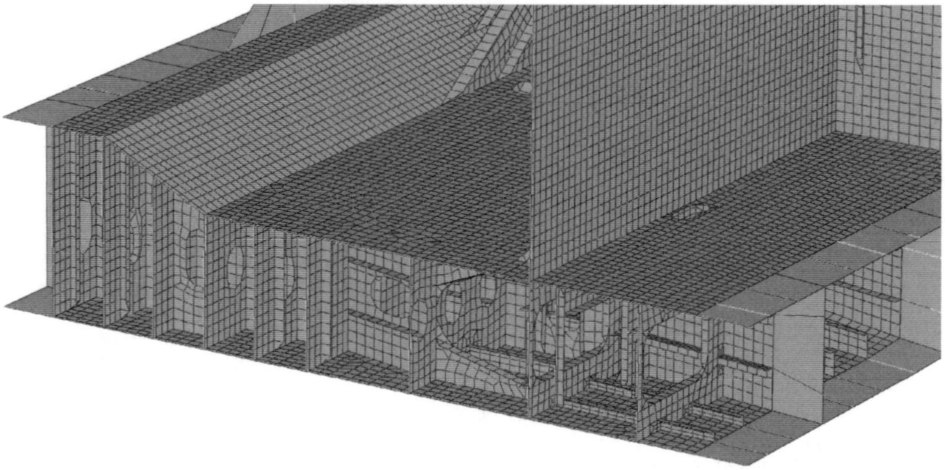

Figure 7: Close up – nonlinear super-element, double bottom area

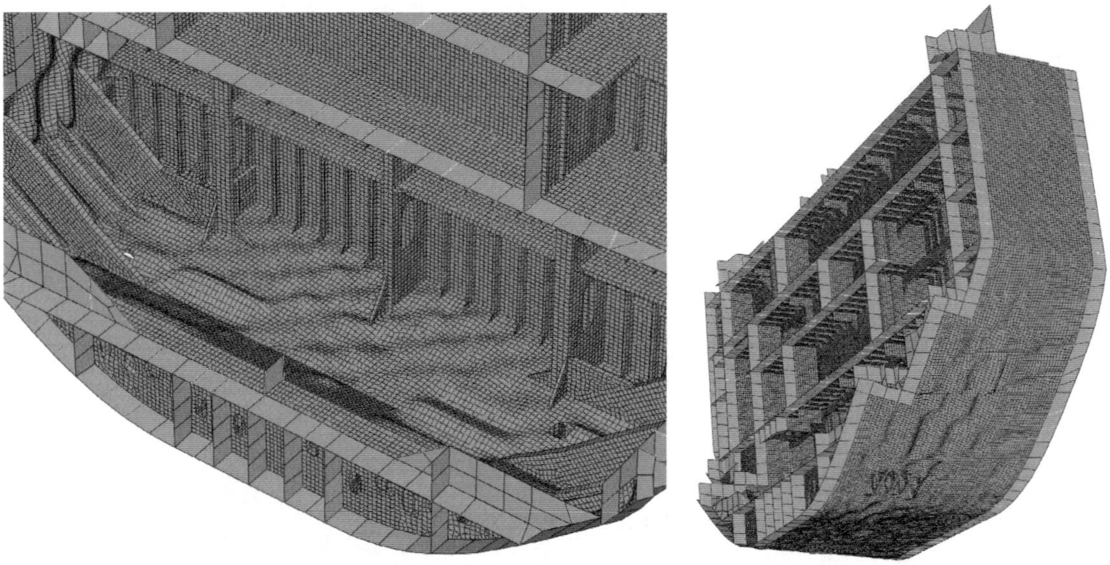

Figure 8: Geometrical imperfections in the non-linear super-element. Imperfection size in figure is scale up 200 times for illustration purposes. Imperfection sizes in the analyses are small reflecting strength consistent with normal fabrication standards

Table 6 Material data and variations in properties

Material	Young's modulus	Poisson ratio	Lower value yield stress	Mean value, yield stress	Upper value, yield stress
Normal Steel	206000	0.3	235	271	307
High strength -32	206000	0.3	315	349	383
High strength -36	206000	0.3	355	394	433

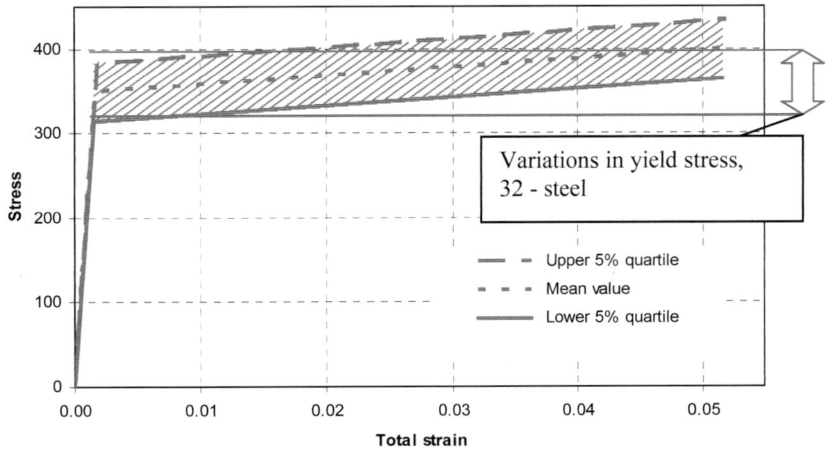

Figure 9: Bi-linear material stress-strain curves for 32 steel with assumed variation range

5.2 HULL CAPACITY RESULTS

The non-linear relation between vertical moment (VBM) at engine room bulkhead (#88) and the vertical deflection at ship bow is shown in Figure 10. The change in slope of the load-displacement curve is due to different load type application, i.e. the first almost linear curve correspond to the still-water loading scaled from zero to the actual level, while the next linear curve part correspond to the scaled wave load distribution up to the final collapse.

For practical purposes it is the maximum point along the moment-deflection curve that is the most interesting parameter. This max point is the capacity of the structure, i.e. here the ultimate collapse moment of the hull girder (Mu), and it is tabulated separately, see Table 7.

Three separate non-linear analyses are run, each corresponding to a selected set of material characteristics according to Table 6, with results as shown in Figure 10.

The capacity range (Mu) is seen to span from 4200 MNm as the lowest level to 4950 MNm as the highest level for engine room bulkhead (#88).

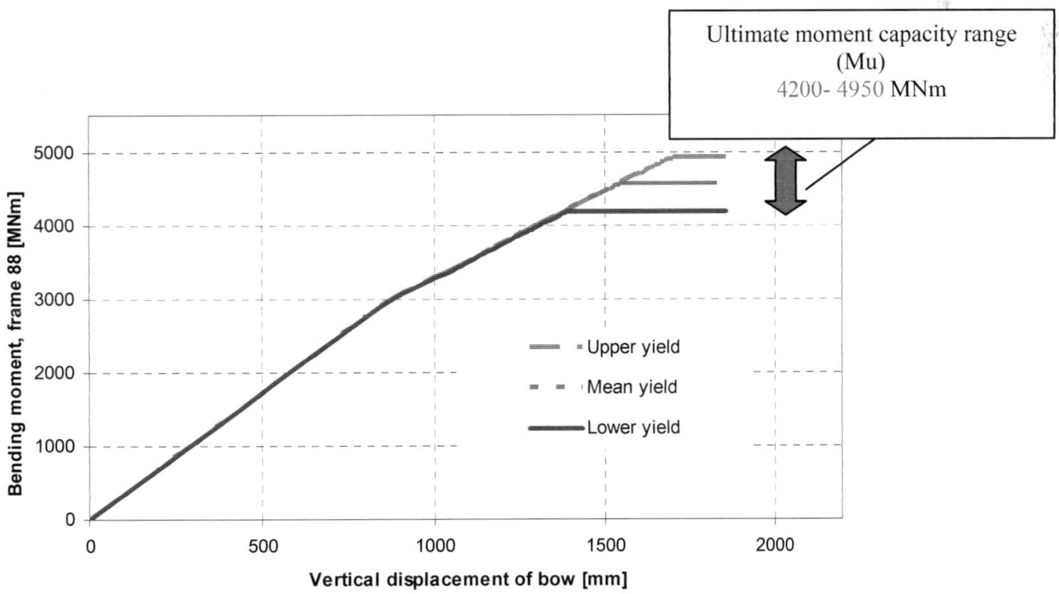

Figure 10: Non-linear moment- displacement relation (Moment at # 88 against vertical bow deflection). Ultimate capacity is the max load indicated for the lower, mean and upper material properties according to Table 6.

Table 7 Results; vertical bending moment (VBM) at # 88 at hull girder collapse

Yield stress used in ABAQUS model acc. to Table 6	**Vertical moment at #88 at collapse (Mu, MNm)**
Lower yield stress	4200
Mean yield stress	4560
Upper yield stress	4950

5.3 HULL GIRDER FAILURE MECHANISM

Associated with the total collapse of the hull girder is the development of a local progressive failure mechanism. The characteristics is that it starts with "elastic" buckling in the hull skin plating in the bilge area in way of # 81 to # 88. From here the buckling gradually develops and progress into the bottom, double bottom and up into the ship side. It finally ends up as a localized collapse pattern in a narrow band across the whole hull section, from bottom to ship side, see Figure 11.

It is also characteristic that local elastic buckling leads to significant stress-redistributions with stress accumulation into stiff areas such as longitudinally girders, stringers, sea-chest area etc. ("hard corners"). The progressive development of the failure mechanism is illustrated in Figure 12, showing buckling deformations and membrane stresses at four different applied load levels (% in relation to ultimate moment Mu); external and internal view.

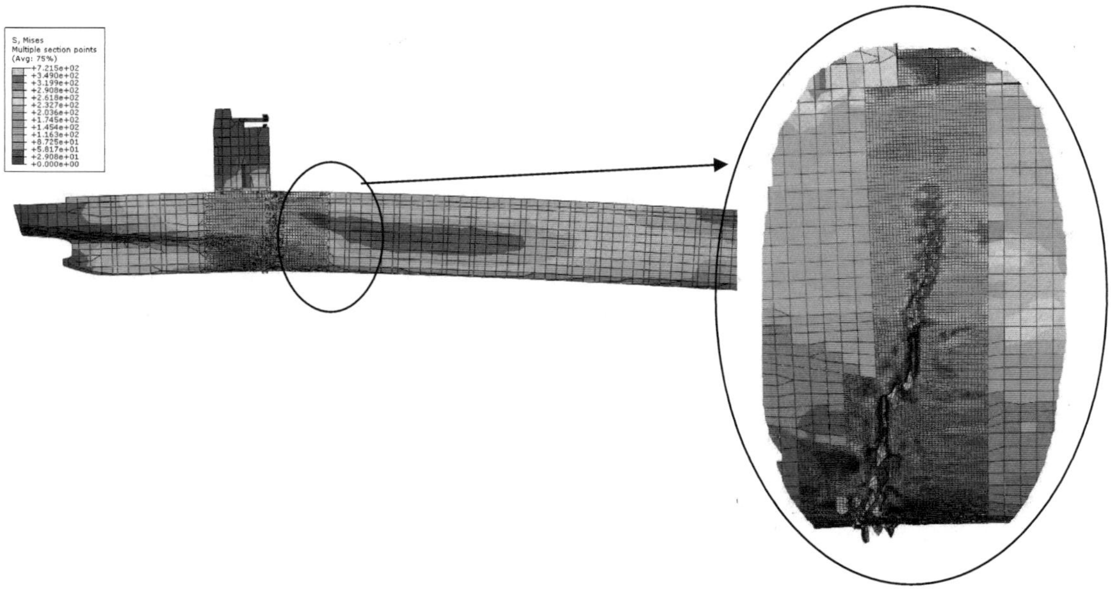

Figure 11: Final failure mechanism as predicted by ABAQUS. Deflections scaled up 5 times for illustration purposes.

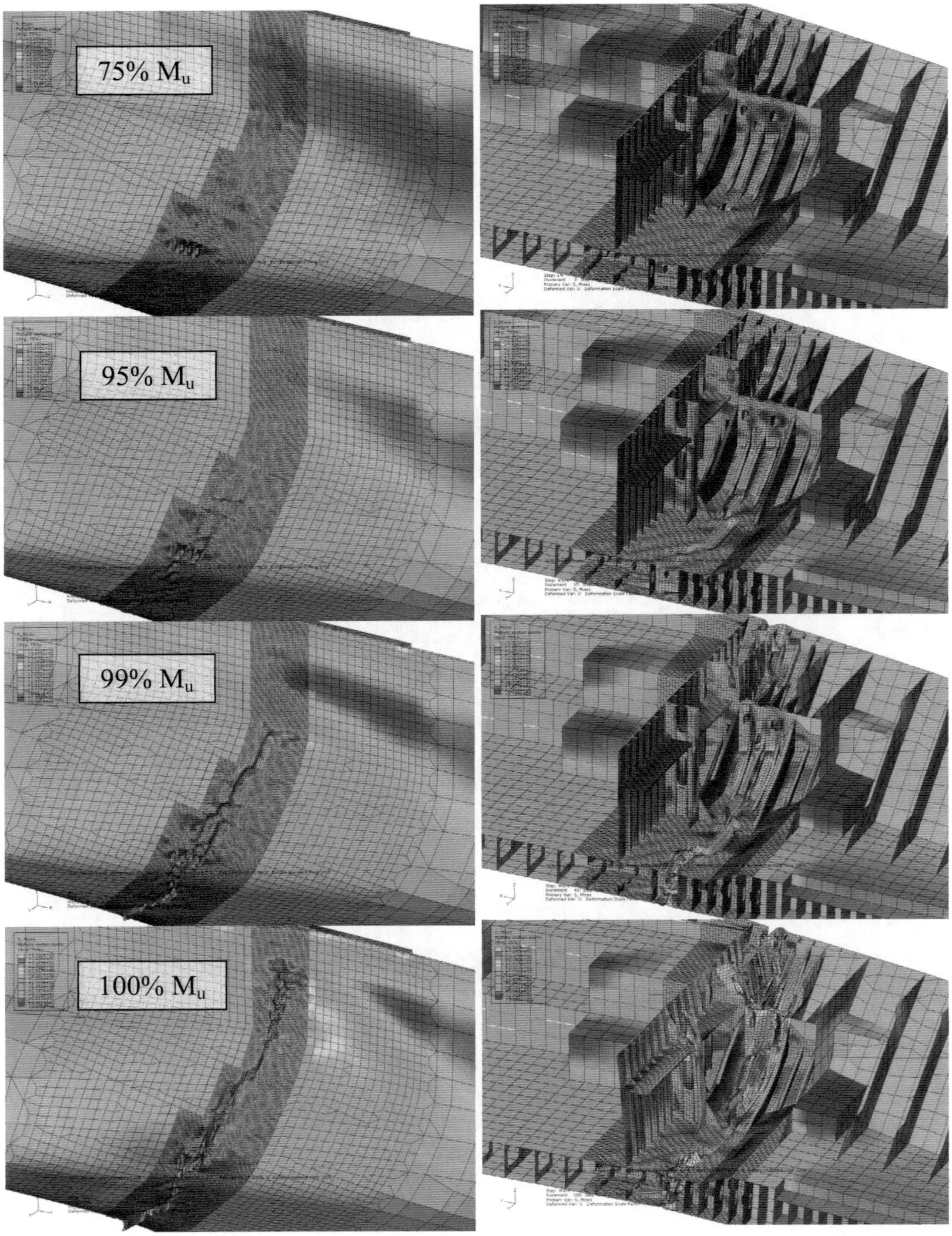

Figure 12: Four stages of progressive collapse, outer (left) and inner structure (right) Deflections scaled up 5 times for illustration purposes.

6. CONCLUSIONS

6.1 GENERAL

The computer analyses demonstrate that a very high vertical hull girder moment, close to or exceeding the corresponding capacity limit, may lead to a type of structural failure between #81 to #88, as is observed on MSC Napoli in open sea. This is evident from a comparison between a photo taken of the vessel's starboard side and a computer simulation image of the same area, see Figure 13. This failure mode is further supported by diving surveys while beached and also later from surveys of the forward part in dry dock.

It is also documented that the overlap in load and capacity is very narrow indicating that such a failure scenario has a very low probability of occurrence for the assumed severe sea states, see Figure 14.

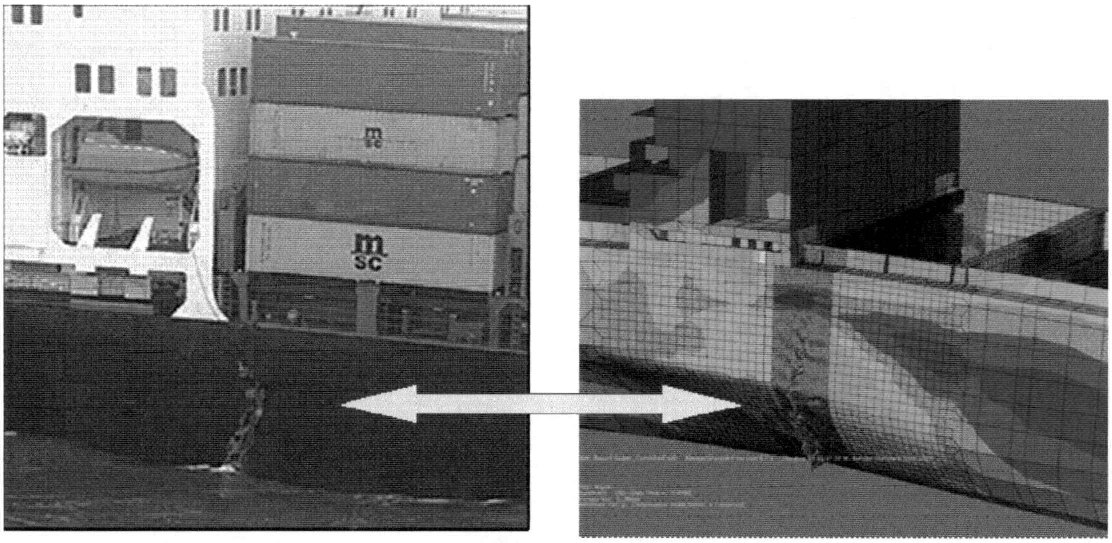

Figure 13: Comparison of real life failure mode with computer simulations (ABAQUS FE model) (Copyright Cargo Law)

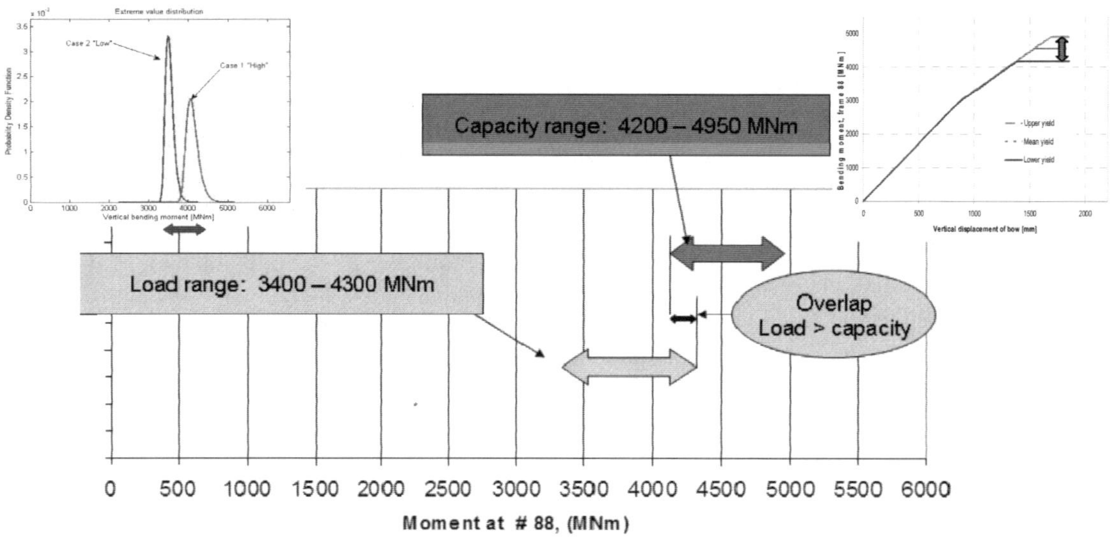

Figure 14: Calculated load and capacity ranges with overlap comparison (no whipping effects). Vertical bending moments (VBM) at engine room bulkhead #88

6.2 DISCUSSIONS

The non-linear FE computer analyses (ABAQUS) document a localized plate buckling mechanism developing between # 81 and # 88 across the whole ship section, i.e. in the bottom structure as well as far up into the shipside. Such large plate bending and folding deformations lead to large plastic strains also in a tension mode. This may lead to sudden ductile plate fracture and cracking with subsequent water ingress and compartment flooding as a consequence.

The exact nature and sequence of the structural failure and plate fracture taking place is difficult to judge, but it is a possibility that not only one severe wave but several subsequent high waves hit the MSC Napoli. This typically will lead to a loading, unloading and reloading again as the ship move through the waves. This again gives locally a plastic straining of the hull plating at one instant, and then in the next moment having a strain relief followed by a new large plastic straining. Such hysteresis behaviour in the large plastic strain region may lead to plate fracture and locally open up large cracks. However, due to the large and localized plate bending deformations initiated from buckling, only one severe wave may be enough to strain the hull plating beyond its fracture limit and thereby generate rupture and large cracks.

A major uncertainty is linked to the wave response characteristics, i.e. the wave response could have had a whipping contribution on top of the calculated wave load, the global 2-noded vibration effect being the result of the vessel slamming into a severe wave train. Measurements on similar vessels have indicated that a whipping contribution in the range of 10-50 % is not unlikely under certain conditions. However, the whipping effect for the MSC Napoli is not assessed explicitly herein as the available analyses tools are not considered fully reliable. Accordingly, whipping effects may have been a triggering factor, but they are of an unknown magnitude and may not have been decisive for the incident.

7. ACKNOWLEDGEMENTS

The analyses carried out and results found are due to the effort of many individuals in DNV. The authors would in particular like to mentioned Michal Moczulski, Thor Hysing, Tom K. Østvold, Eirik Byklum, Torbjørn Lindemark, Ivar Håberg, Håkon Skaret, Gjermund Skailand, Frode Kamsvåg and Rossen K. Panev as key personnel. Also the continuous support from Olav Nortun, technical director in DNV Maritime is highly appreciated.

8. REFERENCES

1. MV MSC Napoli, W05 Spl. TD Voy. 27a - Departure from Antwerp to Sines, Date 16/01/07, MSC. (Onboard Loading Computer) (MAIB 98%)

2. DNV SESAM User Manuals 2007 (FE and WASIM)

3. ABAQUS User Manual 2007, version 6.7-1

4. NAPA User Manual 2006, version 2006.2

9. AUTHORS' BIOGRAPHIES

Eivind Steen holds the current position of Head of Section for Advanced Structural Analyses in the Maritime Technical Consultancy department in DNV. He has 25 years of experience within DNV related to R&D, Rule development and Marine Consultancy within the field of structural analyses and ultimate strength in particular. He has a MSc degree from Norwegian University of Science and Technology (NTNU)/Trondheim in Naval Architecture and a Dr.Scient degree in Structural Mechanics from the University of Oslo.

Gaute Storhaug is a Principle Engineer at the Section for Ship Structures in DNV. His daily work involves trouble shooting, research, consultancy services, and responsibility for full scale measurements and the voluntary class notation HMON (Hull MONitoring). His previous experience includes general hydrodynamic and structural analyses, full scale measurements, model tests and responsibility for fatigue calculations within the Condition Assessment Programme (CAP). Both the MSc and the PhD thesis dealt with the topic of Hydroelasticity. The former was written at MIT and the latter defended at the Centre of Excellence for Ships and Ocean Structures (CeSOS) at the Norwegian University of Science and Technology (NTNU).

Jon Kippenes holds the current position of Project Engineer in the Section for Advanced Structural Analyses in the Maritime Technical Consultancy department in DNV. He has 3 years of experience within DNV related to R&D, Rule development and Consultancy within the field of structural analyses, buckling, fatigue, ultimate strength and non-linear FE analyses in particular. He has a MSc degree in Structural Mechanics from the University of Oslo.

Octavi Sadó holds the current position of Senior Project Engineer in the Section for Hydrodynamics in the Maritime Technical Consultancy department in DNV. He has 5 years of experience within DNV working within the Offshore and Maritime Consultancy marked in Houston, US and at Hovik, Norway. His main field of projects are related to sea keeping, ship response and hydrodynamics in general as well a structural FE analyses. He has a MSc degree in Naval Architecture from the University of New Orleans, US.

Håvard N. Austefjord holds the current position of Project Engineer at the Section for Hydrodynamics in the Maritime Technical Consultancy department in DNV. He has 4 years of experience within DNV working within the Offshore and Maritime Consultancy marked in Hovik, Norway. His main field of projects are related to sea keeping, ship response and hydrodynamics in general. He has a MSc degree in Naval Architecture from the Norwegian University of Science and Technology (NTNU)/Trondheim.

Geir Dugstad holds the current position of Head of Department Class Services & Newbuildings in DNV Maritime. He has 17 years of experience within DNV working as trainee, approval engineer, consultant and surveyor. For the last ten years he has had management positions both related to Ships in Service and Newbuildings. He has a MSc degree in Naval Architecture from the Norwegian University of Science and Technology (NTNU)/Trondheim.

A COMPREHENSIVE AND PRACTICAL STRENGTH ASSESSMENT METHODOLOGY FOR CONTAINER SHIPS TAKING INTO ACCOUNT NON LINEAR AND HYDRO-ELASTIC LOADING

G. de Jong and **M. Huther**, Bureau Veritas, France

SUMMARY

The drive for economy of scale effects has fuelled the strong growth in size and capacity of new container ships. The combination of relatively low hull girder stiffness and high speed (decreased natural frequency of hull girder vibrations and increased wave encounter frequency) may cause phenomena which are of second order for average size vessels to become of high importance for these large vessels. This is particularly the case for vibratory structural response and associated fatigue damage caused by whipping and springing.

Making use of experience feedback with calculations and full scale measurements this paper presents a methodology to estimate the consequences of these phenomena during the design stage of the vessel. Practical application of the methodology has been made possible by further developing the Bureau Veritas in-house hydrodynamic suite HydroSTAR to take into account hydro-elasticity and non linear time domain simulations. The software has been validated against model tests, while a full-scale measurement campaign is ongoing.

Application of the methodology to ultra large container ships confirms that stress levels are indeed increasing. As a consequence, a significant increase both hull girder loads and fatigue damage accumulation is found, which can no longer be ignored when analysing the ship structure. When considering future designs for even larger container ships of 400 m in length and beyond, the effects of whipping and springing will become even more important.

1. INTRODUCTION

Over the past five years the maximum size of Post-Panamax container ships has increased dramatically to achieve economy of scale effects. Together with the rapid growth in trade volume and mileage between the emerging production economies in the Far East and the established economies (mainly Europe), as well as a firm belief that market share is the key for future success, an unprecedented surge in new orders for ultra large container ships (ULCS, 10 000+ TEU) has been created. According to data published by Clarkson, the world orderbook for this segment stood at 187 vessels at the end of the first quarter of 2008. The majority of these giant boxships are expected to be delivered in 2009 (17%), 2010 (29%) and 2011 (43%).

Although the turmoil in the financial markets and uncertainty regarding economic growth (possible slowdown in China, fear for recession in the US, high oil price) have drastically reduced the amount of new orders in the first quarter of 2008, there is no reason to believe that the drive for increasing the vessel size will come to a halt at the present maximum of 13 300 TEU for a ship overall length of about 370 m[1]. Therefore, realizing that a new upheave in container ship ordering would probably produce designs of well over 400 m in length, it is of key importance to reflect on the technical challenges associated with scale enlargement of ultra large boxships.

[1] The only exemption is the E-class series of Maersk, with an estimated capacity of 15 200 TEU and a length of 397 m according to data from AXS Alphaliner.

In fact, if the increase in size is achieved through extrapolation of existing ship dimensions and application of environmental loads estimated from dimensional analysis, it may be expected that phenomena which are of second order for average size vessels can become important for large vessels. This is particularly the case for vibratory structural response and associated fatigue.

Experience feedback shows that two dynamic phenomena affecting the hull girder require specific attention when considering large container ships: whipping (transitory hull girder vibrations caused by hydrodynamic impact at the bow) and springing (excitation of the first natural modes of vibration of the hull girder). Indeed, increasing the size of ships with large deck openings will decrease the natural frequencies of hull girder vibrations, while an increase in service speed will result in a higher wave encounter frequency. Consequently the wave encounter frequency may be in the same range as the lowest hull girder natural frequencies, causing important dynamic amplification effects.

As it is not fully clear to what extent these phenomena are covered by the implicit and explicit safety margins of classical rules and regulations, the effect of the dynamic behaviour of large container ships in a seaway is to be studied by direct calculations.

In this paper the two phenomena and their impact on ship structures are analysed. In addition, the methodology recently developed by Bureau Veritas to assess whipping and springing effects in large container ship is presented, as well as the implementation into the in-house developed hydrodynamic simulation suite HydroSTAR.

2. WHIPPING

Whipping can be characterised as transitory response of the hull girder in the first natural mode caused by hydrodynamic impact at the bow due to re-entry of the foreship into the water (slamming) or violent wave impact at the stem (slapping), see figure 1.

Figure 1: Container ship experiencing slamming

The first natural mode of the hull girder could be simulated by a system with one degree of freedom damped at the corresponding frequency. The associated forces are extremely high but of short time duration, similar to an impulse [1] and consequently infer transitory damped vibrations.

2.1 SHIP STRUCTURAL RESPONSE

Slamming and slapping are known to occur in head seas when the vertical (relative) motions are highest. The deformations associated with the hull girder vibration caused by hydrodynamic impact are superimposed on the wave induced hull girder deformations, see figure 2.

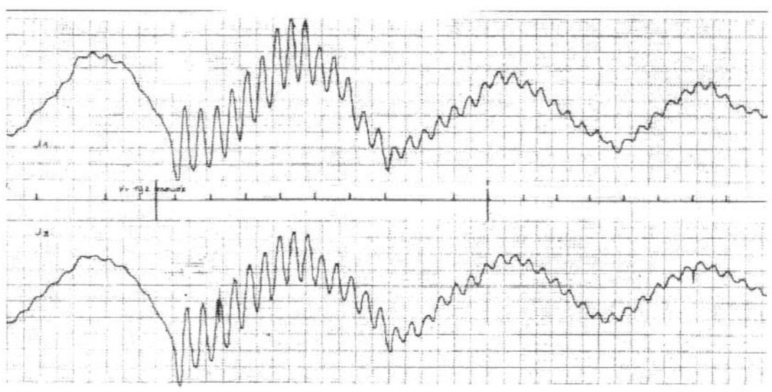

Figure 2: Full scale measurements of stresses in the deck of the vessel Marcel Bayard

It has been known for several years that ships with large deck openings are prone to sustaining damages as a result of whipping. First of all, the design value of the vertical bending moment can be exceeded, causing permanent deformation of the hull girder. Secondly, high frequent stress cycles can be generated, causing premature fatigue cracks in structural details [2, 3].

The understanding and analysis of the phenomenon of exceeding the acceptable bending moment is relatively straightforward, while the prediction of the occurrence of fatigue cracks is much more difficult. However, the risk for fatigue damages has been clearly demonstrated on the vessel Marcel Bayard. In fact, the full scale measurement campaign [2] was initiated after the appearance of fatigue cracks when the vessel was only three years in service. The study showed that, although the vessel length is only 110 m, the cracks could be explained by frequent occurrence of whipping as a result of particular operating conditions for this type of vessel and large deck openings causing high stress concentrations in the deck.

Therefore, it is important to be able to estimate effects of whipping on the stresses in the deck during the design stage. This requires the combined simulation of the behaviour of the vessel under wave loading and hydrodynamic impact caused by slamming and slapping.

2.2 CALCULATION METHODS

Direct calculations on a ship in a seaway in order to determine the extreme hull girder loads and the fatigue life of structural details have been common practice for some decades. The methods used are based on the fast spectral approach and are, after calibration, sufficiently accurate for the classical checks. As whipping is a non linear impulsive phenomenon, the (linear) spectral approach does not longer hold. Instead, time domain methods have been developed which can solve the coupling problem between the 3D hydrodynamic diffraction-radiation calculations and the dynamic deformation of the hull girder modelled as a Timoshenko beam or a 3D finite element model. For this modal approach the motions and deformations of the vessel are represented by a series of six rigid modes and five to ten dry structural modes [4]. It is to be noted that the linear hydrodynamic coefficients in the time step equations are obtained from the classical frequency domain calculations [5, 6].

There are two methods available for the determination of the slamming and slapping loads (dynamically exciting the hull girder), which are essentially 2-dimensional: the 'Generalized Wagner Model' and 'Modified Logvinovich Model'.

The determination of the structural response to whipping is possible by combined application of calculation tools for the rigid ship in the frequency domain and for the elastic ship in the time domain (non linear). This set of tools was successfully verified on vessels of 260 m and 360 m of length [7], see figures 3 and 4.

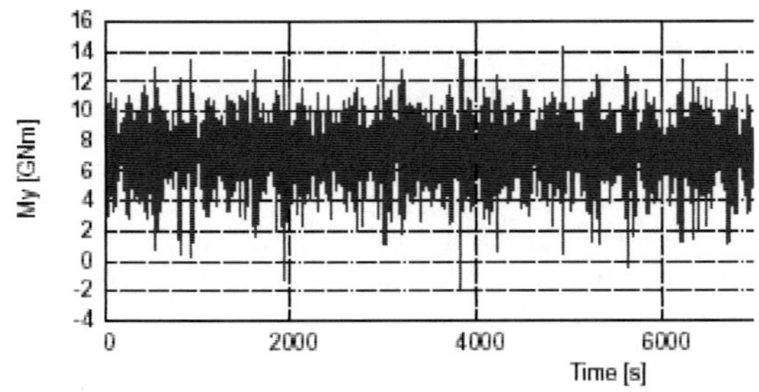

Figure 3: Example of time history of non linear bending moment time including whipping

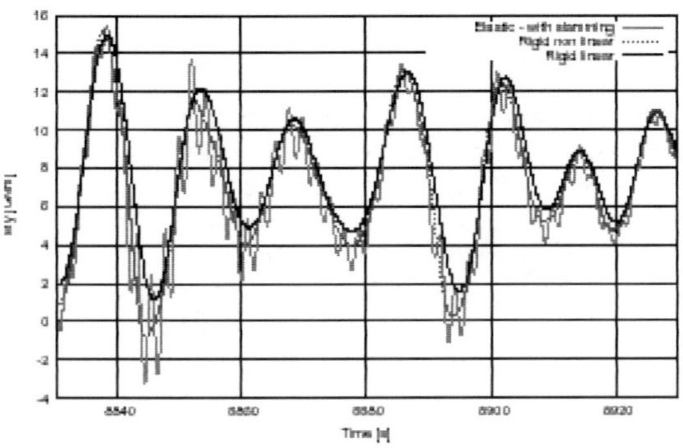

Figure 4: Zoom of figure 3 at the instant of hydro-dynamic impact (response types: elastic non-linear, rigid non linear and rigid linear)

The calculations are performed according to the scheme shown in table 1.

Table 1: Step-by-step procedure for estimating the extreme hull girder loads

1	Selection of the sea states and their probability of occurrence for the relevant operational profile of the vessel (standard: North Atlantic)
2	Calculation of the linear response and determination of the possible extreme navigation conditions for the incident waves (ship speed, wave height)
3	Determination of navigation conditions for which slamming could occur (wave incidence, ship speed, wave height)
4	Calculation of the non linear behaviour for navigation conditions with high risk of slamming

2.3 EXTREME HULL GIRDER LOADS

Observation of the elastic response shows that whipping increases the extreme values of the vertical bending moment. Therefore, it is necessary to take this effect into account for checking the ultimate strength of the hull girder. The performed calculations [7] and full scale measurements [2] show that whipping occurs when the ship is in sagging condition. Therefore, for the most severe navigation conditions the extreme slamming loads and consequently the increase of the sagging wave bending moment are calculated. Full scale measurements show that that the damping of the vibratory response is small (of the order of ξ=1.5%), which means that whipping is still important half a wave period later when the vessel is in hogging condition and that the ultimate strength check is also to be carried out for this condition [2]. This check can be done in analogy with the class rules for the bending moment corrected for the whipping effect as described [8].

Application on a 250 m container ship shows that the increase of the total bending moment can reach 20 to 30% for severe navigation conditions, which is in agreement with full scale measurements [2]. The calculations performed for ultra large container ships show that a similar percentage of increase may be expected for these ships.

2.4 FATIGUE

Observation of the whipping response (figures 3 and 4) reveals two effects regarding fatigue:

- Increase of amplitude of the wave frequent hull girder stresses
- Creation of high frequent damped stress cycles at the first natural frequency of hull girder bending

For welded details it is considered that the residual stresses due to the heat input create a high level of tension in zones where cracks may be expected, resulting in a high R coefficient (minimum stress divided by maximum stress, R > 0.5). For these high R values variation of the average stress level has little influence on the fatigue lifetime.

For details without welding there are no residual stresses and analysis becomes somewhat more delicate. In order to simplify matters, in first instance the same approach as for welded structures is applied. As this means that a high level of residual stress is presupposed, this is a conservative simplification.

Application of Miner's law, which is linear by nature, allows for separating the low frequency (wave) and high frequency (whipping) contributions and add both individual effects to obtain the total result, see figure 5.

[2] Especially since container ships normally operate in hogging condition, thus providing the extreme hull girder bending moment.

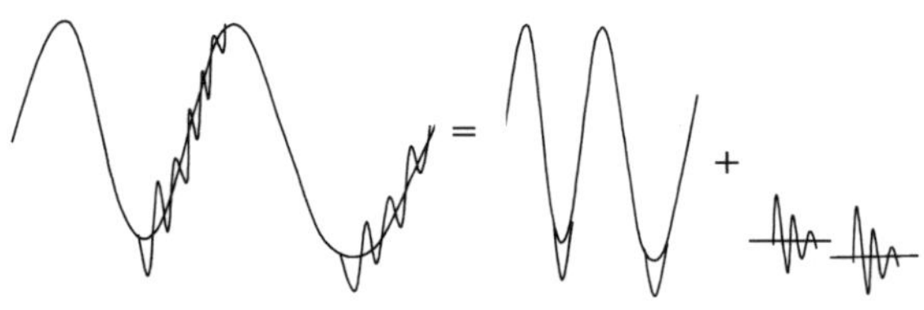

Figure 5: Separation of whipping effects for fatigue calculation

Knowing the stress range of the first whipping cycle and the damping it is possible to determine the distribution of cycles and their associated stress ranges as depicted in figure 6.

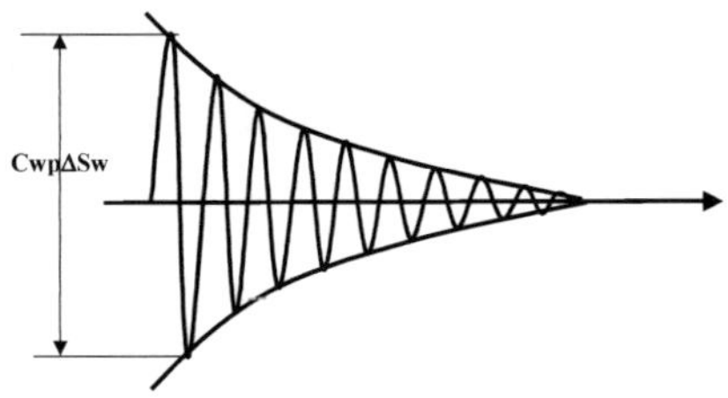

Figure 6: Distribution of high frequent, whipping induced, stress cycles

The associated equations are:

$$\Delta Swp_j = A \exp\left[-\frac{\xi}{\sqrt{1-\xi^2}}(4j-3)\frac{\pi}{2}\right] \qquad (1)$$

with j designating the nth whipping cycle

$$A = \frac{Cwp\Delta Sw}{\exp\left(-\frac{\xi}{\sqrt{1-\xi^2}}\frac{\pi}{2}\right)} \qquad (2)$$

The short term distribution is obtained by noticing that for every single whipping response there is only one cycle at every level and that there will be a whipping response associated to every wave cycle, i.e., at each level the number of cycles is equal to the total number of waves, see figure 7.

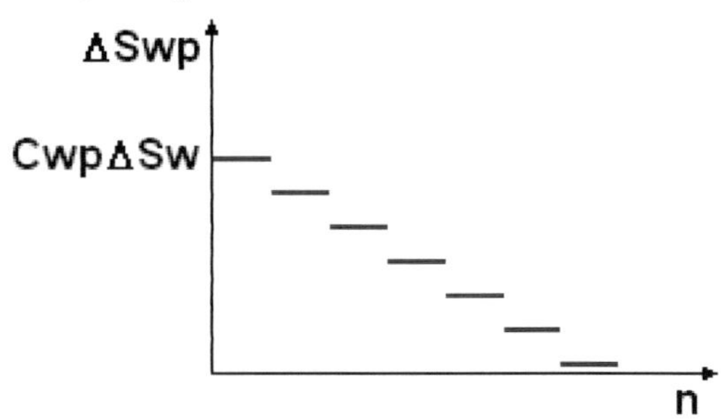

Figure 7: Short term distribution of whipping stresses for a given wave amplitude

Miner's sum can be computed by superimposing the short term distributions of the stress ranges. Every cycle can be considered as a response to a regular wave of constant amplitude, while non linear response calculations to regular waves can be utilised as well.

For a given wave frequency in head seas (being the dominant wave condition for slamming to occur) it is possible to determine the response function for a wave height H and a threshold wave height H_{th} below which there is no risk of slamming:

$$Cwp\Delta Sw = f(H) \qquad H > H_{th} \qquad (3)$$

An overview of the step-by-step calculation method is presented in table 2.

A study performed for an ultra large container ship shows that whipping effectively causes amplification of the hull girder bending stresses, as well as a measurable contribution to the accumulation of fatigue damage. In short term head sea condition the correction of the low frequency stress range can yield a very significant increase of up to 100% for some navigation conditions. Considering that whipping disappears for non-severe sea states, which are far more numerous than severe sea states, and that there is no whipping for transverse and following seas, the total long term increase of the Miner sum due to whipping is estimated to be of the order of 3 to 5%.

Table 2: Step-by-step procedure for estimating the whipping induced fatigue damage

1	Determination of the maximum speed of the vessel as function of the wave height for the considered wave frequencies and incidence angles (maximum bearable navigation condition on the bridge)
2	Calculation of the transfer functions and associated short term response spectra without whipping
3	Determination of the whipping response in head seas as function of the significant wave height according to the wave frequencies of the considered sea states and taking into account the whipping threshold
4	Approximation of a linear rule giving the amplitude of the first whipping cycle as a function of the amplitude of the rigid body response
5	Approximation of the variation on the linear rule as a function of the wave incidence angle
6	Identification of the short term sea states for which whipping is likely to occur
7	Correction of the low frequent (wave) stress range distributions for the whipping effect
8	Calculation of the Miner sum for all short term low frequent distributions

9	Determination of the high frequent whipping responses and associated Miner sums for every short term distribution for which whipping occurs
10	Calculation of total Miner sum taking into account low frequency contributions corrected for whipping and high frequency contributions

Following the analysis it can be concluded that the Miner sum associated to the high frequent part has a relatively small contribution for the following reasons:

- Whipping occurs only in a limited number of sea states (head seas, wave height above threshold value)
- The reduction of the stress amplitudes due to damping puts the majority of the cycles under the changing point of the slope of the S-N curve used (due to the random character of the phenomenon a 2-slope S-N curve is used for calculating the Miner sum).

3. SPRINGING

Springing can be described as enforced wave induced hull girder vibrations. The calculation of the natural frequency of the two node hull girder bending mode for vessels with a length up to 350 m shows values above the frequency of the shortest observed waves. For instance, North Atlantic wave tables give a lower period Tz=3.5 s (0.28 Hz), while the natural frequency including added mass of a 300,000 dwt tanker has been calculated at 0.5 Hz (T=2.0 s). Springing was identified in 1972 during a full scale study on board a 340 m tanker [9], see figure 8. The response of the first natural mode of hull girder vibration, in spite of the high frequency, can be explained by the fact that for a given wave period the energy is distributed over a broad bandwidth. Using a Pierson-Moskowitz spectrum, the energy associated to a wave period of 3.5 s is distributed over a bandwidth of 0.12 to 0.65 Hz, while for a wave period of 8.5 s this is 0.05 to 1.7 Hz.

In addition, for head seas the encounter frequency is increased by the ship speed as follows:

$$\omega_e = \omega_H + \frac{\omega_H^2}{g} V \tag{4}$$

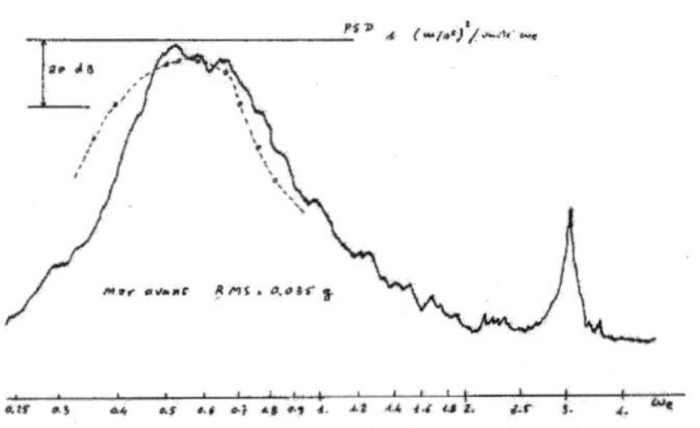

Figure 8: Response spectrum deck longitudinal stresses on a tanker in head seas for rigid mode and including springing [9]

For the referenced tanker, sailing at a speed of 15 kn (7.7 m/s), the energy for a wave period of 3.5 s will be distributed over a bandwidth of 0.19 to 2.73 Hz, while for a wave period of 8.5 s this bandwidth is 0.06 to 16 Hz. These data clearly show that it the natural mode at 0.5 Hz is effectively excited.

On vessels ranging from 300 to 350 m in length, if springing is at all visible from the spectral analysis, the level will be very weak and without effect on the structural resistance. When the length of the vessel increases or when the deck contains large openings the structural rigidity and therefore the first natural frequency mode decrease. When the ship speed increases, the encounter frequency increases as well. In both cases the frequency band associated with springing moves towards the region where the spectral density of wave energy is higher, thus amplifying the springing response (the springing associated peak in figure 8 moves to the left). This effect was noticed when performing design analysis for tankers of 550,000 dwt (414 m in length) and later confirmed during a four year full scale measurement campaign [10, 11].

When considering the characteristics of a typical ultra large container ship, with a length of over 360 m, speed of about 25 kn and large open deck structure, it becomes clear that springing is a phenomenon to be considered during the design of the ship structure.

3.1 SHIP STRUCTURAL RESPONSE

A typical time series of measured longitudinal stress in the deck is presented in figure 9, which shows a high frequent springing signal of nearly constant amplitude superimposed on the low frequent (wave) component.

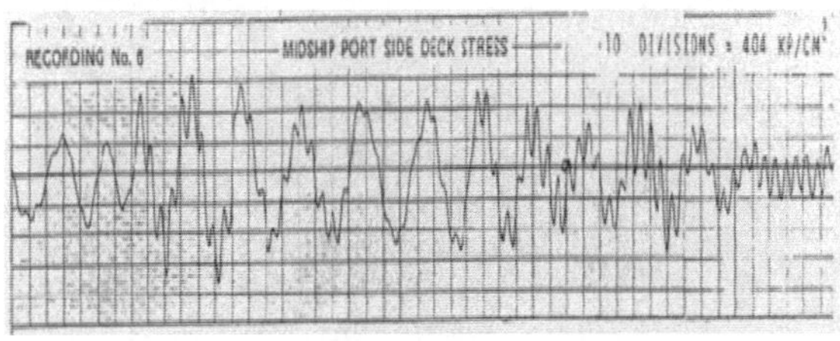

Figure 9: Measurement of deck longitudinal stress with visible whipping effect

Studies and measurements performed on 550,000 dwt tankers show that springing is sensitive for sea states with short wave periods. For these periods the maximum wave height is generally low, see table 3.

Table 3: Maximum significant wave heights for North Atlantic wave environment according to IACS

Tz (s)	3.5	4.5	5.5	6.5	7.5	8.5
Hs (m)	1.5	4.5	9.5	13.5	14.5	14.5

Due to the weak damping, ξ=1.5% obtained from full scale measurements [2, 11], the available energy for the whipping response is limited and the amplitude remains small compared to the wave frequent response. Therefore, the influence of springing on the extreme hull girder loads, corresponding to severe sea states (higher natural period and very high significant wave height), is not significant.

On the other hand, the high frequency and character of the enforced vibration (a quasi permanent phenomenon in sea states with relatively small waves, which are the most frequent during the life of the vessel) yield a non negligible springing induced contribution to the accumulated fatigue damage. This is confirmed by experience feedback of the 550,000 dwt tankers [11].

3.2 CALCULATION METHODS

Taking into consideration that the springing induced deformations are much smaller than the rigid body motions of the hull girder, it can be assumed that the problem can be treated in a linear sense and that the total response can be obtained by summation of the separately calculated rigid and elastic contributions [10]. Therefore it is possible to apply the linear tools developed for hydrodynamic and structural response calculations [13].

First of all, a classical calculation can be performed to determine the linear hydrodynamic coefficients, in particular the added mass and the transfer functions of the rigid body responses (figure 10). Secondly, the natural frequencies for the elastic hull and the associated transfer functions for springing are calculated (figure 11).

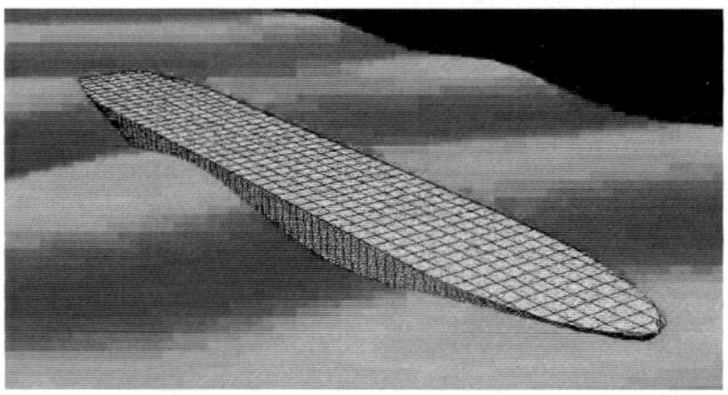

Figure 10: Calculation of rigid body transfer functions

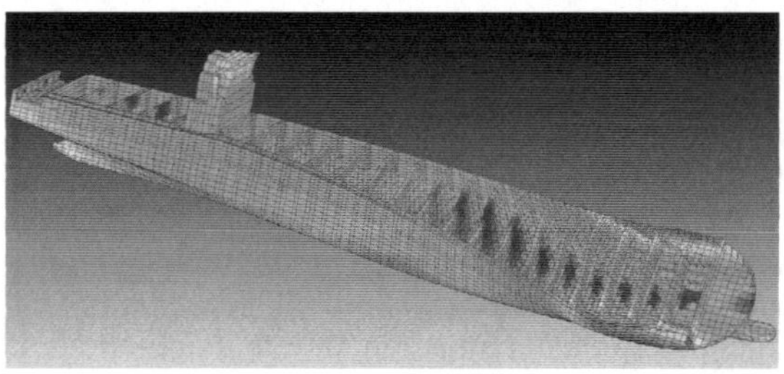

Figure 11: Calculation of transfer function for springing in torsional mode (elastic hull)

In this way the transfer functions for the quasi static, springing and total response are obtained for the selected navigation conditions (figure 12).

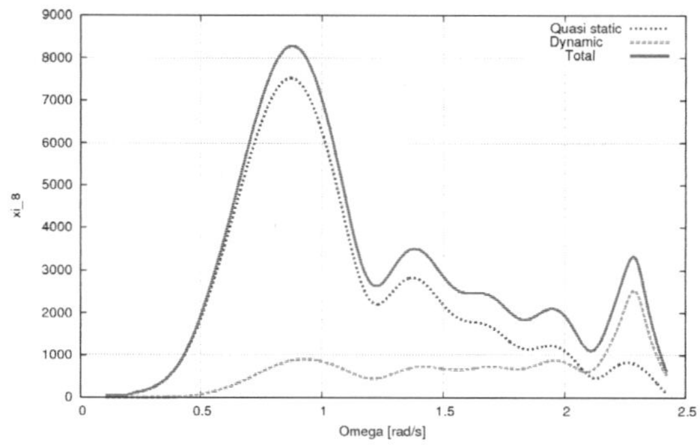

Figure 12: Transfer function of peak stress in hot spot (welded detail) in a container ship (springing, wave and total)

The calculations are performed according to the scheme shown in table 4.

Table 4: Step-by-step procedure for springing analysis.

1	Selection of the sea states and their probability of occurrence for the relevant operational profile of the vessel (standard: North Atlantic)
2	Calculation of the linear response and determination of the possible extreme navigation conditions for the incident waves (ship speed, wave height)
3	Determination of navigation conditions for which springing could occur (wave incidence, ship speed, wave height)
4	Calculation of the linear elastic behaviour for wave incidences where springing occurs

3.3 FATIGUE

Observation of the springing response for a given sea state (figure 9) shows that, on top of the wave frequent response, a time series of stress cycles with nearly constant amplitude at the first natural frequency of the hull girder is generated.

The calculation of the short term fatigue damage accumulation requires the determination of the distribution of the high frequent stress cycles. The exact method for achieving this would be Rainflow counting. In order to do that a time history needs to be available, which is impractical for the spectral approach. By considering the same hypothesis as for the whipping and the using the linearity of Miner's rule, it is possible to separate both the high frequent and low frequent contributions as depicted in figure 13 superimpose them to obtain the total result.

As shown for whipping, the range of the wave frequent stress cycles needs to be corrected for the springing effect as well. As only the navigation conditions with a risk for the occurrence of springing need to be considered, it is important to take into account speed reduction in case the wave height increases, because this can cancel the springing effect altogether.

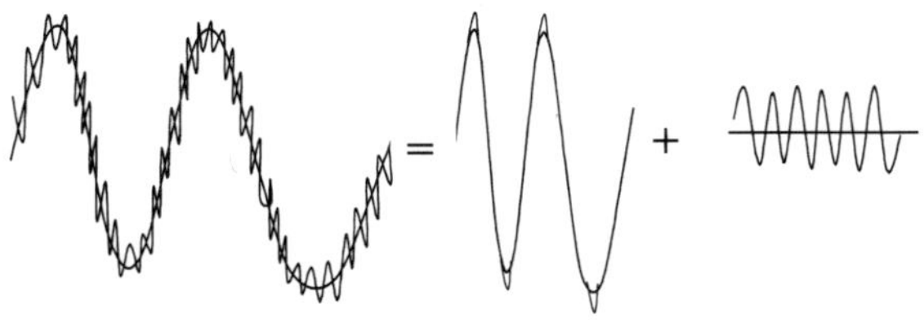

Figure 13: Separation of springing effects for fatigue calculation

For each of the contributions corresponding to a signal with narrow banded spectrum, the fatigue accumulation can be computed according the methodology for wave frequent fatigue assessment [12]. The short term distribution of the stress range ΔS_i can be described by a Rayleigh distribution, where the coefficient m_0 is determined by the area under the response spectrum:

$$\text{prob}(\Delta S > \Delta S_i) = \exp\left[-\frac{\Delta S_i^2}{8m_0}\right] \qquad (5)$$

For a given structural detail with known S-N curve it is possible to calculate the Miner sums separately and add them, taking into account the probability of occurrence of each short term navigation condition (wave height and period, vessel speed and wave incidence angle):

$$D_{total} = D_{wave,corrected} + D_{springing} \qquad (6)$$

An overview of the step-by-step calculation method is presented in table 5.

Table 5: Step-by-step procedure for estimating the springing induced fatigue damage

1	Determination of the maximum speed of the vessel as function of the wave height for the considered wave frequencies and incidence angles (maximum bearable navigation condition on the bridge)
2	Calculation of the transfer functions and associated short term response spectra without springing
3	Determination of the frequency bands with high energy density within the wave spectra for the considered navigation conditions
4	Calculation of the natural frequencies of the hull girder, taking into account the added mass
5	Identification of the short term sea states for which springing is likely to occur
6	Calculation of the transfer functions for springing and associated short term response spectra for the considered sea states
7	Calculation of total Miner sum taking into account low frequency contributions corrected for springing and high frequency contributions

The method has been checked at short term level against the considered exact method, i.e. the Rainflow counting of the stress time history. To do so, a time history has been generated from the short term 2-peak response (see figure 12) using specific software. Then a Rainflow counting according to the AFNOR standard procedure has been performed and the Miner sum calculated.

Two typical cases have been identified. For sea states with Tz above 5 s the low frequent wave stress is dominant over the high frequent springing, while for smaller Tz the situation is reversed, see figure 14.

The comparison between the methods confirms that the proposed procedure provides acceptable short term Miner sum values. However, the correction factor for the low frequent wave stress range has to be adjusted to the typical cases to prevent underestimations.

A study performed for an ultra large container ship shows that, due to the large deck openings, the first natural mode of the hull girder is a torsional mode, which is followed by a number of vertical bending and torsional modes, see table 6.

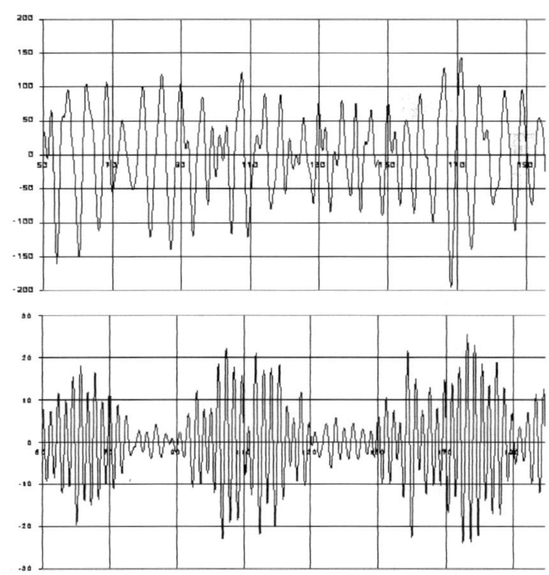

Figure 14: Stress time histories for Tz = 6.5 s (above) and Tz = 3.5 s (below) (First modal period T_1 = 2.15 s)

Table 6: Wet natural hull girder modes and frequencies ultra large container carrier

Hull girder vibration mode	f (Hz)
1-node torsional	0.37
2-node vertical bending	0.47
2-node torsional	0.50
3-node vertical bending	0.90

An investigation of the navigation conditions with a risk for springing occurrence showed that the torsional modes are to be considered for wave incidence angles of 120 and 90 degrees, while the flexural modes are to be considered for wave incidence angles of 180 and 120 degrees (180 degrees corresponding to head seas).

The fatigue lifetime of a structural detail in the deck experiencing high dynamic loading has been calculated according applying North Atlantic scatter diagram, see figure 15.

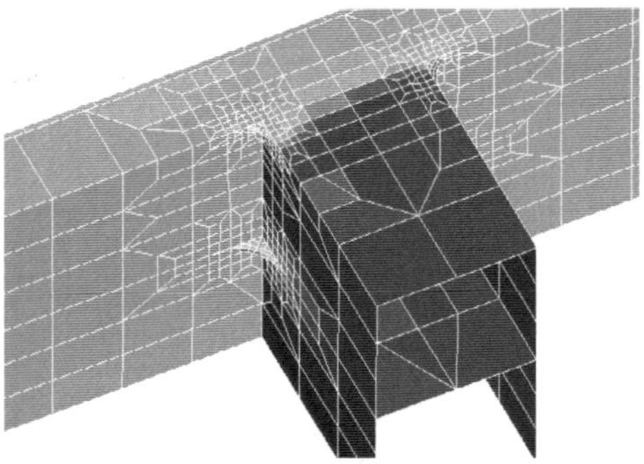

Figure 15: Fatigue analysis of hatch corner detail of an ultra large container ship

The springing response has been calculated for all wave heights and periods of the wave data table until a significant wave height of 5 m. For greater wave heights it is assumed that the vessel speed will be reduced and, consequently, that the shifted encounter frequency will yield only irrelevant springing response.

The calculation of the short term Miner sum including springing shows that, compared to the situation excluding high frequent springing part, an increase in fatigue damage of up to a factor 3 can be reached for small period sea states. However, as small period sea states represent only a limited part of the long term total Miner sum, and taking into account that springing does not occur for transverse and following seas, the total long term increase of the Miner sum due to springing is estimated to be of the order of 4 to 10%.

A sensitivity study shows that if the maximum significant wave height is increased from 5 to 6 m, the springing related Miner sum increases by 20%. However, the effect on the total Miner sum is of second order.

4. HYDROSTAR

The described methodology to assess the effects of whipping and springing during the design stage of ultra large container ships has been made practically applicable within the Bureau Veritas in-house hydrodynamic simulation suite HydroSTAR, which has been developed and validated for more than 20 years to support new technological challenges. Originally started as a linear 3D diffraction-radiation scheme for floating offshore units, today HydroSTAR is a multi disciplinary hydrodynamic package widely used in the marine and offshore industry. Typical issues like irregular frequencies, forward speed effect, roll damping, automatic mesh transfer, second order wave loads and multi body interactions are accounted for.

In order to simulate whipping and springing hydro-elastic and non linear effects are to be accounted for. Hydro-elastic effects are handled using the modal approach, for which the equation of motion can be written as [13]:

$$\left\{-\omega^2([\mathbf{m}]+[\mathbf{A}])-i\omega([\mathbf{B}]+[\mathbf{b}])+([\mathbf{k}]+[\mathbf{C}])\right\}\{\xi\}=\left\{\mathbf{F}^{\mathbf{DI}}\right\} \quad (7)$$

where $[\mathbf{m}]$ is the structural mass, $[\mathbf{b}]$ the structural damping, $[\mathbf{k}]$ the structural stiffness, $[\mathbf{A}]$ the hydrodynamic added mass, $[\mathbf{B}]$ the hydrodynamic damping, $[\mathbf{C}]$ the hydrodynamic restoring, $\{\xi\}$ the modal amplitudes and $\{\mathbf{F}^{\mathbf{DI}}\}$ the modal hydrodynamic excitation. The hydro-elastic coupling is visualised in figure 16.

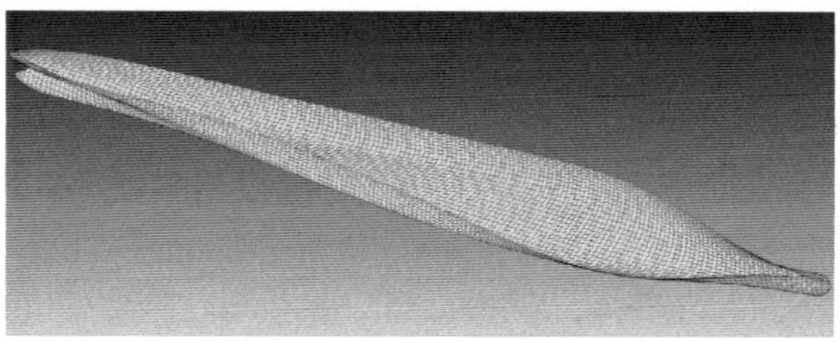

Figure 16: Hydrodynamic mesh following the structural deformation of the hull (first modal period)

The non linear time domain simulations make use of the procedure proposed by Cummins (1962), for which the equation of motion can be written as follows [13]:

$$\begin{aligned}&([\mathbf{m}]+[\mathbf{A}^\infty])\{\ddot{\xi}(t)\}+([\mathbf{k}]+[\mathbf{C}])\{\xi(t)\}\\&+\int_0^t[\mathbf{K}(t-\tau)]\{\dot{\xi}(t)\}d\tau=\{\mathbf{F}(t)+\mathbf{Q}(t)\}\end{aligned} \quad (8)$$

where $[\mathbf{A}^\infty]$ is the infinite frequency added mass matrix and $[\mathbf{K}(t)]$ the matrix of impulse response functions, which can be calculated from the frequency dependent damping coefficients [5, 13]:

$$K_{ij}(t)=\frac{2}{\pi}\int_0^\infty B_{ij}(\omega)\cos\omega t d\omega \quad (9)$$

This method enables the introduction of non linear components in the excitation forces **F**(t) and **Q**(t) [13]. As an example, for the calculation of the extreme hull girder bending moment including whipping effect a (weakly) non linear Froude-Krylov model is applied, see figure 17.

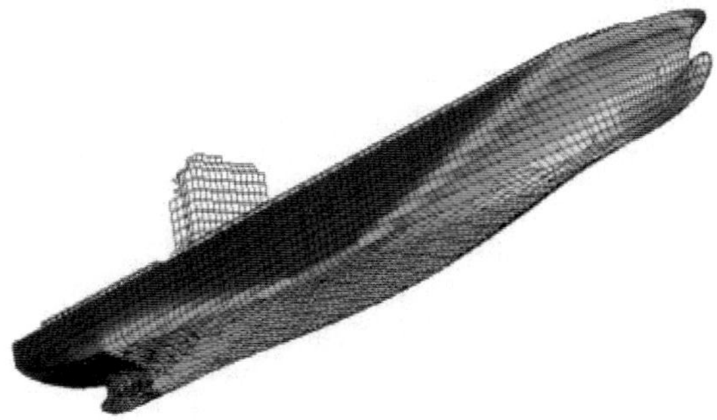

Figure 17: Non linear time domain simulation using Froude-Krylov approach (loads on finite element model)

The resulting overall computation scheme of the software called HydroSTAR++ is depicted in figure 18.

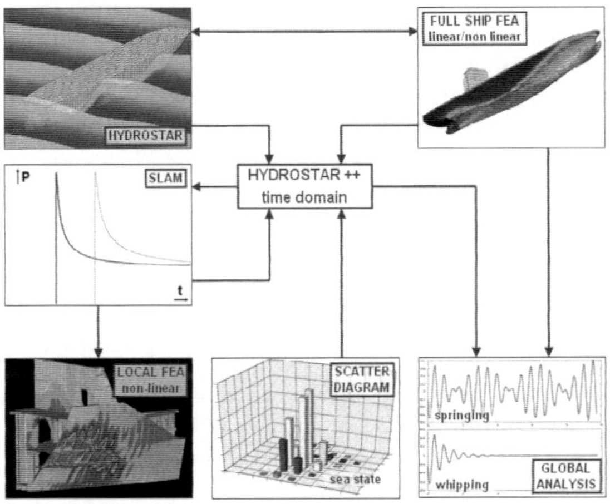

Figure 18: Computation scheme of HydroSTAR++ for hydro-structural simulation of container ships

4.1 MODEL TESTING

In order to validate and calibrate the methodology for the hydro-elastic and time domain simulations, several model tests have been carried out and compared to the calculation results. In figure 19 depicts the torsional response of an elastic barge, showing good agreement between measurements and calculation results [13].

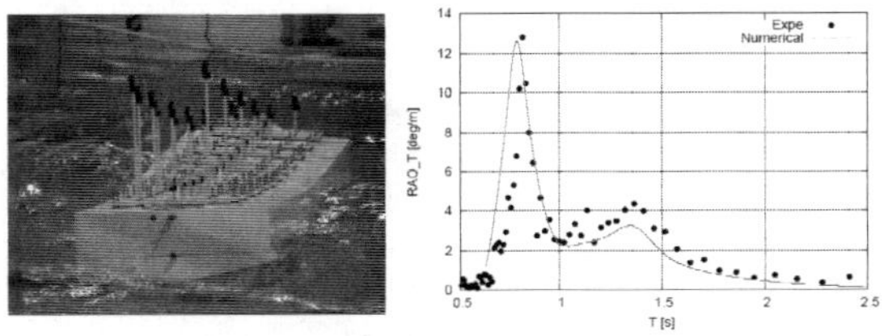

Figure 19: Elastic barge in torsion (impression of test left, RAO of torsional angle right): comparison between experiments and calculation results

In figure 20 the initial deformation and the time history of the vertical displacement after releasing it from the original position are depicted. Again, there is good agreement between calculations and experiments.

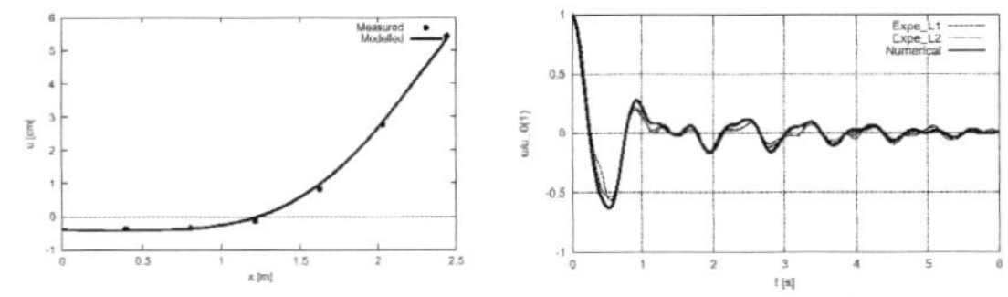

Figure 20: Elastic decay test (initial position left, time history right): comparison between experiments and calculation results

4.2 FULL SCALE MEASUREMENTS

Within the scope of the Joint Industry Project *Lashing@Sea* full scale measurements are carried out on board the BV class 9,400 TEU container ship CMA-CGM Rigoletto. Motion and acceleration sensors have been distributed over the hull and two cross sections have been equipped with strain sensors to record the dynamic response in terms of accelerations and internal hull girder loads.

In order to study the behaviour of stacked containers and lashing equipment a dedicated measurement container has been installed above the aft deck. An example of a time series of measured accelerations in the container is shown in figure 21.

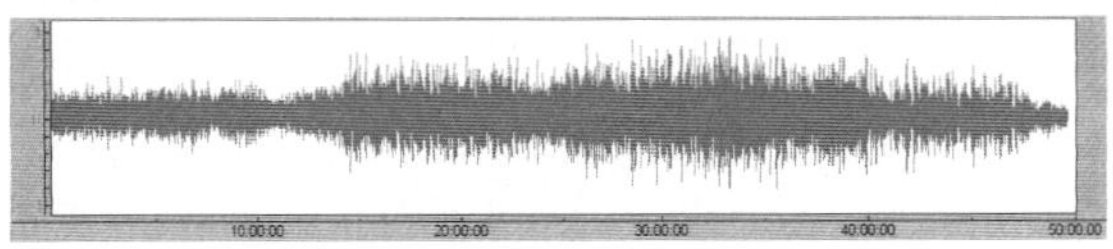

Figure 21: Time series of full scale measurement in transverse accelerations in container (normalised)

The results of the measurement campaign can be used to study the 'real life' behaviour of large container ships and to further validate and calibrate the developed numerical tools.

5. CONCLUSIONS

An increase in the size and speed of container ships importantly changes the behaviour in a seaway. Phenomena already known for smaller size vessels can be amplified to reach levels high enough to have an impact on the design of the hull. This is particularly true for the cases of whipping and springing.

The springing phenomenon was first studied during the construction of the 550,000 dwt tankers in 1975, for which time domain calculations have been performed. At that time the whipping phenomenon was also known, specifically in relation to ships with large deck openings, but computer simulation was not yet feasible. Recent developments in computer computation power and efficient algorithms allow performing non linear hydro-structural calculations in the time domain.

A methodology has been proposed to estimate the effects of whipping and springing on the extreme hull girder loads and fatigue lifetime of structural details. In order to be able to study the fatigue related effects during the design stage of a vessel, it is assumed that the low frequent and high frequent contributions can be dealt with separately. This simplification importantly reduces the required amount of non linear time domain simulations.

The application of the methodology has been made possible by further development of the Bureau Veritas in-house hydrodynamic simulation suite HydroSTAR to solve hydro-elastic and non linear time domain problems. Extensive validation has been done by model testing, while the ongoing full scale measurement campaign will provide additional valuable feedback.

Application of the methodology to ultra large container ships confirms that the stress levels are indeed increasing. As a consequence, a significant increase both hull girder loads and fatigue damage accumulation is found, which can no longer be ignored when analysing the ship structure.

Whipping effects can cause an increase of about 20% in tot total vertical bending moment, for which the hull girder ultimate strength is to be checked.

For current container ship designs of about 350 m with extremely large open deck structure, whipping can cause an increase in short term fatigue damage accumulation by a factor of up to 100% for some head sea navigation conditions. The increasing effect on the long term whipping induced fatigue damage accumulation is estimated to be in the range of 3 to 5%.

With regard to springing, as a consequence of the flexibility of the hull girder and high vessel speed, current designs of about 350 m in length show an increase in short term fatigue damage accumulation by a factor of up to 3 for small period sea states, which is superior to the referenced tankers of 414 m. The increasing effect on the long term springing induced fatigue damage accumulation is estimated to be in the range of 4 to 10%.

It is expected that full scale measurements results will provide valuable insight into the real dynamic behaviour of large container ships, as well as the accuracy of the calculations. In addition, further development of the methodology and hydro-structural computation schemes are ongoing and are expected to further enhance prediction capabilities.

When considering future designs for even larger container ships of 400 m in length and beyond, it is certain that with similar hull structural design the increase of fatigue damage accumulation due to whipping and springing will become even more important. By realizing that springing is a forced vibration phenomenon with little damping and that the response is of non linear character, the dynamic phenomena and associated consequences will have to be analysed with greater attention.

6. ACKNOWLEDGEMENTS

The models and results presented in this paper require strong interaction of many technical competencies. Therefore, the authors would like to thank all the people involved in the research and application of the developed methodologies, in particular Š. Malenica, F. Mauduit, N. Germain, F. Bigot, F.X. Sireta, S. Maherault, E. Stumpf, V. Bouitillier, J. Henry and G. Parmentier.

7. REFERENCES

1. Huther M, Ket N, 'Water impact risk estimation during ship design', Ship Behaviour at Sea, International Symposium on Ship Hydrodynamics and Energy Saving, ISSHES, Madrid, September 1983

2. Osouf J, 'Etude expérimentale du comportement dynamique dur houle du navire câblier Marcel Bayard', Revue NTM, 1973 (in French)

3. Drummen I, Moan T, Storhaug G, Moe E, 'Experimental and full scale investigation of the importance of fatigue damage due to wave-induced vibration stress in a container vessel', Design & Operation of Container Ships, RINA, London, November 2006

4. Malenica Š, Molin B, Remy F, Senjanovic I, 'Hydroelastic response of a barge to impulsive wave loads', 3rd International Conference on Hydroelasticity, Oxford, 2003

5. Cummins WE, 'The impulsive response function and ship motions', Schiffstechnik, 1962

6. Ogilvie TF, 'Recent progress toward the understanding and prediction of ship motions', 9th International Conference on Numerical Ship Hydrodynamics, Ann Arbor (USA), 2007

7. Tuitman J, Malenica Š, 'Some aspects of whipping response of container ships', 23rd IWWWFV, Jegu (Korea), 2008

8. Bureau Veritas, 'Rules for the Classification of Steel Ships', November 2007

9. Planeix JM, 'Wave loads, A correlation between calculations and measurements', Revue International Shipbuilding Progress, August 1972

10. d'Hautefeuille B, Huther M, Baudin M, Osouf J, Calcul de springing par équations intégrales', Revue NTM, 1977

11. Huther M, Osouf J, 'SEFACO, Four years of experience in ship hull girder stress monitoring', MARINTEC, Shanghai, October 1983

12. Huther M, Beghin D, Mahérault S, 'A fatigue strength guidance note for welded ship structure assessments', IIW/IIS doc XIII-111742-98/XV-992-98, 1998

13. Malenica Š, 'Hydro structure interaction in seakeeping, International Workshop on Coupled methods in numerical dynamics IUC, Dubrovnik, September 2007

8. AUTHORS' BIOGRAPHIES

Gijsbert de Jong holds the current position of product manager at Bureau Veritas. He is responsible for the international business development in the field of container ships. Gijsbert joined Bureau Veritas in 2001 after obtaining an MSc in Naval Architecture & Marine Engineering from Delft University of Technology. Before moving to the sales and marketing management team in Paris, he has worked as hull surveyor and department manager in the plan approval office in Rotterdam.

Michel Huther is a graduate Mechanical Engineer (Ecole Centrale de Lyon, 1965) and Naval Architect & Marine Engineer (Ingénieur Civil du Génie Maritime, 1967). Michel joined Bureau Veritas in 1969 and has worked as research engineer and department head for the ship research and rule development department, with many contributions to European R&D projects. In 2000 he was promoted deputy to the director of the marine technical management. After his retirement in 2003 Michel is holding the position of technical advisor to the Bureau Veritas marine technical management. He is a member of SNAME, ATMA & SF2M and has published over 200 scientific and technical papers on ship structural behaviour and safety.

HYDROELASTICITY OF VERY LARGE CONTAINER SHIPS

I. Senjanović, **S. Tomašević**, **S. Rudan**, **M. Tomić**, **N. Vladimir**, University of Zagreb, Croatia
Š. Malenica, Bureau Veritas, France

SUMMARY

Sea transport is rapidly increasing and Very Large Container Ships (VLCS) are built. They are more flexible and their natural frequencies can fall into the range of the encounter frequencies in an ordinary sea spectrum. Such conditions are not covered by the present Classification Rules for ship design and construction. So, structure design of VLCS is at the margin of the rules. Therefore, the wave induced hydroelastic response of VLCS becomes an imperative issue for improving the Classification Rules and ensuring ship safety.

This paper is dealing with the methodology of hydroelasticity investigation, based on mathematical model which includes structural, hydrostatic and hydrodynamic submodels assembled in hydroelastic one. Special attention is paid to restoring stiffness in hydrostatic submodel. The modal superposition method is used, and ship natural modes are determined by the advanced beam theory which includes shear influence on torsion. Contribution of transverse bulkheads to hull stiffness is analysed. 1D FEM model is verified by correlating dry natural vibrations with those obtained from 3D FEM model. Validation of the hydroelastic mathematical model is done by comparison with results from model test of a flexible segmented barge. Application of the developed numerical procedure is illustrated in case of a very large container ship.

NOMENCLATURE

A	wave amplitude
A_{ij}	added mass coefficient
A_s	shear area
B_{ij}	damping coefficient
B_w	warping bimoment
C_{ij}	restoring coefficient
E	Young's modulus
F	force
G	shear modulus
g	gravity constant
H_s	significant wave height
I_b	moment of inertia of cross-section
I_t	torsional modulus
I_w	warping modulus
i	imaginary unit
M	bending moment
S	surface
T	torque
T_z	zero wave crossing period
U	strain energy
u	axial displacement
$\overline{u}$	warping function
V	ship speed
w	deflection
x, y, z	Cartesian coordinates
Z	coordinate of wetted surface
$\mathbf{A}$	modal added mass
$\mathbf{B}$	damping
$\mathbf{C}$	restoring stiffness
$\mathbf{H}$	deformation vector
$\mathbf{h}$	mode vector
$\mathbf{i}, \mathbf{j}, \mathbf{k}$	unit vectors
$\mathbf{K}$	stiffness matrix
$\mathbf{k}$	modal stiffness
$\mathbf{M}$	mass matrix
$\mathbf{m}$	modal mass
$\mathbf{N}$	normal vector
$\mathbf{n}$	unit normal vector
$\boldsymbol{\delta}$	displacement vector
δ	variation
ϑ	variation of twist angle
ν	Poisson's coefficient
ξ	modal coefficient
ρ	density
φ	velocity potential, rotation of cross-section
χ	heading angle
ψ	twist angle
Ω	dry natural frequency
ω	wave frequency
∇	Hamilton operator

1. INTRODUCTION

Sea transport is rapidly increasing and tendency to build large container ship becomes reality [1, 2]. They are relatively more flexible and their structural natural frequencies can fall into the range of the encounter frequencies in an ordinary sea spectrum [3]. Therefore, the wave induced hydroelastic response of large container ships becomes an important issue especially for improving the classification rules and ensuring ship safety [4].

For ships with closed cross-section and ordinary hatch openings such as tankers, bulk carriers, general cargo vessels etc., the lowest natural frequencies are usually associated with the vertical bending. On the other hand, for ships with open cross-section, such as container ships, the lowest elastic natural modes are those of coupled horizontal and torsional vibrations. This coupling is highly pronounced due to the fact that the torsional (shear) centre is below the keel [5, 6].

The classical approach to determine ship motions and wave loads is based on the assumption that the ship hull acts as a rigid body [7]. The obtained wave load is then imposed to the elastic 3D FEM model of ship structure in order to analyse global longitudinal and transverse strength, as well as local strength with stress concentrations related to fatigue analysis [8].

The above approach is not reliable enough for ultra large container ships due to mutual influence of the wave load and structure response [9, 10, 11]. Therefore, a reliable solution requires analysis of wave load and ship vibration as a coupled hydroelastic problem [12]. This is especially important for impulsive loads such as ship slamming that causes whipping [13].

The methodology of hydroelastic analysis is shown in Figure 1 according to [14, 15]. It includes definition of the structural model, ship and cargo mass distributions, and geometrical model of ship surface. First, dry natural vibrations are calculated. Then modal hydrostatic stiffness, modal added mass, damping and modal wave load are determined. Finally, wet natural vibrations are obtained as well as the transfer functions (RAO-response amplitude operator) for determining ship structural response to wave excitation.

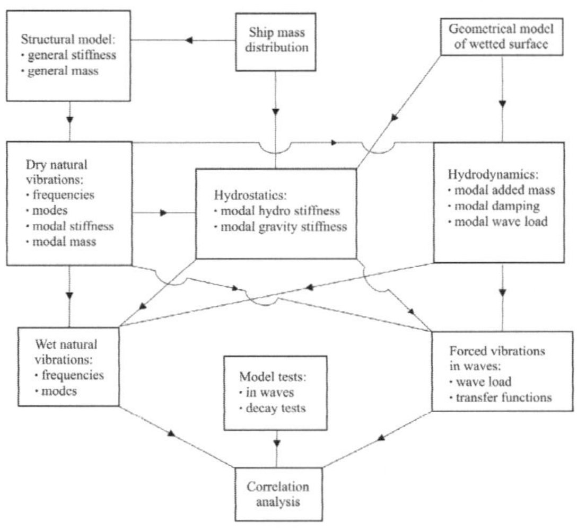

Figure 1 Methodology of the hydroelastic analysis

2. STRUCTURAL MODEL

The hydroelastic problem can be solved at different levels of complexity and accuracy. The best, but highly time-consuming way is to consider 3D FEM structural model and 3D hydrodynamic model based on the radiation-diffraction theory [16]. Such an approach is recommended only for the final strength analysis since it is time consuming. However, in the preliminary strength analysis it is more rational and convenient to couple 1D FEM model of ship hull [6] with 3D hydrodynamic model. Much better results are obtained if the beam model is based on the advanced thin-walled girder theory, Appendix A.

In both cases of the FEM approach the governing matrix equation of dry natural vibrations yields [17]

$$\left(\mathbf{K} - \Omega^2 \mathbf{M}\right)\boldsymbol{\delta} = \mathbf{0}, \tag{1}$$

where **K** is stiffness matrix, **M** is mass matrix, Ω is dry natural frequency and $\boldsymbol{\delta}$ is dry natural mode.

As solution of the eigenvalue problem (*1*) Ω_i and $\boldsymbol{\delta}_i$ are obtained for each the i-th dry mode, where $i = 1,2...N$, N is total number of degrees of freedom. Now natural modes matrix can be constituted

$$\boldsymbol{\delta} = \left[\boldsymbol{\delta}_1, \boldsymbol{\delta}_2 ... \boldsymbol{\delta}_i ... \boldsymbol{\delta}_N\right] \tag{2}$$

and the modal stiffness and mass can be determined [18]

$$\mathbf{k} = \boldsymbol{\delta}^T \mathbf{K} \boldsymbol{\delta}, \quad \mathbf{m} = \boldsymbol{\delta}^T \mathbf{M} \boldsymbol{\delta}. \tag{3}$$

Since the dry natural vectors are mutually orthogonal, matrices **k** and **m** are diagonal. Terms k_i and $\Omega_i^2 m_i$ represent strain and kinetic energy of the i-th mode respectively.

Note that generally the first six natural frequencies Ω_i are zero with corresponding eigenvectors representing the rigid body modes. As a result, the first six diagonal elements of **k** are also zero, while the first three elements in **m** are equal to structure mass, the same in all directions x, y, z, and the next three elements represent the mass moment of inertia around the corresponding coordinate axes.

3. GEOMETRICAL MODEL OF WETTED SURFACE

For determining pressure forces acting on the wetted surface it is necessary to specify panels and their position in space. Wet surface is given by offsets of waterline ordinates at body plan stations. If the strip method is used for pressure calculation, then the panels bounded with two close stations can be used.

If 3D radiation-diffraction theory is used for hydrodynamic pressure determination, the wetted surface mesh can be created in such a way that panels of more regular shape and refined subdivision in the area of the free surface are achieved [19]. Such a rational mesh is shown in Figure 2, where the panel rows follow the ship body diagonals similarly to the structural elements of ship outer shell. In this way, efficiency of the hydrodynamic calculation is increased.

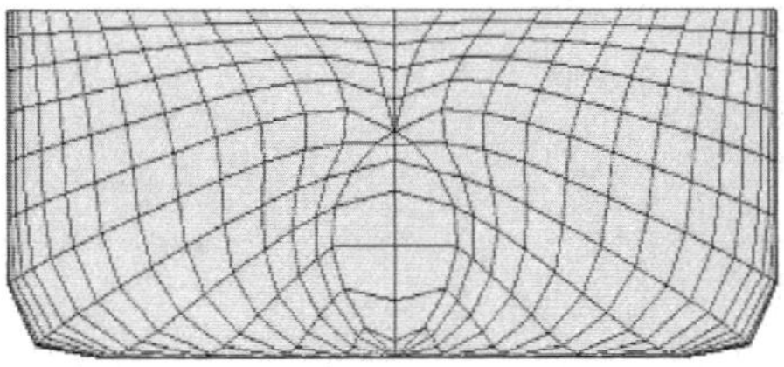

Figure 2 Rational mesh of wetted surface

4. DRY MODES OF WETTED SURFACE

As mentioned in Section 2, structural dry modes can be determined by 1D or 3D FEM analysis. If 1D analysis is used, the beam modes are spread to the ship wetted surface as follows:

vertical vibration

$$\mathbf{h}_i = -\frac{\mathrm{d}\, w_{vi}}{\mathrm{d}\, x}(z - z_N)\mathbf{i} + w_{vi}\mathbf{k}\,, \tag{4}$$

horizontal vibration

$$\mathbf{h}_i = -\frac{\mathrm{d}\, w_{hi}}{\mathrm{d}\, x} y\,\mathbf{i} + w_{hi}\,\mathbf{j}\,, \tag{5}$$

torsional vibration

$$\mathbf{h}_i = \psi_i (z - z_S)\,\mathbf{j} - \psi_i\, y\,\mathbf{k}\,, \tag{6}$$

where w is hull deflection, ψ is twist angle, y and z are coordinates of the point on ship surface, and z_N and z_S are coordinates of neutral line and shear centre respectively.

If strong coupling between horizontal and torsional vibration occurs, as in the case of container ships, the coupled mode yields

$$\mathbf{h}_i = \left(-\frac{\mathrm{d}\, w_{hi}}{\mathrm{d}\, x} y + \frac{\mathrm{d}\psi_i}{\mathrm{d}\, x}\overline{u} \right)\mathbf{i} + \left[w_{hi} + \psi_i (z - z_S) \right]\mathbf{j} - \psi_i\, y\,\mathbf{k}\,, \tag{7}$$

where $\overline{u} = \overline{u}(x, y, z)$ is the cross-section warping function reduced to the wetted surface [20, 21].

5. HYDRODYNAMIC MODEL

5.1 GENERAL

The fully consistent and efficient seakeeping model with forward speed, even for rigid body, does not exist yet and only the approximate models are employed for the time being. These approximated models spread from the so called 2D strip theories, over so called encounter frequency approximation (main effect is put on the change of frequency due to speed) and different 3D approximations based on the exact forward speed Green function with or without coupling with the ship steady flow, to non-linear time domain seakeeping models. However, according to the authors' knowledge, none of the above methods has been validated enough, and the problem is still open.

The most widely used method, the one based on the encounter frequency approximation, is applied in this study, even if the method utilizing the exact forward speed Green function (together with so called Neuman-Kelvin approximation) is also implemented in the software used [22].

The choice of hydrodynamic model is not likely to change the general conclusions of this paper, because the experience shows that, in the case of a rigid body, the encounter frequency method agrees quite well with more complicated methods, at least for the global ship loadings. This, however, does not mean that we should forget this important assumption, and that point should be kept in mind when more detailed correlations with experiments and full scale results will be undertaken.

Anyhow, the coupling procedure does not depend on the used hydrodynamic model, and is therefore described here for the zero speed case, as the simplest one.

5.2 THEORETICAL OUTLINE

Harmonic hydroelastic problem is considered in frequency domain and therefore we operate with amplitudes of forces and displacement. In order to perform coupling of the structural and hydrodynamic models, it is necessary to express the external pressure forces in a convenient manner [13]. First, the total hydrodynamic force F^h has to be split into two parts: the first part F^R depending on the structural deformations, and the second one F^{DI} representing the pure excitation

$$F^h = F^R + F^{DI}\,. \tag{8}$$

Furthermore, the modal superposition method can be used. Vector of the wetted surface deformations $\mathbf{H}(x, y, z)$ can be presented as a series of dry natural modes $\mathbf{h}_i(x, y, z)$

$$\begin{aligned}\mathbf{H}(x,y,z) &= \sum_{i=1}^{N} \xi_i\, \mathbf{h}_i(x,y,z) \\ &= \sum_{i=1}^{N} \xi_i \left[h_x^i(x,y,z)\mathbf{i} + h_y^i(x,y,z)\mathbf{j} + h_z^i(x,y,z)\mathbf{k} \right],\end{aligned} \tag{9}$$

where ξ_i are unknown coefficients. Vectors $\mathbf{h}_i(x, y, z)$ related to wetted surface are obtained from the structural dry modes as explained in the previous section.

The potential theory assumptions are adopted for the hydrodynamic part of the problem. Within this theory the total velocity potential φ, in the case of no forward speed, is defined with the Laplace differential equation and the given boundary values

$$\Delta\varphi = 0 \qquad \text{within the fluid}$$

$$-\nu\varphi + \frac{\partial\varphi}{\partial z} = 0 \qquad \text{at the free surface, } z = 0 \qquad (10)$$

$$\frac{\partial\varphi}{\partial n} = -i\omega \mathrm{H\,n} \qquad \text{on the wetted body surface, } S,$$

where ν is the wave number, $\nu = \omega^2/g$, ω is wave frequency, **n** is the wetted surface normal vector, and i is the imaginary unit.

Furthermore, the linear wave theory enables the following decomposition of the total potential (Salvesen et al., 1970)

$$\varphi = \varphi_I + \varphi_D - i\omega\sum_{j=1}^{N}\xi_j\,\varphi_{Rj}\,, \qquad (11)$$

where:

$$\varphi_I = -i\frac{gA}{\omega}\mathrm{e}^{\nu(z+ix)} \qquad (12)$$

φ_I - incident wave potential

φ_D - diffraction potential

φ_{Rj} - radiation potential

A - wave amplitude.

Now, the body boundary conditions (10) can be deduced for each potential

$$\frac{\partial\varphi_D}{\partial n} = -\frac{\partial\varphi_I}{\partial n}, \quad \frac{\partial\varphi_{Rj}}{\partial n} = \mathrm{h}_j\mathrm{n}\,. \qquad (13)$$

It is necessary to point out that the diffraction and radiation potentials should also satisfy the radiation condition at infinity.

Once the potentials are determined, the modal hydrodynamic forces are calculated by pressure work integration over the wetted surface. The total linearised pressure can be found from Bernoulli's equation

$$p = i\omega\rho\varphi - \rho gz\,. \qquad (14)$$

First, the term associated with the velocity potential φ is considered and subdivided into excitation and radiation parts *(11)*

$$F_i^{DI} = i\omega\rho\iint_S(\varphi_I + \varphi_D)\mathbf{h}_i\mathbf{n}\,\mathrm{d}S \qquad (15)$$

$$F_i^R = \rho\omega^2\sum_{j=1}^{N}\xi_j\iint_S\varphi_{Rj}\mathbf{h}_i\mathbf{n}\,\mathrm{d}S\,. \qquad (16)$$

Thus, (15) represents the modal pressure excitation. Now one can decompose (16) into the modal inertia force and damping force associated with acceleration and velocity respectively

$$F_i^a = \mathrm{Re}(F_i^R) = \omega^2\sum_{j=1}^{N}\xi_jA_{ij}\,, \quad A_{ij} = \rho\,\mathrm{Re}\iint_S\varphi_{Rj}\,\mathbf{h}_i\mathbf{n}\,\mathrm{d}S \qquad (17)$$

$$F_i^v = \mathrm{Im}(F_i^R) = \omega\sum_{j=1}^{N}\xi_jB_{ij}\,, \quad B_{ij} = \rho\omega\,\mathrm{Im}\iint_S\varphi_{Rj}\,\mathbf{h}_i\mathbf{n}\,\mathrm{d}S \qquad (18)$$

where A_{ij} and B_{ij} are elements of added mass and damping matrices respectively.

Determination of added mass and damping for rigid body modes is a well-known procedure in ship hydrodynamics [7]. Now the same procedure is extended to the calculation of these quantities for elastic modes.

The hydrostatic part of the total pressure, $-\rho gz$ in (14), is considered within the hydrostatic model.

6. HYDROSTATIC MODEL

6.1 GENERAL

Hydroelastic analysis is performed by the modal superposition method. Modal forces represent work of actual static and dynamic forces on rigid body and elastic mode displacements. Thus, modal restoring forces consist of time dependent modal pressure forces and modal gravity forces.

6.2 PRESSURE FORCES

In order to specify contribution of the hydrostatic part of the total pressure, $-\rho gz$ in (14), to hydrostatic stiffness it is necessary to determine the change of the modal hydrostatic force as the difference between its instantaneous value and the initial value for the vibration mode $\mathbf{h}_i$ of the body wetted surface $Z = Z(x, y)$ [23]

$$\delta F_i^H = -\rho g\iint_{\tilde{S}}\tilde{Z}\,\tilde{\mathbf{h}}_i\,\tilde{\mathbf{n}}\,\mathrm{d}\tilde{S} + \rho g\iint_S Z\,\mathbf{h}_i\,\mathbf{n}\,\mathrm{d}S\,. \qquad (19)$$

Each of the above quantities can be presented in the form

$$\widetilde{(...)} = (...) + \delta(...)$$

Where δ denotes the variation. By neglecting small terms of higher order, one can write for the modal hydrostatic force *(19)*.

$$\delta F_i^H = -\rho g\iint_S(\delta Z\,\mathbf{h}_i\,\mathbf{n} + Z\,\delta\mathbf{h}_i\,\mathbf{n} + Z\,\mathbf{h}_i\,\delta\mathbf{n})\mathrm{d}S\,. \qquad (20)$$

Variation of the particular quantity can be determined by applying the notion of directional derivative $\mathbf{h}_j\nabla$, where ∇ is Hamilton differential operator

$$\mathbf{h}_j\nabla = h_x^j \frac{\partial}{\partial x} + h_y^j \frac{\partial}{\partial y} + h_z^j \frac{\partial}{\partial z}. \qquad (21)$$

As a result

$$\delta Z = (\mathbf{h}_j\nabla)Z = \mathbf{h}_j\mathbf{k}, \quad \delta\mathbf{h}_i = (\mathbf{h}_j\nabla)\mathbf{h}_i, \quad \delta\mathbf{n} = (\mathbf{h}_j\nabla)\mathbf{n}. \qquad (22)$$

Determining the variation of the body surface normal vector, $\delta\mathbf{n}$, according to (22) is a rather difficult task, Figure 3. Therefore, a relatively simpler procedure, taken from [23], is applied.

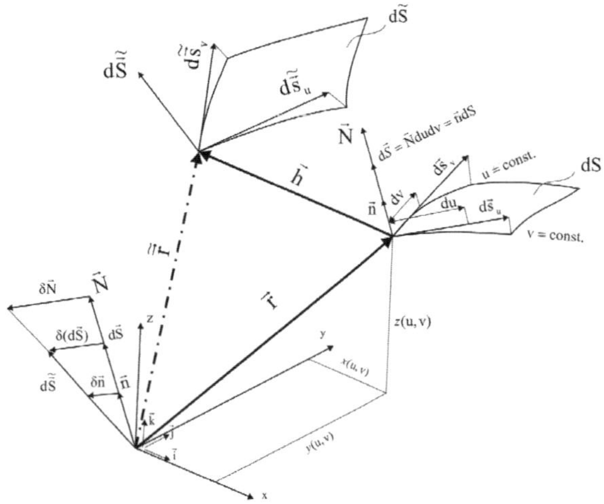

Figure 3 Definition of normal vector variation $\delta\mathbf{n}$ due to modal deformation $\mathbf{h}$

Since the variations (22) depend on all modes, the modal force is expressed by series:

$$F_i^H = \sum_{j=1}^{N} \xi_j \delta F_{ij}^H = -\sum_{j=1}^{N} \xi_j C_{ij}^H, \qquad (23)$$

where:

$$C_{ij}^H = C_{ij}^{Hp} + C_{ij}^{Hh} + C_{ij}^{Hn} \qquad (24)$$

Is the i,j-th element of the hydrostatic stiffness matrix, composed of static pressure, surface mode and normal vector contributions, respectively:

$$C_{ij}^{Hp} = \rho g \iint_S h_z^j (h_x^i n_x + h_y^i n_y + h_z^i n_z) \mathrm{d}S \qquad (25)$$

$$C_{ij}^{Hh} = \rho g \iint_S Z \Bigg[\left(h_x^j \frac{\partial h_x^i}{\partial x} + h_y^j \frac{\partial h_x^i}{\partial y} + h_z^j \frac{\partial h_x^i}{\partial z} \right) n_x + \left(h_x^j \frac{\partial h_y^i}{\partial x} + h_y^j \frac{\partial h_y^i}{\partial y} + h_z^j \frac{\partial h_y^i}{\partial z} \right) n_y + \left(h_x^j \frac{\partial h_z^i}{\partial x} + h_y^j \frac{\partial h_z^i}{\partial y} + h_z^j \frac{\partial h_z^i}{\partial z} \right) n_z \Bigg] \mathrm{d}S \qquad (26)$$

$$C_{ij}^{Hn} = \rho g \iint_S Z \Bigg\{ \left[\left(\frac{\partial h_y^j}{\partial y} + \frac{\partial h_z^j}{\partial z} \right) h_x^i - \frac{\partial h_x^j}{\partial y} h_y^i - \frac{\partial h_x^j}{\partial z} h_z^i \right] n_x + \left[-\frac{\partial h_y^j}{\partial x} h_x^i + \left(\frac{\partial h_x^j}{\partial x} + \frac{\partial h_z^j}{\partial z} \right) h_y^i - \frac{\partial h_y^j}{\partial z} h_z^i \right] n_y + \left[-\frac{\partial h_z^j}{\partial x} h_x^i - \frac{\partial h_z^j}{\partial y} h_y^i + \left(\frac{\partial h_x^j}{\partial x} + \frac{\partial h_y^j}{\partial y} \right) h_z^i \right] n_z \Bigg\} dS. \qquad (27)$$

Note that the wetted surface coordinate Z is measured from the waterplane. Based on the constitution of the above coefficients, it is evident that the hydrostatic stiffness matrix is not diagonal.

6.3 GRAVITY FORCES

The above expressions represent only the action of the hydrostatic pressure, and the gravity part has to be added in order to complete the total restoring coefficients. Similarly to the pressure part, Eqs. (19) and (20), a change of the generalised modal gravity force associated with a particular mode yields [23]

$$\delta F_i^m = -g \iiint_V \delta \mathbf{h}_i \mathbf{k} \, \mathrm{d}m, \qquad (28)$$

where V is body volume and dm is differential mass.

By employing (22) and further (21) one can write

$$F_i^m = \sum_{j=1}^{N} \xi_j \delta F_{ij}^m = -\sum_{j=1}^{N} \xi_j C_{ij}^m, \qquad (29)$$

where

$$C_{ij}^m = g \iiint_V \left(h_x^j \frac{\partial h_z^i}{\partial x} + h_y^j \frac{\partial h_z^i}{\partial y} + h_z^j \frac{\partial h_z^i}{\partial z} \right) \mathrm{d}m. \qquad (30)$$

6.4 RESTORING STIFFNESS

Finally, the complete restoring coefficients read

$$C_{ij} = C_{ij}^H + C_{ij}^m. \qquad (31)$$

It is important to point out that the above expressions for the hydrostatic and gravity coefficients are general and therefore valid not only for the elastic modes but also for the rigid body modes as well as for their coupling. Since the rigid body modes are a special case of elastic modes, the corresponding modal restoring stiffness is reduced to the hydrostatic coefficients [7].

In the literature various definitions of hydrostatic stiffness matrix can be found, Appendix B. The advantage of the present method is that the derived formulae are general and applicable for a complex body as well as for its parts. This is important for determining the local hydrostatic action as internal loads, transfer of load to a FEM structural model, etc.

7. HYDROELASTIC MODEL

After the structural, hydrostatic and hydrodynamic models have been determined, the hydroelastic model can be constituted. For that purpose, let us impose modal hydrodynamic forces (16), (17) and (18) and hydrostatic and gravity forces (23), (28) to the modal structural model, Section 2

$$\left(\mathbf{k}-\omega^2\mathbf{m}\right)\boldsymbol{\xi}=\mathbf{F}^{DI}+\mathbf{F}^{a}+\mathbf{F}^{v}+\mathbf{F}^{H}+\mathbf{F}^{m}. \qquad (32)$$

Furthermore, all terms dependent on unknown modal amplitudes, ξ_i, can be separated on the left hand side. Thus, the governing matrix differential equation for ship motions and vibrations is deduced

$$\left[\mathbf{k}+\mathbf{C}-i\omega\left(\mathbf{d}+\mathbf{B}(\omega)\right)-\omega^2\left(\mathbf{m}+\mathbf{A}(\omega)\right)\right]\boldsymbol{\xi}=\mathbf{F}, \qquad (33)$$

where $\mathbf{k}$ is structural stiffness, $\mathbf{d}$ is structural damping, $\mathbf{m}$ is structural mass, $\mathbf{C}$ is restoring stiffness, $\mathbf{B}(\omega)$ is hydrodynamic damping, $\mathbf{A}(\omega)$ is added mass, $\boldsymbol{\xi}$ is modal amplitudes, $\mathbf{F}$ is wave excitation and ω is encounter frequency. All quantities, except ω and $\boldsymbol{\xi}$, are related to the dry modes.

Structural damping can be given as a percentage of the critical value based on experience. As it is well-known, added mass and hydrodynamic damping depends on the frequency. The solution of (33) gives the modal amplitudes ξ_i and displacement of any point of the structure obtained by retracking to (9).

The wet natural modes can also be determined by solving the eigenvalue problem extracted from (33)

$$\left[\mathbf{k}+\mathbf{C}-\omega^2\left(\mathbf{m}+\mathbf{A}(\omega)\right)\right]\boldsymbol{\xi}=\mathbf{0}. \qquad (34)$$

Now damping is neglected since its influence on the eigenpair is very small. The solution of (34) gives natural frequencies of ship motion and vibration in water and the corresponding so-called wet natural modes. Since added mass is a frequency dependent function, it is evident that an iteration procedure has to be employed to solve (34). Therefore, the wet modes are not orthogonal. Also, there are no more zero natural frequencies and pure rigid body modes due to their coupling with the elastic modes [13].

8. VERIFICATION OF NUMERICAL PROCEDURE

Based on the presented theory computer software has been developed. Both theory and code are checked by correlation analysis of the simulation results and the measured ones for a flexible segmented barge consisting of 12 pontoons, for which test results are available [13, 14], Figures 4 and 5. For illustration, Figures 6 and 7 show correlation of the calculated and the measured transfer function of horizontal bending moment and torque, respectively, as function of wave period T, [3, 13]. Since quite good agreement between the results is achieved, the developed procedure can be used for ship hydroelasticity analysis

Also, convergence of the applied modal superposition method in hydroelastic analysis is confirmed in the case of the same flexible barge with no forward speed [4]. Figure 8 shows absolute amplitude values ξ_i of normalized modes. The first three modes are related to sway, roll and yaw, while the remaining modes are elastic. All diagrams are determined for heading angle χ=120°, and each of them is a result of single wave length in the range λ/L=0.5-2.

Figure 4 Barge test in waves

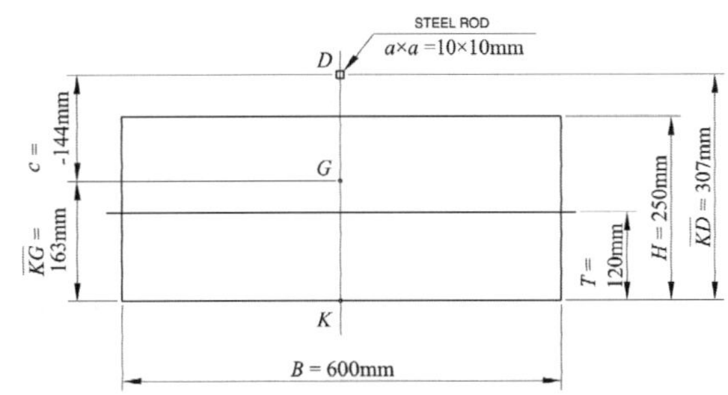

Figure 5 Barge cross-section

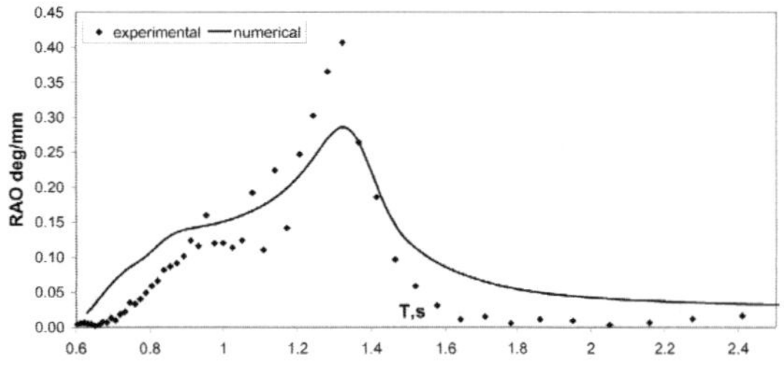

Figure 6 Transfer function of barge horizontal bending moment, $\chi = 60°$

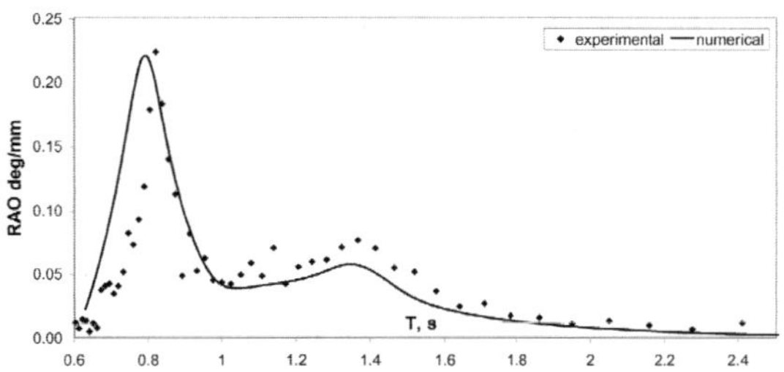

Figure 7 Transfer function of barge torque, $\chi = 60°$

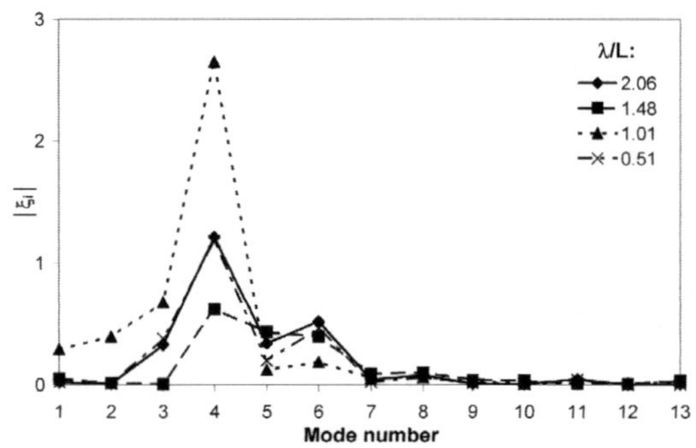

Figure 8 Modal amplitudes of coupled horizontal and torsional vibrations of flexible barge, $V = 0$ kn, $\chi = 120°$

9. HYDROELASTICITY ANALYSIS OF CONTAINER SHIP

9.1 SHIP PARTICULARS

The application of developed theory is illustrated in case of a 7800 TEU VLCS (Very Large Container Ship), Figure 9. The main vessel particulars are the following:

Length overall	$L_{oa} = 334$ m
Length between perpendiculars	$L_{pp} = 319$ m
Breadth	$B = 42.8$ m
Depth	$H = 24.6$ m
Draught	$T = 14.5$ m
Displacement, full load	$\Delta_f = 135336$ t
Displacement, ballast	$\Delta_b = 68387$ t
Engine power	$P = 69620$ kW
Ship speed	$v = 25.4$ kn.

The midship section, which shows a double skin structure with the web frames and longitudinals, is presented in Figure 10. Rows and tiers of containers at the midship section are indicated in Figure 11. Vertical positions of neutral line, deformation (shear, torsional) centre, and centre of gravity are also marked in the figure. Large distance between gravity centre and deformation centre causes high coupling of horizontal and torsional vibrations.

The ship hull stiffness properties are calculated by program STIFF, based on the theory of thin-walled girders [20, 21]. The geometrical properties rapidly change values in the engine and superstructure area due to closed ship cross-section. This is especially pronounced in case of torsional modulus, which takes quite small values for open cross-section and rather high for the closed one [4].

Influence of the transverse bulkheads is taken into account by using the equivalent torsional modulus for the open cross-sections instead of the actual values, i.e. $I_t^* = 2.4 I_t$. Physical explanation for this step is given in Appendix C.

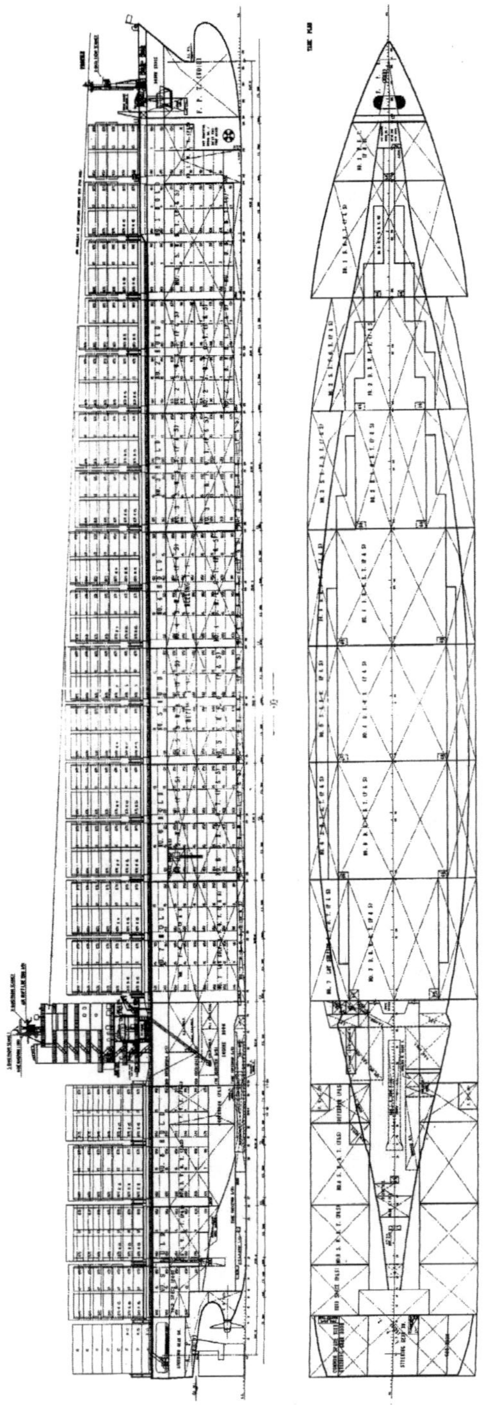

Figure 9 7800 TEU Container Ship

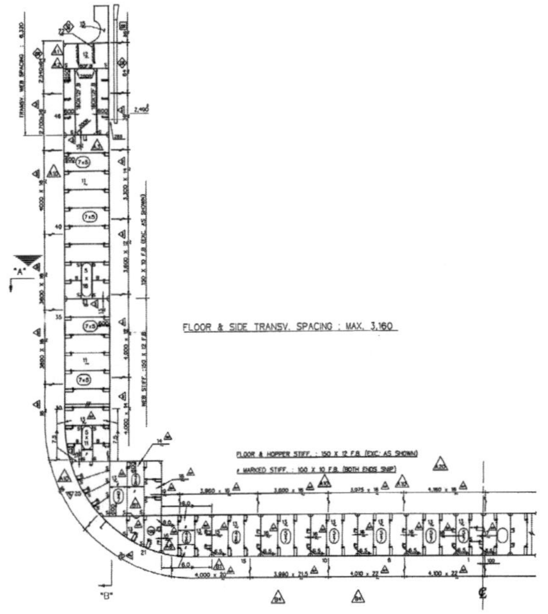

Figure 10 Midship section

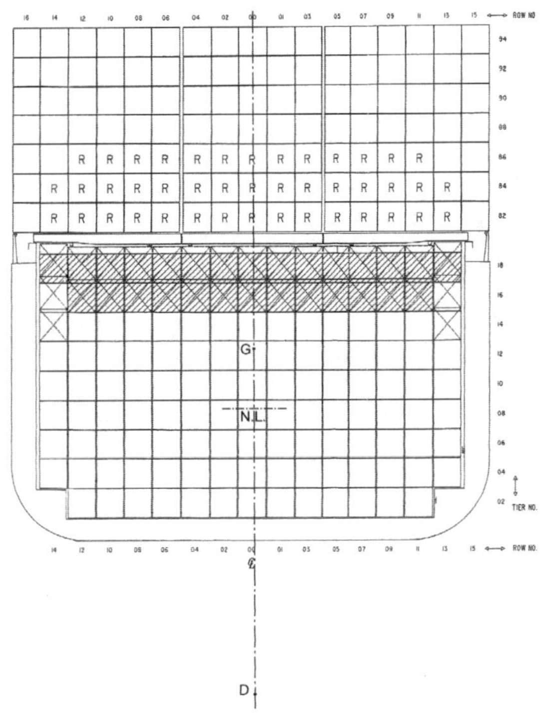

Figure 11 Container distribution at midship section

9.2 NATURAL VIBRATIONS

Dry natural vibrations of fully loaded ship are calculated by program DYANA [4]. The ship hull is divided into 50 beam finite elements. Finite elements of closed cross-section (6 d.o.f.) are used in the ship bow, ship aft and in the engine room area. Dry natural frequencies for vertical vibrations, and coupled horizontal and torsional vibrations for full load and ballast conditions are listed in Table 1. Their values for ballast condition are higher due to the lower mass. The lowest frequency value, which plays the main role in wave excitation, is detected for coupled vibrations. As a result of rather low torsional stiffness, this value belongs to primarily torsional mode.

Table 1 Dry natural frequencies, ω_i [rad/s]

Mode no.	Full load		Ballast	
	Vertical	Horizontal + torsional	Vertical	Horizontal + torsional
1	3.99	2.12	5.58	3.15
2	8.39	3.58	11.56	5.27
3	13.18	5.84	17.79	9.20
4	18.02	7.44	24.92	11.62
5	22.97	10.01	32.37	15.48
6	27.99	12.92	39.00	17.08
7	32.61	13.65	46.01	20.79
8	36.77	16.22	52.92	25.12
9	38.94	18.71	60.03	26.63
10	41.64	21.04	66.45	30.49

Once the dry natural modes of ship hull are determined it is possible to transfer the beam node displacements to the ship wetted surface for the hydrodynamic calculation, Eqs. (4) and (7). In addition, the values of natural frequencies of wet ship in full load condition are listed in Table 2. By comparing them to those of dry ship in Table 1, one sees that vertical vibration frequencies are more reduced than those of coupled horizontal and torsional vibrations.

Table 2 Wet natural frequencies, ω_i [rad/s]

Mode no.	Full load	
	Vertical	Horizontal + torsional
1	3.06	1.98
2	6.27	3.14
3	9.81	5.34
4	14.50	6.69
5	18.71	9.09

9.3 SHIP RESPONSE

Numerical calculation of ship response to waves is performed for full load and ballast condition, unit harmonic wave amplitude, and set of heading angles, ship speed and wave length [4]. Here, only some selected results for full load and ship speed of $V = 25$ kn are presented. Vertical response is shown for head sea, $\chi = 180°$, and coupled horizontal and torsional response for quartering sea, $\chi = 120°$. In all figures hydroelastic response is correlated to rigid body response determined by program HYDROSTAR [22], where the so called encounter frequency approximation was used for the solution of the seakeeping problem with forward speed. Due to the lack of measured data on large container vessels, modal damping that is usually used in global vibration calculations of merchant ships, i.e. 2% of the critical value, is applied in the analysis [24].

Figure 12 shows the transfer function of the vertical bending moment at the midship section. Heave resonance occurs at accounter frequency $\omega_e = 0.83$ rad/s, while bending resonance is achieved at $\omega_e = 3.01$ rad/s. In the former case, bending moments determined by hydroelasticity and rigid body motion are almost the same, Figure 13. In the latter case, the hydroelastic bending moment takes very high values compared to the rigid body one, Figure 14.

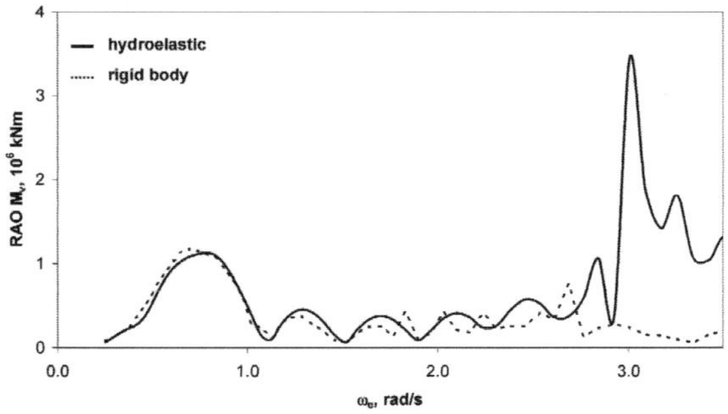

Figure 12 Transfer function of vertical bending moment, $x = 161.35$ m, $\chi = 180°$, $V = 25$ kn

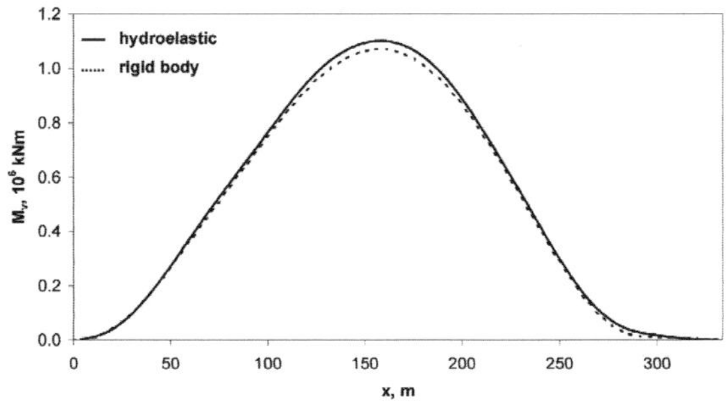

Figure 13 Vertical bending moment, $\omega_e = 0.83$ rad/s, $\lambda = 246.55$ m, $\chi = 180°$, $V = 25$ kn

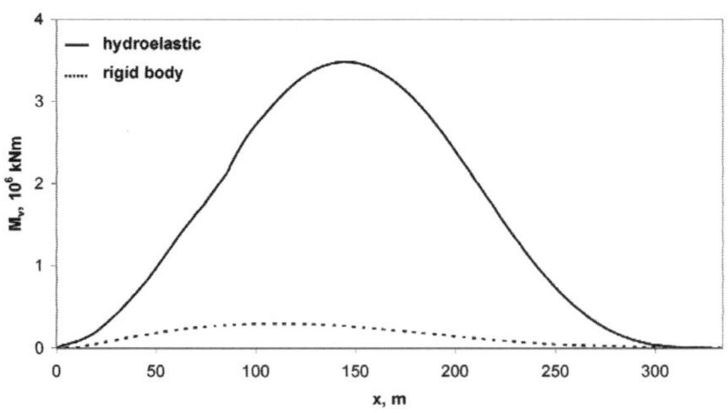

Figure 14 Vertical bending moment, $\omega_e = 3.01$ rad/s, $\lambda = 44.27$ m, $\chi = 180°$, $V = 25$ kn

The results of coupled horizontal and torsional ship response are presented in Figures 15 to 20. Figures 15 and 16 show transfer functions of horizontal bending moment and torque referred to the shear (torsional) centre, at the midship section, respectively. Hydroelastic effect is more pronounced in torsional than in bending response.

Large relative discrepancies between the hydroelastic and rigid body torque in low frequency domain in Figure 16 might be the result of ill-conditioning of the former. In hydro-rigid-body analysis, namely, sectional forces are determined by pressure integration over wetted surface, while in hydroelastic analysis sectional forces are obtained by derivation of hull displacements, which are predominantly rigid body ones in low frequency domain. Looking through the modal decomposition, derivatives of rigid body modes are zero, while those of elastic body are quite small with decreased accuracy.

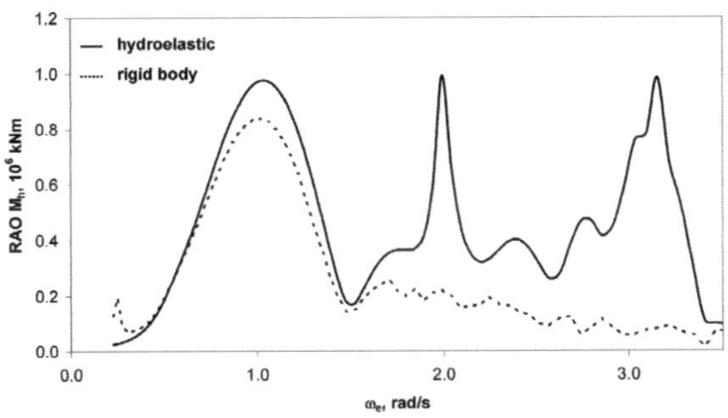

Figure 15 Transfer function of horizontal bending moment, $x = 161.35$ m, $\chi = 120°$, $V = 25$ kn

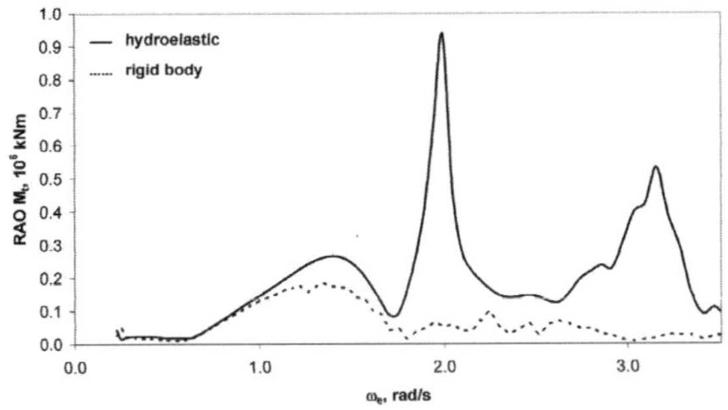

Figure 16 Transfer function of torque $x = 161.35$ m, $\chi = 120°$, $V = 25$ kn

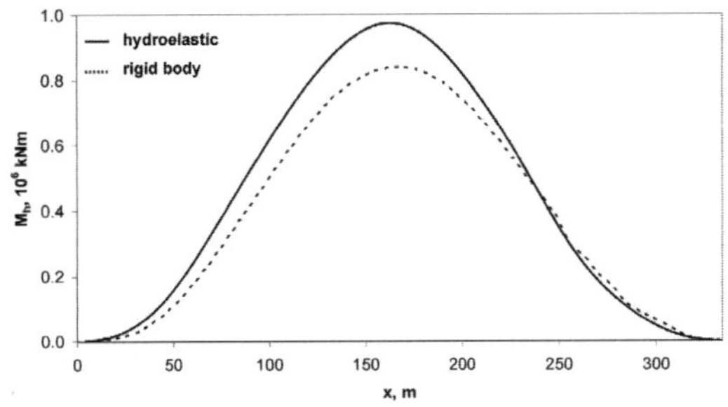

Figure 17 Horizontal bending moment, $\omega_e = 1.02$ rad/s, $\lambda = 125.79$ m, $\chi = 120°$, $V = 25$ kn

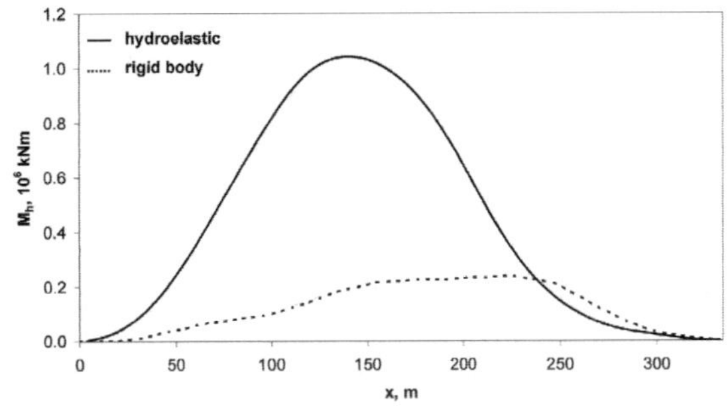

Figure 18 Horizontal bending moment, $\omega_e = 1.99$ rad/s, $\lambda = 47.43$ m, $\chi = 120°$, $V = 25$ kn

Distributions of bending moment in case of rigid body and elastic resonance are shown in Figures 17 and 18. In the former case hydroelastic and rigid body bending moments are quite close, while in the latter case the elastic bending moment is much higher than the rigid body one. Similar situation occurs with torques, Figures 19 and 20. When elastic response is in resonance with wave excitation the hydroelastic torque takes very high values.

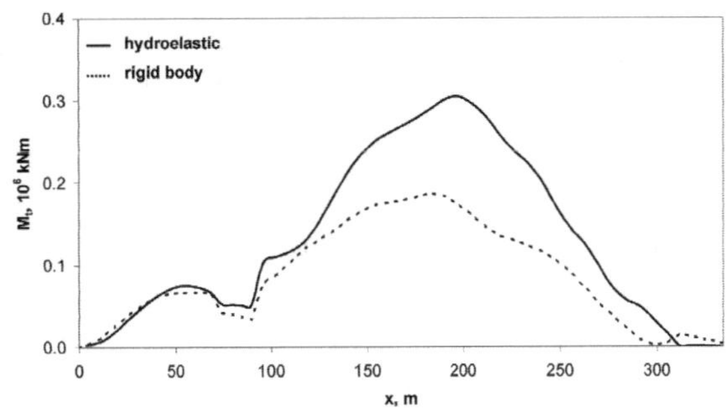

Figure 19 Torque, $\omega_e = 1.39$ rad/s, $\lambda = 79.60$ m, $\chi = 120°$, $V = 25$ kn

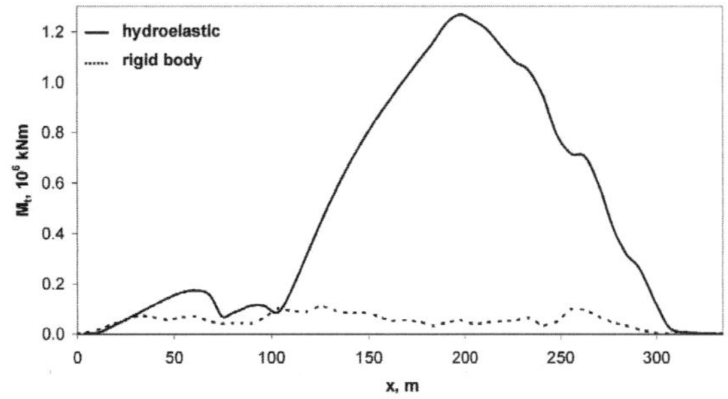

Figure 20 Torque, $\omega_e = 1.99$ rad/s, $\lambda = 47.43$ m, $\chi = 120°$, $V = 25$ kn

While diagram of vertical wave bending moment is generally a positive or a negative one-peak curve with zero boundary values, the realisation of wave torque takes different shapes, depending on ship parameters and sea conditions [11]. However, the envelope of realisations is a positive two-peak curve [10]. Therefore, such a diagram is used for the design torsional moment in classification rules [9].

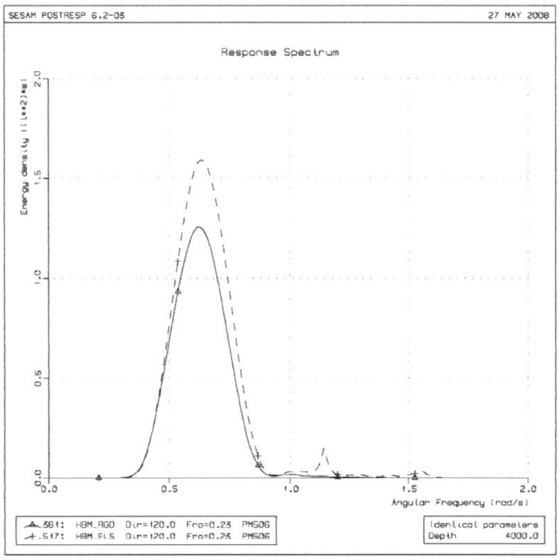

Figure 21 Energy spectrum of midship horizontal bending moment v.s. wave frequency: $H_S = 4.5$ m, $T_Z = 9.5$ s, $\chi = 120°$, $V = 25$ kn

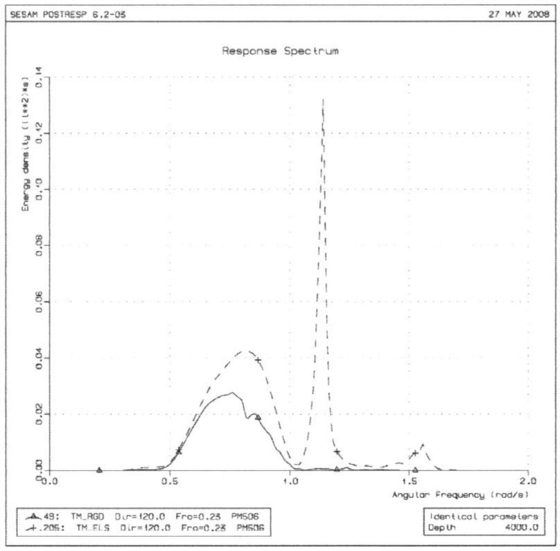

Figure 22 Energy spectrum of midship torque v.s. wave frequency: $H_S = 4.5$ m, $T_Z = 9.5$ s, $\chi = 120°$, $V = 25$ kn

Considerable difference between elastic and rigid body transfer functions results in large difference of the corresponding response energy spectrum. For example, Figures 21 and 22 represent the energy spectra of horizontal bending moment and torque amidships as a function of the wave frequency, respectively, determined for a sea state characterized by a Pierson-Moskowitz spectrum with the significant wave height $H_S = 4.5$ m and zero crossing period $T_Z = 9.5$ s. The ship sails in quartering sea, $\chi = 120°$, with a forward speed $V = 25$ kn [25].

In addition, the long term distribution of horizontal bending moment and torque amidships, based on DnV's World Wide trade scatter diagram, are shown in Figures 23 and 24, respectively. Only simplified analysis, based on the assumption of constant maximum ship's speed, was conducted. Quite large differences between the elastic and rigid body responses for the torque are noticed. As a result the hydroelastic approach leads to shorter return period of a considered torque. Also, the rules values according to IACS requirements, $M = 3444$ MNm and $T = 900$ MNm, are included in the figures for orientation purpose [26]. Another aspect of the analysis is the large exceedance of the IACS horizontal bending moment and especially torque with probability of 10^{-8}. Taking reduction of ship speed in severe seas into account, would result in more realistic long-term distributions. The above very important facts should be taken into account both in hull girder ultimate strength analysis and fatigue consideration of structural details.

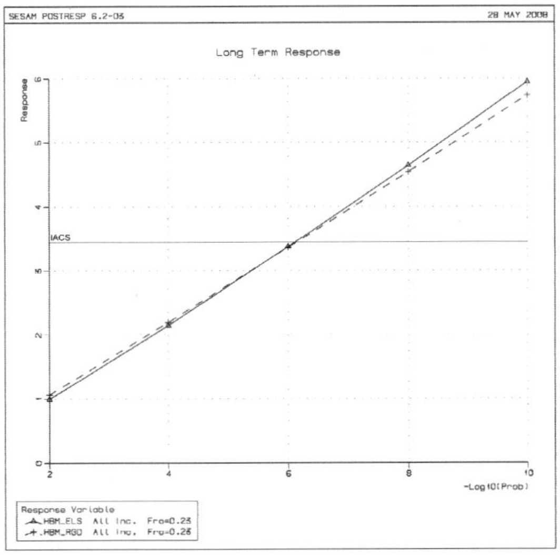

Figure 23 Long-term distribution of midship horizontal bending moment, $V = 25$ kn

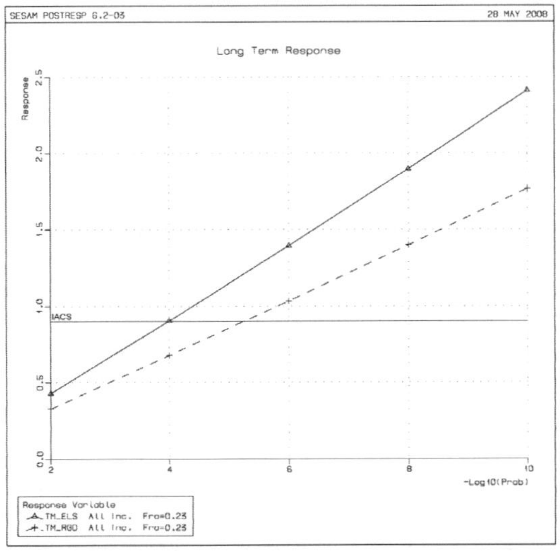

Figure 24 Long-term distribution of midship torque, $V = 25$ kn

9.4 VALIDATION OF 1D FEM MODEL

The reliability of 1D FEM analysis is verified by 3D FEM analysis of the considered ship. For this purpose, the light weight loading condition of dry ship with displacement Δ=33692 t is taken into account. There is no visual difference between the warping function of the open cross-section determined by 2D strip model and 3D FEM calculation, Figure 25. The lateral and bird view of the first dominantly torsional mode and the second dominantly horizontal flexural mode of the wetted surface, determined by 1D model, are shown in Figures 26 and 27. The first two 3D dry coupled natural modes of the complete ship structure are shown in Figures 28 and 29, where Y and Z are vertical and transversal axis, respectively. They are similar to those of 1D analysis for the wetted surface. Warping of the transverse bulkheads, which increases the hull torsional stiffness, is evident. The first four corresponding natural frequencies obtained by 1D and 3D analyses are compared in Table 3. Quite good agreement is achieved. Values of natural frequencies for higher modes are more difficult to correlate, since strong coupling occurs between global hull modes and local substructure modes of 3D analysis. Each uncoupled global natural frequency is split into two frequencies, one with a higher, and another with a lower value [6].

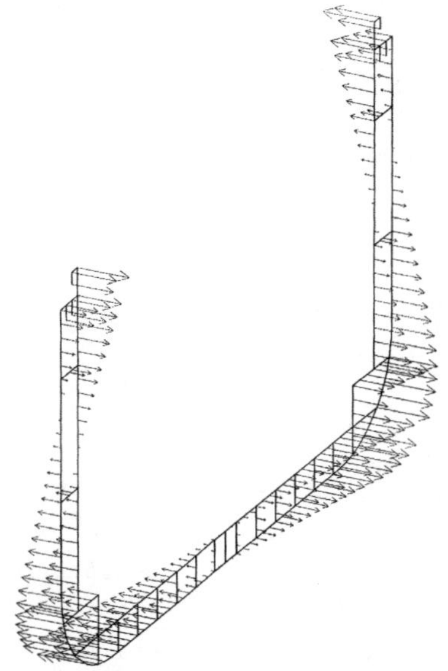

Figure 25 Warping function of open cross-section

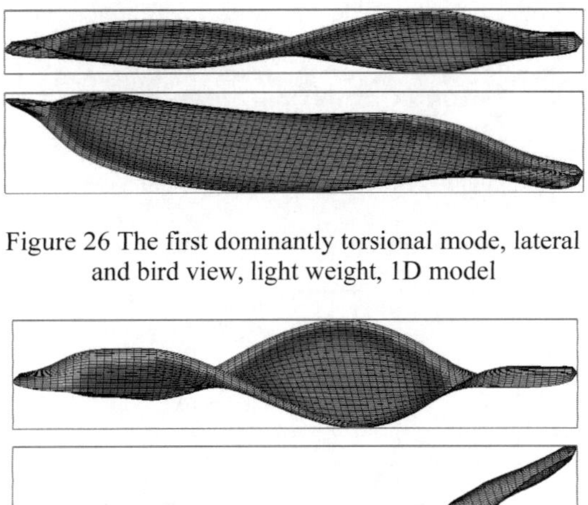

Figure 26 The first dominantly torsional mode, lateral and bird view, light weight, 1D model

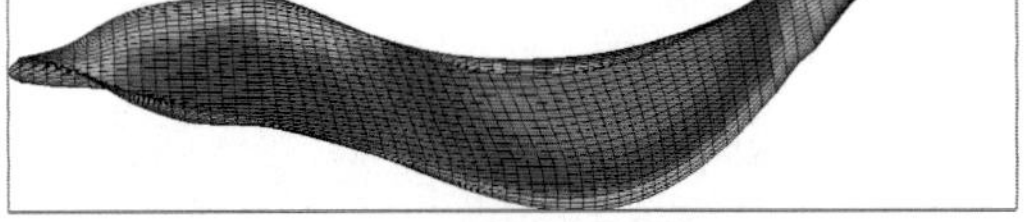

Figure 27 The second dominantly horizontal mode, lateral and bird view, light weight, 1D model

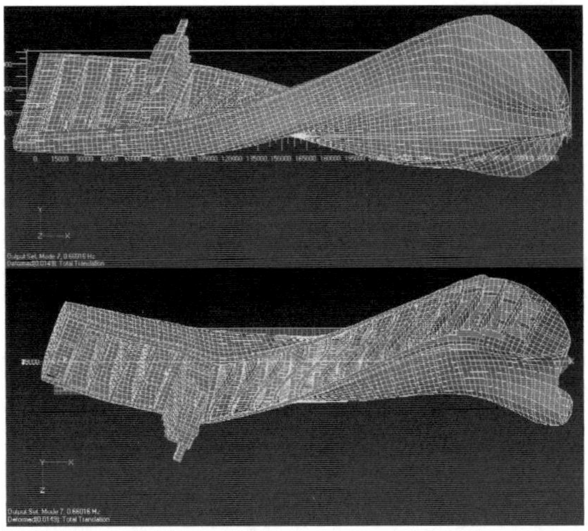

Figure 28 The first dominantly torsional mode, lateral and bird view, light weight, 3D model

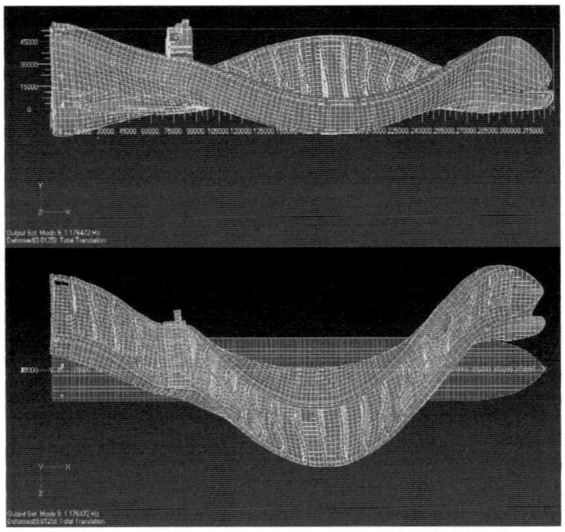

Figure 29 The second dominantly horizontal mode, lateral and bird view, light weight, 3D model

Table 3 Dry natural frequencies, light weight, ω_i [rad/s]

Mode no.	Vertical		Horizontal + torsional		Mode no.
	1D	3D	1D	3D	
1	7.35	7.33	4.17	4.15	1(H0 + T1)
2	15.00	14.95	7.34	7.40	2(H1 + T2)
3	24.04	22.99	12.22	12.09	3(H2 + T3)
4	35.08	34.21	15.02	16.22	4(H3 + T4)

10. CONCLUSION

Ultra large container ships are quite flexible and stretch the bounds of present classification rules for reliable structure design. Therefore, hydroelastic analysis has to be performed. One of the basic steps is dry natural vibration analysis of ship hull. Vertical vibration calculation is performed in a standard way, while the coupled horizontal and torsional vibration is rather complex.

The performed vibration analysis of the 7800 TEU Container ship shows that the developed hydroelasticity theory, utilizing 1D FEM structural model and 3D hydrodynamic model, is an efficient tool for application in ship hydroelastic analyses. The obtained results point out that the transfer functions of hull sectional forces in case of resonant vibration (springing) are much higher than in resonant ship motion. This is the main issue of the paper, which requires further investigation.

In order to complete hydroelastic analysis of container ships and confirm its importance for ship safety, it is necessary to proceed further to ship motion calculation in irregular waves for different sea states, based on the known transfer functions. This includes determination of global wave loads, i.e. bending and torsional moments and their conversion into stresses, stress concentration in critical areas of ship structures, especially in hatch corners due to suspended warping, and fatigue of structural details. The same numerical procedures, which are used in hydro-rigid-body analysis, are at disposal and are applicable in this case.

The used beam model of ship hull is a reasonable choice for determining wave load. However, stress concentration in hatch corners calculated directly by the beam model is underestimated [5, 27]. This problem can be overcome by applying substructure approach, i.e. 3D FEM model of substructure with imposed boundary conditions from beam response. In any case, 3D FEM model of complete ship is preferable from the viewpoint of determining stress concentration.

Another aspect of hydroelasticity is slamming and whipping, which are related to vertical bending. This impulsive load is considered in time domain and is outside the scope of this paper. However, it must be considered within general container ship problematics.

At the end of a complete investigation, which also has to include model tests and full scale measurements, it will be possible to decide on the extent of the revision of Classification Rules for the design and construction of ultra large container ships. The main reason is indication on the large exceedance of the IACS bending moments and torque. Also, operational conditions for this type of ships, i.e. reduction of ship speed and change of heading angle for some sea states due to resonant flexural and torsional response, as well as parametric rolling have to be reconsidered.

APPENDIX A: ADVANCED THIN-WALLED GIRDER THEORY

A.1 DISPLACEMENTS AND FORCES

Referring to the flexural beam theory [6], the total beam deflection, w, consists of the bending deflection, w_b, and the shear deflection, w_s, i.e., Figure A1

$$w = w_b + w_s \,. \tag{A1}$$

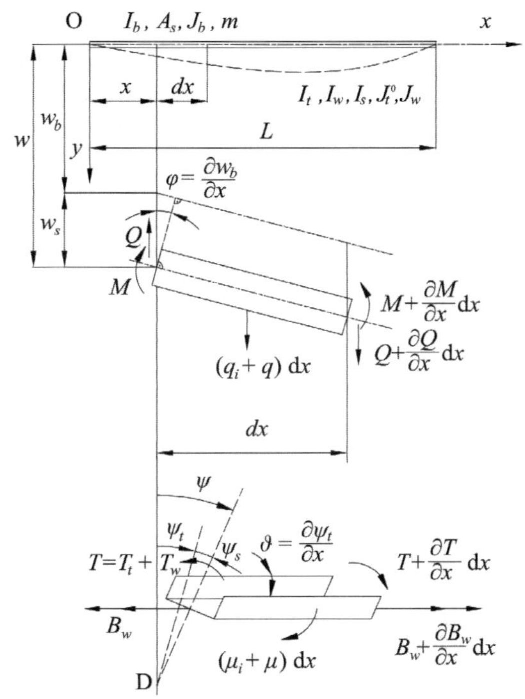

Figure A1 Beam bending and torsion

The shear deflection is a function of w_b

$$w_s = -\frac{EI_b}{GA_s}\frac{\partial^2 w_b}{\partial x^2}\,, \tag{A2}$$

where E and G are the Young's and shear modulus, respectively, while I_b and A_s are the moment of inertia of cross-section and shear area, respectively. The angle of cross-section rotation is caused by the bending deflection

$$\varphi = \frac{\partial w_b}{\partial x}\,. \tag{A3}$$

The cross-sectional forces are the bending moment

$$M = -EI_b \frac{\partial^2 w_b}{\partial x^2} \tag{A4}$$

and the shear force

$$Q = GA_s \frac{\partial w_s}{\partial x} = -EI_b \frac{\partial^3 w_b}{\partial x^3}\,. \tag{A5}$$

Concerning torsion, the total twist angle, ψ, consists of the pure twist angle, ψ_t, and the shear contribution, ψ_s, i.e., Figure A1 [28]

$$\psi = \psi_t + \psi_s \,. \tag{A6}$$

Referring to the analogy of torsion and bending [29], the shear angle depends on the twist angle, similarly to Eq. (*A2*)

$$\psi_s = -\frac{EI_w}{GI_s}\frac{\partial^2 \psi_t}{\partial x^2}\,, \tag{A7}$$

where I_w is the warping modulus and I_s is the shear inertia modulus. The second beam displacement, which causes warping of cross-section (similarly to the cross-section rotation due to bending) is a variation of the pure twist angle

$$\vartheta = \frac{\partial \psi_t}{\partial x}\,. \tag{A8}$$

The sectional forces include the total torque, T, consisting of the pure torsional torque, T_t, and the warping torque T_w [28] i.e.

$$T = T_t + T_w\,, \tag{A9}$$

where

$$T_t = GI_t \frac{\partial \psi_t}{\partial x} \tag{A10}$$

$$T_w = GI_s \frac{\partial \psi_s}{\partial x} = -EI_w \frac{\partial^3 \psi_t}{\partial x^3} \tag{A11}$$

and the bimoment given by

$$B_w = EI_w \frac{\partial^2 \psi_t}{\partial x^2}\,. \tag{A12}$$

A.2 FINITE ELEMENT FORMULATION

1D FEM procedure for vertical ship hull vibrations is well known in literature. Coupled horizontal and torsional vibrations are a more complex problem. Due to analogy between bending and torsion the same shape functions, represented by Hermitian polynomials, are used. The matrix finite element equation for coupled natural vibrations yields [18]

$$\mathbf{f}^e = \mathbf{k}^e\boldsymbol{\delta}^e + \mathbf{m}^e\ddot{\boldsymbol{\delta}}^e, \qquad (A13)$$

where $\mathbf{f}^e$ is nodal forces vector, $\boldsymbol{\delta}^e$ is nodal displacements vector, $\mathbf{k}^e$ is stiffness matrix, and $\mathbf{m}^e$ is mass matrix.

These quantities consist of flexural and torsional parts

$$\mathbf{f}^e = \begin{Bmatrix}\mathbf{P}\\ \mathbf{R}\end{Bmatrix}, \quad \boldsymbol{\delta}^e = \begin{Bmatrix}\mathbf{U}\\ \mathbf{V}\end{Bmatrix} \qquad (A14)$$

$$\mathbf{k}^e = \begin{bmatrix}\mathbf{k}_{bs} & 0\\ 0 & \mathbf{k}_{wt}\end{bmatrix}, \quad \mathbf{m}^e = \begin{bmatrix}\mathbf{m}_{sb} & \mathbf{m}_{st}\\ \mathbf{m}_{ts} & \mathbf{m}_{tw}\end{bmatrix}. \qquad (A15)$$

Vectors of nodal forces and displacements are

$$\mathbf{P} = \begin{Bmatrix}-Q(0)\\ M(0)\\ Q(l)\\ -M(l)\end{Bmatrix}, \quad \mathbf{R} = \begin{Bmatrix}-T(0)\\ -B_w(0)\\ T(l)\\ B_w(l)\end{Bmatrix} \qquad (A16)$$

$$\mathbf{U} = \begin{Bmatrix}w(0)\\ \varphi(0)\\ w(l)\\ \varphi(l)\end{Bmatrix}, \quad \mathbf{V} = \begin{Bmatrix}\psi(0)\\ \vartheta(0)\\ \psi(l)\\ \vartheta(l)\end{Bmatrix}. \qquad (A17)$$

In the above formulae symbols Q, M, T and B_w denote shear force, bending moment, torque and warping bimoment, respectively. Also, w, φ, ψ and ϑ are deflection, rotation of cross-section, twist angle and its variation, respectively. The submatrices, which are specified in Ref. [30], have the following meaning:

$\mathbf{k}_{bs}$ – bending - shear stiffness matrix
$\mathbf{k}_{wt}$ – warping - torsion stiffness matrix
$\mathbf{m}_{sb}$ – shear - bending mass matrix
$\mathbf{m}_{tw}$ – torsion - warping mass matrix
$\mathbf{m}_{st} = \mathbf{m}_{ts}^T$ – shear - torsion mass matrix.

It is evident that coupling between horizontal and torsional vibrations is realised through the mass matrix due to eccentricity of the centre of gravity and shear centre.

Before assembling of finite elements it is necessary to transform Eq. (A13) in such a way that all the nodal forces as well as nodal displacement, Eqs. (A16) and (A17), are related to the first and then to the second node. Furthermore, Eq. (A13) has to be transformed from local to global coordinate system. The origin of the former is located at the shear centre, and of the latter at the base line.

APPENDIX B: DIFFERENT FORMULATION OF RESTORING STIFFNESS

It seems that the determination of the ship restoring stiffness is a more difficult problem than one would expect. Basically, there are two approaches to solve the problem, a hydromechanical one and a structural one. Within the former approach, there are three well-known formulations for the hydrostatic stiffness of a flexible structure. They are presented here in the index (tensor) notation.

Price and Wu [31]:

$$C_{ij} = -\rho g \iint_S h_3^j h_k^i n_k dS\,, \qquad (B1)$$

Newman [32]:

$$C_{ij} = -\rho g \iint_S \left(h_3^i + Z h_{l,l}^i\right) h_k^j n_k dS\,, \qquad (B2)$$

Riggs [33]:

$$C_{ij} = -\rho g \iint_S \left(h_3^i + Z h_{l,l}^i\right) h_k^j n_k dS + g \iiint_V \rho_S h_{3,k}^i h_k^j dV\,, \qquad (B3)$$

where:

C_{ij} –element of the modal restoring stiffness matrix
h – natural mode (rigid and elastic)
X, Y, Z– global coordinates
S – wetted surface
n – unit normal vector to the wetted surface
V – volume of the structure
ρ, ρ_S – fluid and structure mass density
g – gravity constant.

In the first formulation, Eq. (B1), only the basic hydrostatic pressure term is considered. Eq. (B2) extends the first one, giving the correct hydrostatic pressure coefficients. However, neither first formulation, nor the second one give the complete restoring stiffness coefficients for the rigid body motion because the gravity part is missing. The necessity of the gravity contribution was commented in [32]. This is realized in the third formulation, Eq. (B3). All three formulations result in an asymmetric stiffness matrix.

The formulation of Huang and Riggs [34], as a more structural approach, results in the following definition of the complete restoring stiffness, also including the structure internal stresses and structure gravity as a special case

$$C_{ij} = -\rho g \iint_S h_k^i \left(h_3^j + Z h_{l,l}^j\right) n_k dS + \rho g \iint_S Z h_l^i h_{k,l}^j n_k dS + \iiint_V \sigma_{lm} h_{k,l}^i h_{k,m}^j dV. \qquad (B4a)$$

In that particular paper, the importance of internal stresses, as a way to complete restoring stiffness, is pointed out. Moreover, the authors claim that the restoring stiffness matrix formulated in such a way has the important symmetry property, as an obvious advantage in the hydroelasticity analysis.

The fifth formulation of restoring stiffness, presented by Malenica [23] and applied in [15], is based on the variational principle and can be presented in the index notation like the others

$$C_{ij} = \rho g \iint_S h_3^j h_k^i n_k dS + \rho g \iint_S Z h_l^j h_{k,l}^i n_k dS \\ + \rho g \iint_S Z \left[h_{l,l}^j n_k - h_{l,k}^j n_l \right] h_k^i dS + g \iiint_V \rho_S h_{3,k}^i h_k^j dV. \tag{B5}$$

Molin [35] derived the restoring coefficients by employing the vector differential and integral calculus [36], in the following form:

$$C_{ij} = \rho g \iint_S \left\{ \begin{matrix} \left(h_z^j + Z \nabla \cdot \mathbf{h}^j \right) \mathbf{h}^i \mathbf{n} \\ + Z \left[\left(\mathbf{h}^j \nabla \right) \mathbf{h}^i - \left(\mathbf{h}^i \nabla \right) \mathbf{h}^j \right] \mathbf{n} \end{matrix} \right\} dS \\ + g \iiint_V \rho_s \left(\mathbf{h}^j \nabla \right) h_z^i dV. \tag{B6}$$

It can be proved that formulations (B5) and (B6) are identical.

The above formulae show the development of the restoring stiffness from the basic hydrostatic part, through an extension by the gravity contribution up to taking the stress distribution into account.

In formulation (B4a) the orientation of the normal vector is opposite to that in (B5). Due to easier correlation of these two formulations, let us change the sign of the normal vector in (B4a) and present it in the form similar to Eq. (B5)

$$C_{ij} = \rho g \iint_S h_3^j h_k^i n_k dS \\ + \rho g \iint_S Z \left(h_{l,l}^j h_k^i - h_{k,l}^j h_l^i \right) n_k dS + \iiint_V \sigma_{lm} h_{k,l}^i h_{k,m}^j dV \ . \tag{B4b}$$

By comparing (B4b) and (B5), it is obvious that the first integrals (pressure variation) are identical. The second integral (mode variation) of (B5) is not taken into account in (B4b). The third integral in (B5) and the second integral in (B4b) are related to the normal vector variation. Their first terms are identical, while there are some differences in the second terms. The volume integral in (B5) (gravity force variation) is equal to that in (B3), and to the rigid body contribution of the stress integral in (B4b). Differences between (B4b) and (B5) are discussed in Ref. [37].

APPENDIX C: CONTRIBUTION OF TRANSVERSE BULKHEADS TO HULL STIFFNESS

This problem for container ships is extensively analysed in [38], where torsional modulus of ship cross-section is increased proportionally to the bulkhead strain energy. The bulkhead is considered as an orthotropic plate with very strong stool [39]. The bulkhead strain energy is determined for the given warping of cross-section as a boundary condition. The warping causes bulkhead screwing and bending. Here, only the review of the final results is presented.

The bulkhead deflection (axial displacement) is given by the following formula, Figure C1:

$$u(y,z) = -y \left\{ (z-d) + \left[1 - \left(\frac{y}{b} \right)^2 \right] \frac{z^2}{H} \left(2 - \frac{z}{H} \right) \right\} \psi', \tag{C1}$$

where H is the ship height, b is one half of bulkhead breadth, d is the distance of warping centre from double bottom neutral line, y and z are transverse and vertical coordinates, respectively, and ψ' is the variation of twist angle.

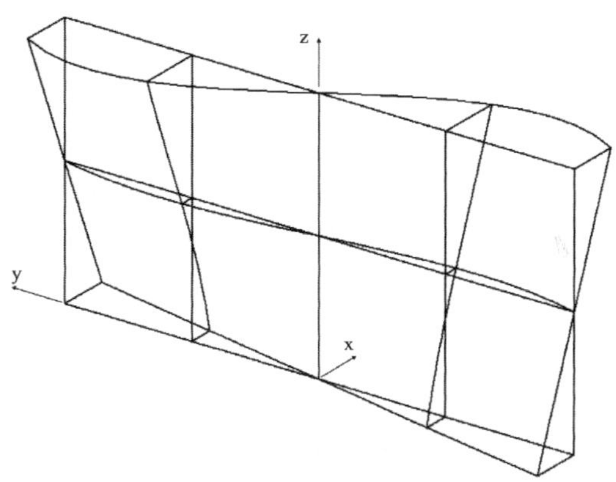

Figure C1 Shape of bulkhead deformation

The bulkhead grillage strain energy includes vertical and horizontal bending with contraction, and torsion [38].

$$U_g = \frac{1}{1-\nu^2} \left[\frac{116H^3}{35b} i_y + \frac{32b^3}{105H} i_z + \frac{8Hb}{75} \nu \left(i_y + i_z \right) \right. \\ \left. + \frac{143Hb}{75} (1-\nu) i_t \right] E \psi'^2 \tag{C2}$$

where i_y, i_z and i_t are the average moments of inertia of cross-section and torsional modulus per unit breadth, respectively.

The stool strain energy is comprised of the bending, shear and torsional contributions

$$U_s = \left[\frac{12h^2 I_{sb}}{b} + 72(1+\nu)\frac{h^2}{b^3}\frac{I_{sb}^2}{A_s} + \frac{9bI_{st}}{10(1+\nu)} \right] E\psi'^2, \quad \text{(C3)}$$

where I_{sb}, A_s and I_{st} are the moment of inertia of cross-section, shear area and torsional modulus, respectively. Quantity h is the stool distance from the inner bottom, Figure C2.

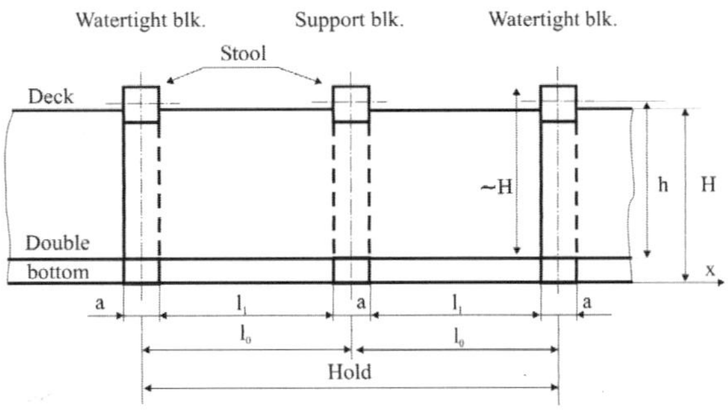

Figure C2 Longitudinal section of container ship hold

The equivalent torsional modulus yields, Figure C2

$$I_t^* = \left[1 + \frac{a}{l_1} + \frac{4(1+\nu)C}{I_t l_0} \right] I_t, \quad \text{(C4)}$$

where a is the web height of bulkhead girders (frame spacing), l_0 is the bulkhead spacing, $l_1 = l_0 - a$ is the net length, and C is the energy coefficient

$$C = \frac{U_g + U_s}{E\psi'^2}. \quad \text{(C5)}$$

The second term in (C4) is the main contribution of the bulkhead as the closed cross-section segment of ship hull, and the third one comprises the bulkhead strain energy.

Large container ships are designed with alternate watertight and support bulkheads, Figure C2. These bulkheads for the considered ship are shown in Figures C3 and C4. The stiffness parameters of the bulkhead girders are listed in Tables C1 and C2, while the stool parameters are given in Table C3. The bulkhead dimensions are the following:

$$H = 24.60 \text{ m}, \; b = 19.17 \text{ m}, \; l_0 = 14.44 \text{ m}, \; a = 1.81 \text{ m}.$$

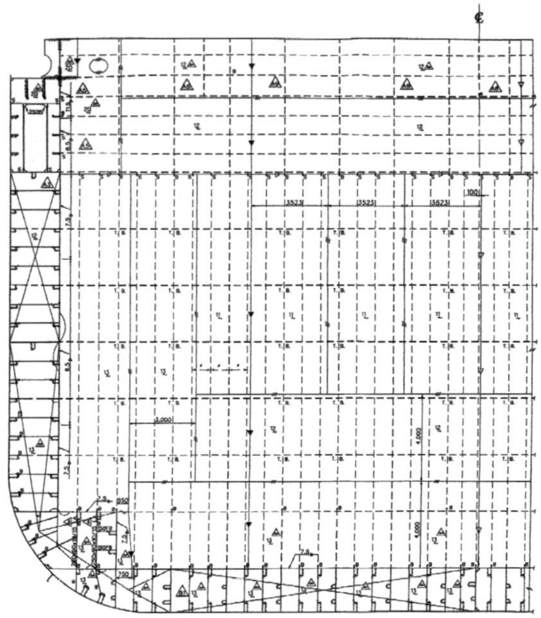

Figure C3 Watertight bulkhead

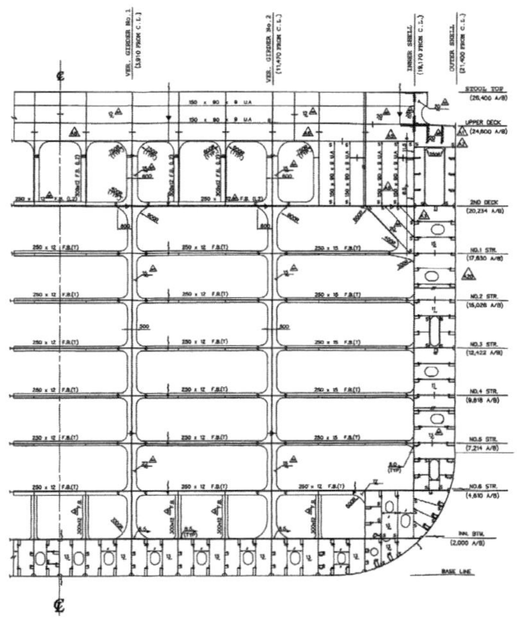

Figure C4 Support bulkhead

The bulkhead strain energy, determined according to (C2) and (C3), is summarized in Table C4, where also the energy coefficient is calculated as the average value of the watertight and support bulkhead strain energies. Most of the hull induced energy is absorbed by the stool. Thus, the equivalent torsional modulus for midship section yields $I_t^* = 2.4I_t$. This value is applied for all ship cross-sections as the first approximation.

Table C1 Stiffness parameters of watertight bulkhead

Girder	Moment of inertia	Torsional modulus	Girder spacing	Moment of inertia per unit breadth	Torsional modulus per unit breadth
	I (m^4)	I_t (m^4)	c (m)	i (m^3)	i_t (m^3)
Horizontal	0.02356	0.01555	2.6	0.00906	0.00493
Vertical	0.04196	0.03205	7.9	0.00531	

Table C2 Stiffness parameters of support bulkhead

Girder	Moment of inertia	Torsional modulus	Girder spacing	Moment of inertia per unit breadth	Torsional modulus per unit breadth
	I (m^4)	I_t (m^4)	c (m)	i (m^3)	i_t (m^3)
Horizontal	0.00972	0.00486	2.6	0.00374	0.001696
Vertical	0.01944	0.01215	7.9	0.00246	

Table C3 Stool stiffness parameters

Shear area	Moment of inertia	Torsional modulus
A_s (m^2)	I_s (m^4)	I_{ts} (m^4)
0.45	0.07804	0.131

Table C4 Bulkhead strain energy, $U/\left(E\psi'^2\right)$

Watertight bulkhead		Support bulkhead		Energy coefficient
Grillage	Stool	Grillage	Stool	C, Eq. (C5)
(1)	(2)	(3)	(4)	(5) = [(1)+ (2)+ (3)+ (4)]/2
29.691	28.872	12.051	28.872	49.743

ACKNOWLEDGMENT

This research was supported by the Ministry of Science, Education and Sport of the Republic of Croatia (Project No. 120-1201703-1704).

REFERENCES

1. Proceedings of International Conference Design & Operation of Container Ships, RINA, London, 2006.

2. PAYER, H. 'Technological an Economic Implications of Mega-Container Carriers', SNAME Transactions, Vol.109 (2001), 101-120.

3. MALENICA, Š., SENJANOVIĆ, I., TOMAŠEVIĆ, S., STUMPF, E., 'Some aspects of hydroelastic issues in the design of ultra large container ships', The 22nd IWWWFB, Plitvice Lakes, Croatia, 2007.

4. TOMAŠEVIĆ, S. 'Hydroelastic model of dynamic response of Container ships in waves', Doctoral Thesis, University of Zagreb, Zagreb, 2007.

5. PEDERSEN, P.T., 'Torsional response of container ships', Journal of Ship Research, Vol.29, (1985) 3, 194-205.

6. SENJANOVIĆ, I., GRUBIŠIĆ, R., 'Coupled horizontal and torsional vibration of ship hull with large hatch openings' Computers & Structures, Vol.41, (1991) 2, 213-226.

7. 'Principles of Naval Architectures', SNAME, Jersey City, 1988.

8. HUGHES, O.F., 'Ship Structural Design', SNAME, Jersey City, 1988.

9. VOLSGARD, S., SVENSEN, T.E., THORKILDSEN, H., 'A computational method for analysis of container vessels', SNAME Transactions, Vol.103 (1995), 371-393.

10. IIJIMA, K., SHIGEMI, T., MIYAKE, R., KUMANO, A., 'A practical method for torsional strength assessment of container ship structures', Marine Structures, (2004) 17, 355-384.

11. SHI, B., LIU, D., WIERNICKI, C., 'Dynamic loading approach for structural evaluation of ultra large container carriers', SNAME Transactions, Vol.113 (2005), 402-417.

12. BISHOP, R.E.D., PRICE, W.G., 'Hydroelasticity of Ships', Cambridge University Press, 1979.

13. MALENICA, Š., MOLIN, B., REMY, F., SENJANOVIĆ, I., 'Hydroelastic response of a barge to impulsive and non-impulsive wave load', Hydroelasticity in Marine Technology, Oxford, UK, (2003), 107-115.

14. REMY, F., MOLIN, B., LEDOUX, A., 'Experimental and numerical study of the wave response of a flexible barge', Hydroelasticity in Marine Technology, Wuxi, China, (2006), 255-264.

15. SENJANOVIĆ, I., MALENICA, Š., TOMAŠEVIĆ, S., 'Investigation of ship hydroelasticity', Ocean Engineering, Vol.35, (2008) 5-6, 523-535.

16. SALVESEN, N., TUCK, E.O., FALTINSEN, O.M. 'Ship motion and sea loads', SNAME Transactions, Vol.70, (1970), 250-287.

17. BATHE, K.J., 'Finite Element Procedures', Prentice Hall, 1996.

18. SENJANOVIĆ, I., 'Finite Element Method in Ship Structures', University of Zagreb, Zagreb, 1998.

19. MALENICA, Š., SENJANOVIĆ, I., CHEN, X.B., 'Automatic mesh generation for naval and offshore hydrodynamic simulations', SORTA'04, Plitvice Lakes, Croatia, 2004.

20. SENJANOVIĆ, I., FAN, Y., 'A higher-order theory of thin-walled girders with application to ship structures', Computers & Structures, Vol.43, (1992) 1, 31-52.

21. SENJANOVIĆ, I., FAN, Y., 'A finite element formulation of ship cross-sectional stiffness parameters', Brodogradnja 41 (1993) 1, 27-36.

22. Hydrostar, User's manual, Bureau Veritas, Paris, 2006.

23. MALENICA, Š., 'Some aspects of hydrostatic calculations in linear seakeeping', The 14th NAV Conference, Palermo, Italy, 2003.

24. KUMAI, T., 'Damping factors in the higher modes of ship vibrations', European Shipbuilding, Vol.7, (1958) 1, 29-34.

25. SESAM,User's manual, Det norske Veritas, Høvik, 2008.

26. PARUNOV, J., SENJANOVIĆ, I., SOARES, C.G., 'Hull-girder reliability of new generation oil tankers', Marine Structures, 20 (2007), 49-70.

27. SENJANOVIĆ, I., FAN, Y., 'Pontoon torsional strength analysis related to ship with large deck openings', Journal of Ship Research, Vol.35, (1991) 4, 339-351.

28. SENJANOVIĆ, I., ĆATIPOVIĆ, I., TOMAŠEVIĆ, S., 'Coupled flexural and torsional vibrations of ship-like structures', Thin-Walled Structures, Vol.45 (2007), 1002-1021.

29. PAVAZZA, R., 'Torsion of thin-walled beams of open cross-sections with influence of shear', International Journal of Mechanical Sciences, 47, (2005), 1099-1122.

30. SENJANOVIĆ, I., TOMAŠEVIĆ, S., VLADIMIR, N., 'An advanced theory of thin-walled girders with application to ship vibrations', Thin-Walled Structures, (submitted).

31. PRICE, W.G., WU, Y., 'Hydroelasticity of Marine Structures', Theoretical and Applied Mechanics, Eds. NIORSON, F.I. and OLHOFF, N., Elsevier Science Publishers B.V., North Holland, 1985.

32. NEWMAN, J.N., 'Wave effects on deformable bodies', Applied Ocean Research, 16, (1994), 47-59.

33. RIGGS, H.R., 'Hydrostatic stiffness of flexible floating structures', Proceedings of the International Workshop on Very Large Floating Structures, Hayama, Japan, 1996, 229-234.

34. HUANG, L.L., RIGGS, H.R., 'The hydrostatic stiffness of flexible floating structure for linear hydroelasticity', Marine Structures, 13, (2000), 91-106.

35. MOLIN, B., 'Hydrostatique d'un corps deformable', Technical note, Ecole Superieure de Marseille, France, 2003.

36. KREYSZIG, E., 'Advanced Engineering Mathematics', John Wiley & Sons, 1993.

37. SENJANOVIĆ, I., TOMIĆ, M., TOMAŠEVIĆ, S., 'An explicit formulation for restoring stiffness and its performance in hydroelasticity', Ocean Engineering, (submitted).

38. SENJANOVIĆ, I., TOMAŠEVIĆ, S., RUDAN, S., SENJANOVIĆ, T., 'Role of transverse bulkheads in hull stiffness of large container ships', Engineering Structures (2008), doi:10.1016/j.engstruct.2008.01.016.

39. SZILARD, R., 'Theories and Applications of Plate Analysis', New York, John Wiley & Sons, 2004.

AUTHOR'S BIOGRAPHIES

Ivo Senjanović, D.Sc. Professor of Naval Architecture at the Faculty of Mechanical Engineering and Naval Architecture, University of Zagreb. Teaching several courses on strength and vibration of ships and offshore structures, submarines etc. Investigation fields are ship stability, launching, numerical methods, numerical simulations, non-linear dynamics, ship strength and vibrations, shell theory and design of pressure vessels. Over 120 papers in scientific journals and conference proceedings. Book "Theory of Plates and Shells", "Finite Element Method in Ship Structure Analysis", "Ship Vibrations". ISSC member and fellow of the Croatian Academy of Sciences and Arts.
E-mail: ivo.senjanovic@fsb.hr
http://mahazu.hazu.hr/Akademici/ISenjanovic.html

Stipe Tomašević, D.Sc. Lecturer at the Faculty of Mechanical Engineering and Naval Architecture, University of Zagreb. Support in education on the finite element method and software application. Investigation of ship strength, vibration and fatigue.
E-mail: stipe.tomasevic@fsb.hr

Smiljko Rudan, D.Sc. Lecturer at the Faculty of Mechanical Engineering and Naval Architecture, University of Zagreb. Support in education on the finite element method and software application. Investigation of ship strength, vibration and fatigue.

Marko Tomić, Dipl. Ing, assistant, Faculty of Mechanical Engineering and Naval Architecture, University of Zagreb.

Nikola Vladimir, Dipl. Ing, assistant, Faculty of Mechanical Engineering and Naval Architecture, University of Zagreb.

Šime Malenica, graduated 1990 in Naval Architecture at the Faculty of Mechanical Engineering and Naval Architecture, University of Zagreb. He obtained his PhD degree in 1994, at the University Paris VI, on the problem of nonlinear water wave diffraction. He has published more than 70 papers both in scientific journals and at the specialized conferences. He is the actual head of hydro-structure section in the Bureau Veritas Research Department leading several national and international research projects. His main fields of research concern the hydro-structural interactions due to the sea waves (linear, nonlinear, frequency domain, time domain, ships and offshore…). Actual topics concern the hydroelasticity issues for ships (springing, whipping, slamming, sloshing…).
E-mail: sime.malenica@bureauveritas.com

FULL SCALE MEASUREMENTS OF STRESS AND DEFLECTIONS OF POST-PANAMAX CONTAINER SHIP

M. Toyoda, and **T. Okada**, IHI Marine United Inc., Japan
T. Maeda, IHI Corporation, Japan
T. Matsumoto, Nippon Kaiji Kyokai, Japan

SUMMARY

Strong demand on efficient container trade has accelerated the development of large container ships. The increase of size brings concern about not only strength but deformation of large hatch opening, such as interferences between hull structure and deck fittings.

To cope with such problems, full-scale measurements on two post panamax container ships had been conducted separately. Procedures to derive components of deformations from selected measurement point were developed and actual values were calculated based on actual measurement. Comparison between both ships shows that design assumption is generally appropriate.

Furthermore, whipping stresses were observed during measurement. Magnitude of effect on fatigue strength by whipping was investigated based on actual measurement, and it is confirmed that appropriate fatigue design has sufficient fatigue strength against whipping.

1. INTRODUCTION

Large container ship has required designers to take account of constraints such as hull girder strength due to torsional, vertical and horizontal bending moment, fatigue strength of hatch corner and hatch opening deflection.

On the other hand, rapid enlargement of rows and tiers of container ships raises large hatch opening deflection, and accurate estimation of deflection is now more important. Interferences between fittings such as hatch covers, lashing bridges and other container securing instruments need to be addressed when new size of ship is explored.

The deflection has three components of torsional deflection, cross deck fore-aft deflection due to container inertia force, and hatch longitudinal deflection due to hull girder vertical bending. However, there was no clear method how to combine the deflection from these three components.

Therefore, the authors conducted full scale measurement of stresses and deflections of Ship B between 2002 and 2004 as well as Ship A between 1998 and 2001, which revealed more accurate correlation among the three components.

Furthermore, whipping due to wave impact load were observed in measurements, it brought a concern about fatigue strength of hull girder. Therefore the authors evaluate the effects on fatigue strength of hull girder.

This paper shows results of full scale measurements of ships in service, and summarizes the two ship measurements. And examples of design application are introduced.

2. FULL SCALE MEASUREMENT

2.1 PRINCIPAL DIMENSIONS

The measurements were performed on two post-panamax container ships. Those ships have made a round trip, displayed in the route map in Figure 1, between Japan to Europe through the Suez Canal. Principal dimensions are shown in Table 1. The ship A and B were delivered in 1998 and 2002, respectively.

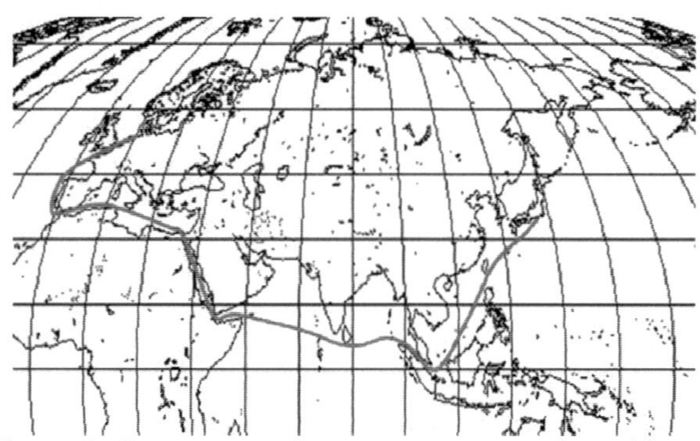

Figure 1: Route map of post-panamax container ship Japan-Europe

Table 1: Principal Dimensions

	Ship A	Ship B
Length, between perpendiculars	283.8m	281.4m
Breadth	42.8m	40.0m
Depth	24.4m	24.2m
Design draught	13.5m	12.5m
Scantling draught	14.0m	14.0m
Service speed	24.5kt.	25.6kt

2.2 MEASURING SYSTEM

The measuring system on the ship A is presented in detail in [1]. The following items were measured on the ship B [2].

1. 6 degrees of freedom of ship motion measured by the gyroscope at engine room
2. Wave height measured by the microwave sensor at the front end of forecastle deck
3. 3 accelerations at bow
4. Longitudinal stresses of upper deck and hold bottom measured by seven optical strain gauges (OSMOS).
5. Wave impact pressure at fore peak tank and bow thrusters room
6. Relative displacement between the lateral deformation of the cross deck and the steel wire, presenting lateral deformation of the transverse bulkhead measured by two laser gap sensors.
7. Relative displacement in longitudinal direction between the cross deck and the hatch cover measured by three laser gap sensors.
8. Vessel speed, wind speed, wind direction and ship position were obtained from ship's equipments.

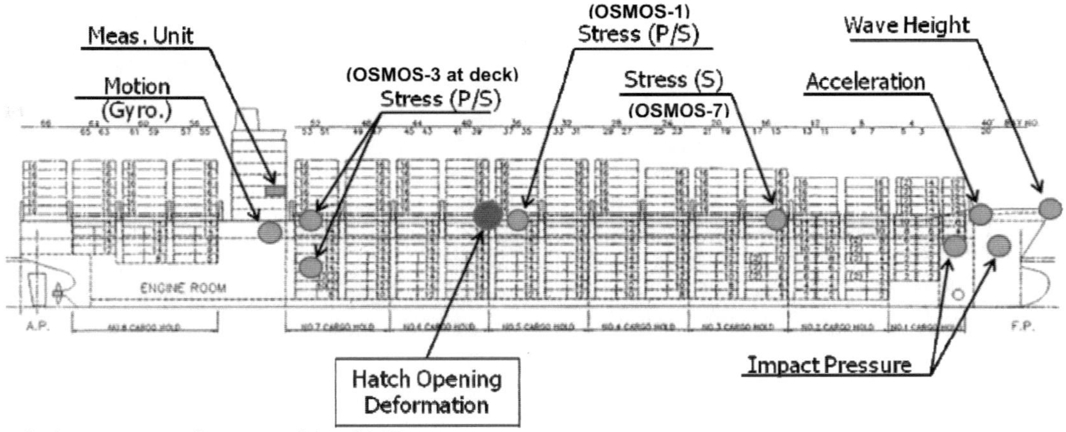

Figure 2: Arrangement of sensors of the ship B

Measurements were made for 20 minutes in every 2 hours automatically. Data sampling rate was 10 Hz, except bow flare impact pressure, considering that the frequency of the stresses and deformation due to wave load is low. The ship arrives at Japan port once in 2 months, therefore, measured data stored in measuring unit was taken. And sea and operation condition on the log book and stability report submitted to the authors by the ship master. This measurement had been continued for two and a half years.

3. HATCH OPENING DEFORMATION

3.1 THREE COMPONENTS OF HATCH OPENING DEFORMATION

Although the hatch opening of a container ship deforms in various shape due to wave loads, total deformation can be presented as a summation of the following three components of deformation [1, 3] as illustrated in Figure 3.

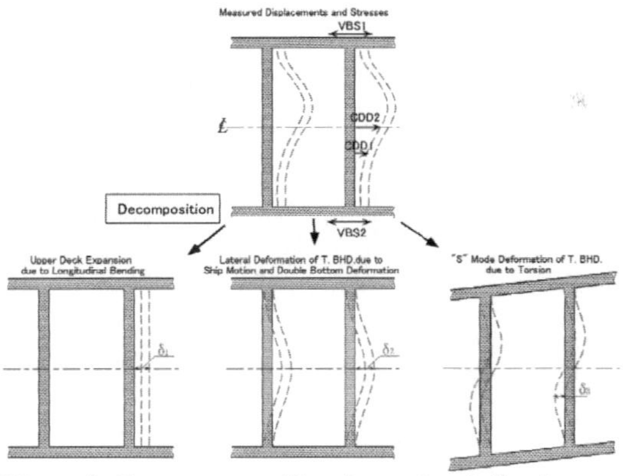

Figure 3: Components of hatch opening deflection

3.1 (a) Upper Deck expansion due to longitudinal bending: δ_1

Upper deck causes extension and shrinkage due to longitudinal bending induced by wave or static load as shown in Figure 4. This expansion results in the relative displacement between the two transverse bulkheads.

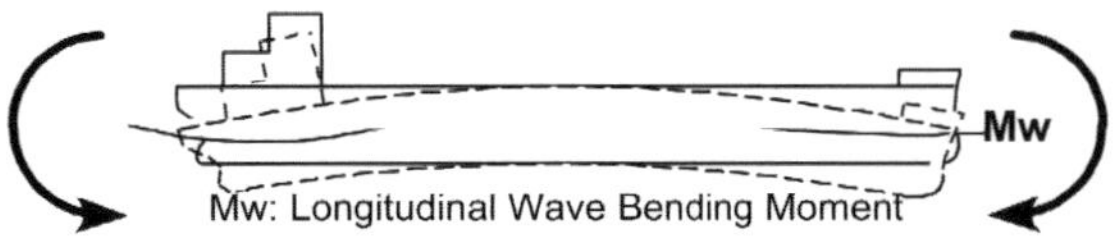

Figure 4: Upper deck expansion due to longitudinal wave bending moment δ_1

3.1 (b) Lateral Deformation of T.BHD due to Ship Motion: δ_2

Lateral deformation of transverse bulkhead is caused by the longitudinal inertia force of in-hold containers and of on-deck containers due to ship motion, such as surging or pitching. An example of this deformation due to ship motion is demonstrated in Figure 5, which is calculated by a half of finite element model of fore cargo hatches of a post panamax container ship.

In addition to this deformation, another lateral deformation of transverse bulkhead is superposed; vertical deformation of the double bottom due to hydrodynamic pressure causes lateral deformation of transverse bulkhead as illustrated in Figure 6. This deformation is considered to be caused by unsymmetrical structural arrangement in fore-aft direction, but finite element analysis revealed that this effect is relatively small compared to deformation due to longitudinal accelerations.

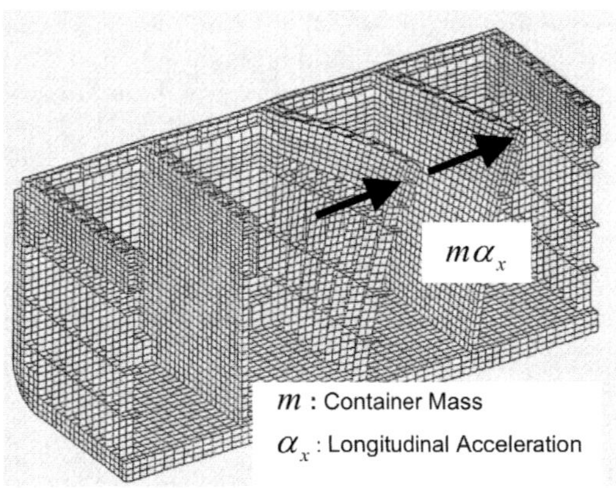

Figure 5: Lateral deformation of transverse bulkhead due to ship motion δ_2

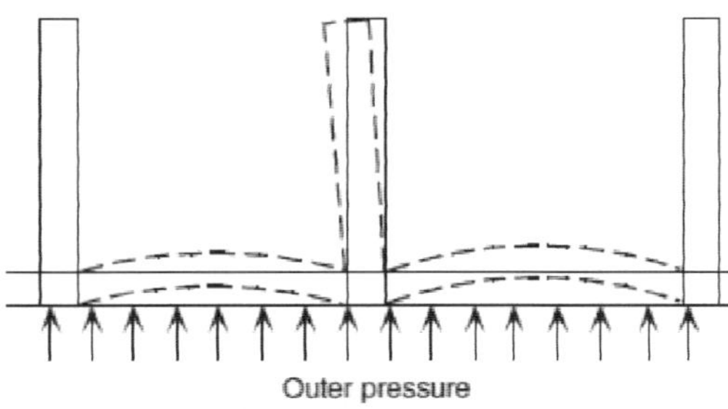

Figure 6: Lateral deformation of transverse bulkhead due to double bottom deformation

3.1 (c) "S" Mode Deformation of T.BHD due to Torsion: δ_3

In oblique wave, the vessel is twisted due to torsional moment of wave as illustrated in Figure 7. T.BHD can no longer maintain the plane but causes lateral distortion. Figure 8 presents the shape of deformation in way of hatch coaming obtained by finite element analysis of whole container ship under torsion. According to Figure 8, it is found that the deformation shape of T.BHD is not in straight form but in "S" form.

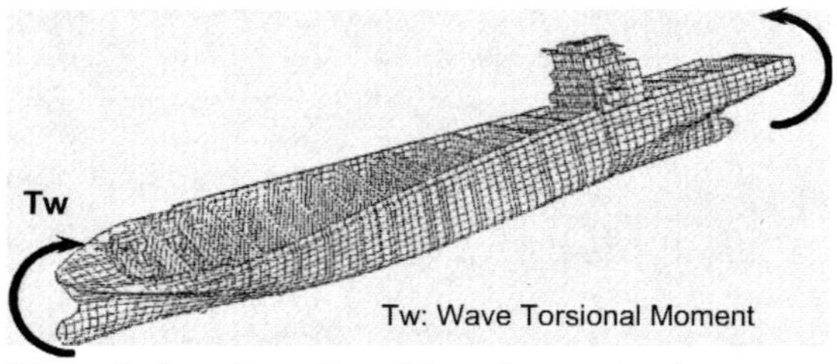

Figure 7: Container ship subjected to torsional moment

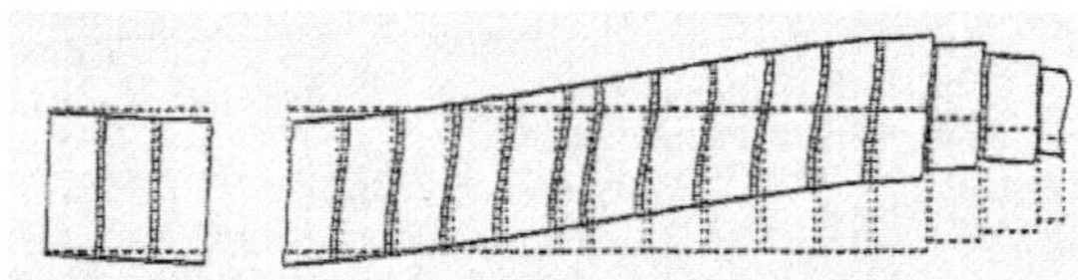

Figure 8: Hatch opening deformation caused by torsion δ_3

3.2 MEASUREMENT SYSTEM OF HATCH OPENING DEFORMATION

These deformation components are difficult to measure directly, however they can be obtained by decomposing the lateral deformation of T.BHD and the longitudinal deck stresses. Figure 9 shows the arrangement of sensors for measuring hatch opening deformation. The concept of this decomposition is explained in Figure 3. δ_1 can be calculated by measured longitudinal deck stresses at midship section.

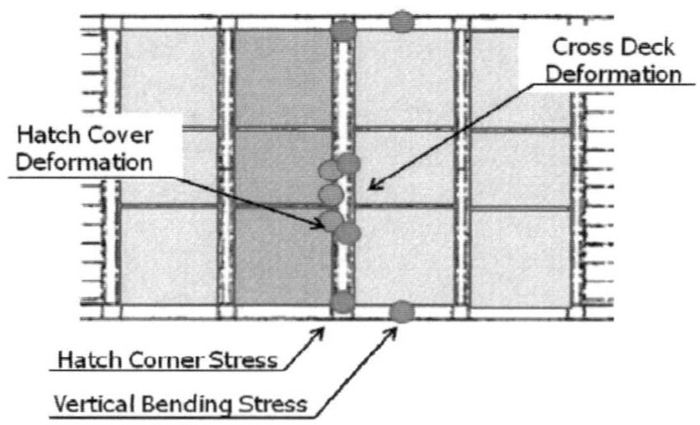

Figure 9: Arrangement of Sensors for Hatch Opening Deformation

3.2 (a) Measurement System of Hatch Opening Deformation by Steel Wires

δ_2 and δ_3 can be obtained by the lateral deformation of transverse bulkhead using measurement system as shown in Figure 9 and 10. In Figure 9, lateral deformation of T.BHD can be obtained by measuring relative displacement between the bulkhead and the target on steel wire by using laser displacement sensor installed on the bulkhead, since steel wire stretched from one side of the vessel to the opposite side provides an immovable point.

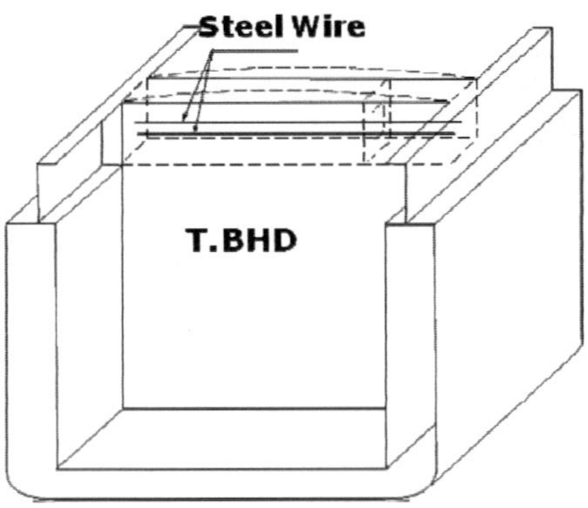

Figure 9: Measurement System of Lateral Deformation of Transverse Bulkhead by Steel Wire

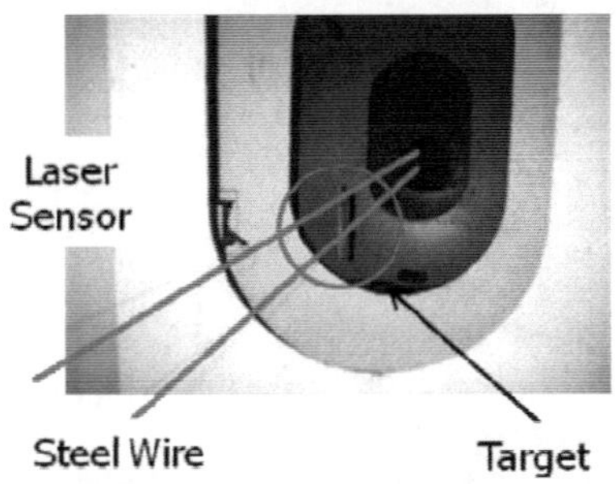

Figure 10: Steel Wires and Laser Sensor for Measuring the Lateral Deformation of Transverse Bulkhead

3.2 (b) Measurement System of Hatch Opening Deformation by Hatch Corner Movement

Although hatch opening deformation can be measured by the steel wire stretched from side to side as explained in 3.2. (a), this deformation can be also obtained by an alternative procedure; i.e., by measuring the relative movement of hatch covers. As illustrated in Figure 11, the hatch covers of this vessel are fixed in the cross-deck structure by the pin stopper at their aft end, therefore, relative longitudinal displacement is caused by torsional deformation of opening between the centre hatch cover (HCD2) and the side hath cover (HCD1). Hence, diagonal displacement of hatch opening can be estimated by measuring the relative motion between HCD1 and HCD2.

In this case, the deviation between HCD1 and HCD2 is also affected by the fore-aft deflection of the cross deck strip δ_2. To obtain accurate deformation due to pure torsion, it is necessary to subtract the component due to the fore-aft deflection obtained from the procedure stated in 3.2. (a).

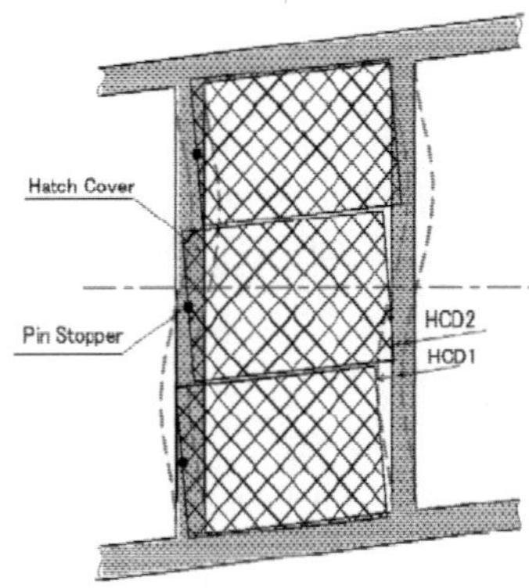

Figure 11: Measurement of Hatch Cover Displacement

3.2 MEASURED DATA

Comparisons between the 10^{-8} extreme value of the deformation components predicted based on Weibull long-term distributions and design assumptions are tabled in Table 2.

Table 2: Comparison between predicted value and design assumption (Predicted/Design)

SHIP	δ_1	δ_2	δ_3
A	78%	30%	100%
B	66%	13%	108%

From Table 2, it is observed that the predicted extreme values related to fore-aft deflection of cross deck strip are generally smaller than the design assumption. One reason for this can be found in the stack load analysis. In case of ship A [1], actual stack load was about 30% of the design maximum stack load in average, and about 42% in long voyages. For Ship B, averaged stack load of container was about 24% of the design maximum stack load. Therefore, fore-aft deflection actually became much smaller. Another reason may be caused by friction forces between containers, hatch covers and hull structure.

3.3 CORRELATION FACTORS FOR HATCH CORNER STRENGTH

When Z is composed of two components X and Y, correlation factor ρ is defined in the following equation:

$$\rho = \frac{Z^2 - \left(X^2 + Y^2\right)}{2XY} \tag{1}$$

Thus ρ is ranged between -1 and $+1$, and can be interpreted to have the following physical meaning:

$$\begin{cases} \rho = 1: & Z = X + Y & (full\ conbination) \\ 0 < \rho < 1: & Z > \sqrt{X^2 + Y^2} & \\ \rho = 0: & Z = \sqrt{X^2 + Y^2} & (no\ correlation) \\ -1 < \rho < 0: & Z < \sqrt{X^2 + Y^2} & \\ \rho = -1 & Z = X - Y & (cancellation) \end{cases} \quad (2)$$

Correlation factors of Ship B calculated for each 20 minutes short term duration are plotted in Figure 12. Figure 12 shows correlation factors between two components, with horizontal axis of the standard deviation of component per significant wave height. From Figure 12 (c), we can observe that the correlation factors converge to 0.0 when the response per unit significant wave height increase, thus indicating little correlation among these three responses. Ship A also shows same feature [1].

Table 3 summarizes correlation factors of both Ship A and Ship B between δ_1 and δ_2, δ_2 and δ_3, δ_3 and δ_1, based on long-term extreme value from the measurement.

Table 3: Correlation Factors among δ_1, δ_2 and δ_3

Correlation Factor	Ship A	Ship B
ρ_{12} ($\delta_1 + \delta_2$)	-0.313	-0.418
ρ_{23} ($\delta_2 + \delta_3$)	-0.244	0.166
ρ_{31} ($\delta_3 + \delta_1$)	-0.242	0.235

Regarding δ_1 and δ_2, negative correlation to some degree was observed in the results of the both ships, therefore full combination of δ_1 and δ_2 at design consideration, i.e., correlation factor of 1.0, is considered to be too sever.

Although negative correlation is observed, positive correlation is derived at the opposite side of hatch or cross deck strip. Therefore some moderate positive value of correlation factor may be assumed at design stage. No other apparent correlation could be observed in both ships.

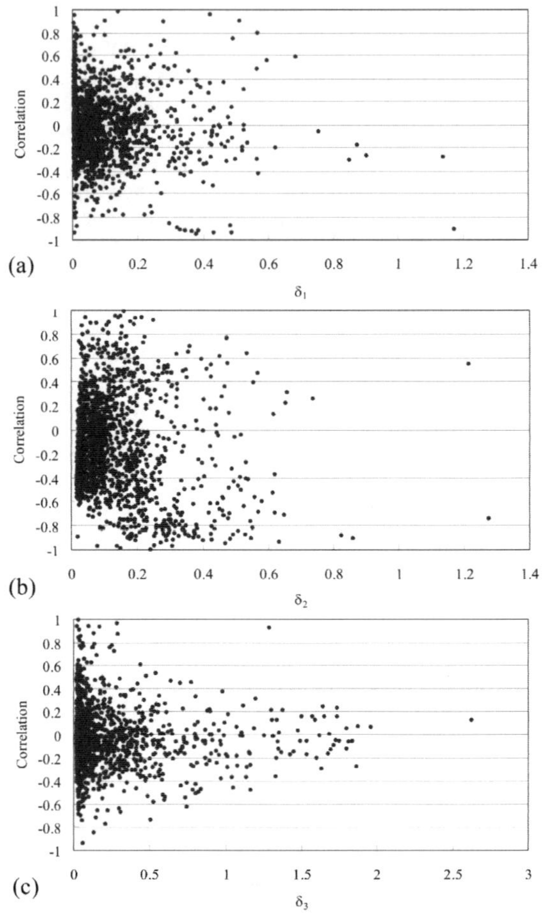

Figure 12: Measured Correlation Factors of Ship B

3.4 CORRELATION FACTORS FOR DECK FITTING DESIGN

In the deck fitting design, there seems to be no explicit regulations to specify the hatch opening deformation. Hence, maximum deformations of hatch opening are required to conduct the deck fitting design, such as hatch covers and lashing bridges. On the other hand, total deformation of hatch opening is assumed to be the summation of three components of deformation; i.e., δ_1, δ_2 and δ_3, already obtained by measurement.

Therefore, maximum deformations are investigated by considering the correlation factors. The correlation factor means the phase relationship, where the maximum deformation of each component in 20 minutes is normalized by the maximum value in all measured period as expressed by the following formulae:

$$\begin{aligned} \delta_{total} &= \delta_1 + \delta_2 + \delta_3 \\ &= a_1 \times \delta_1^* + a_2 \times \delta_2^* + a_3 \times \delta_3^* \end{aligned} \quad (3)$$

Where δ_1^*, δ_2^*, δ_3^* is the maximum value in all measured period, a_1, a_2, a_3 the correlation factor;

$$a_1 = \frac{\delta_1}{\delta_1^*},\ a_2 = \frac{\delta_2}{\delta_2^*},\ a_3 = \frac{\delta_3}{\delta_3^*} \quad (4)$$

In order to investigate the correlation factors, three sets of correlation factors are considered at the instance when the three components δ_1, δ_2, δ_3 became the largest in 20 minutes of measuring time. The maximum value of each component does not always appear simultaneously, therefore, three sets of correlation factors $(a_1^{\max}, a_2, a_3)$, $(a_1, a_2^{\max}, a_3)$, $(a_1, a_2, a_2^{\max})$ can be calculated in one series of measurement in 20 minutes. All sets of correlation factor through all measuring period are plotted in Figure 13.

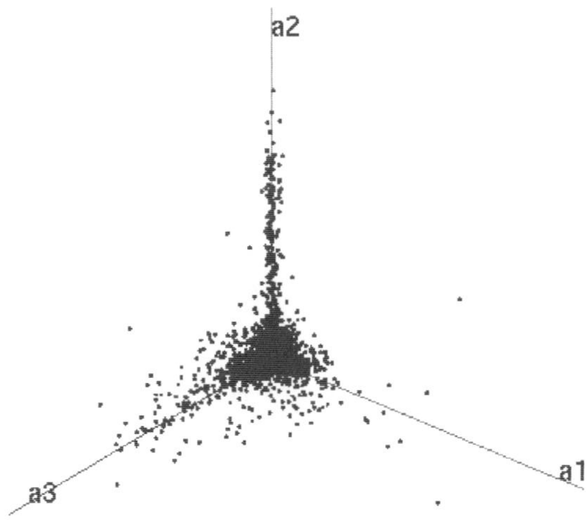

Figure 13: Plotting of all Measured Correlation Factor of Ship B

By assuming the correlation factor adequately, the extreme deformation can be estimated in good accuracy. One of the examples of assumed correlation factors of Ship B is represented by the following formulae:

$$\begin{aligned} Case\ A: \delta_{total} &= 1.0 \times \delta_1 + 0.6 \times \delta_2 + 0.4 \times \delta_3 \\ Case\ B: \delta_{total} &= 0.2 \times \delta_1 + 1.0 \times \delta_2 + 0.3 \times \delta_3 \\ Case\ C: \delta_{total} &= 0.45 \times \delta_1 + 0.55 \times \delta_2 + 1.0 \times \delta_3 \end{aligned} \quad (5)$$

Assumed correlation factors can be verified by confirming that all measured data are within the criteria specified by the above cases, namely, at least one of the cases covers any one of the measured values. Validation process is illustrated in Figure 14, 15 and 16.

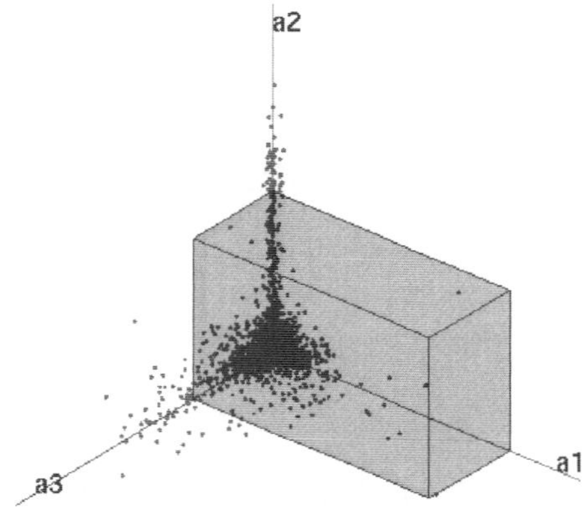

Figure 14: Specified Data by Criterion Case A

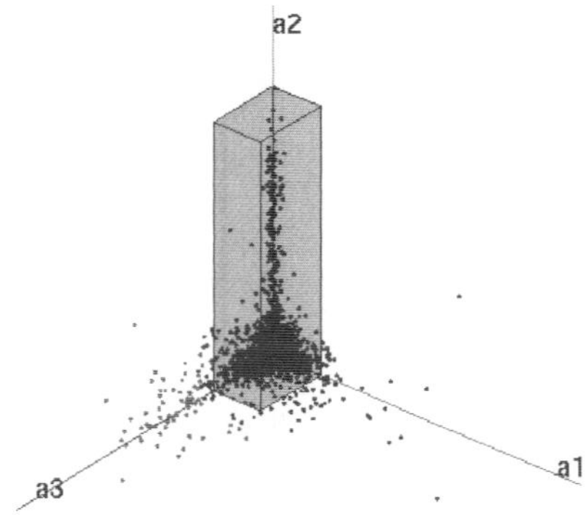

Figure 15: Specified Data by Criterion Case B

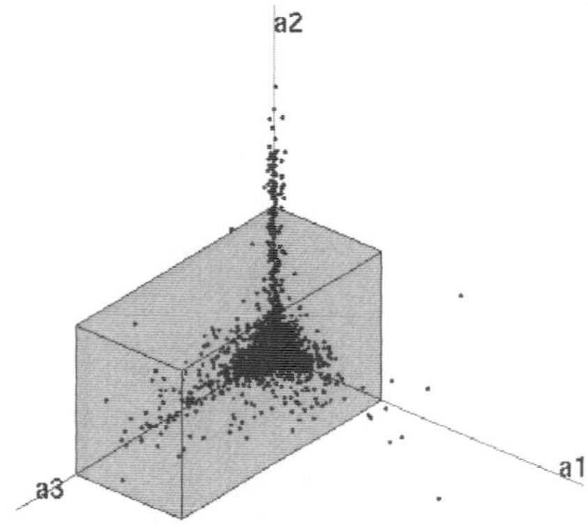

Figure 16: Specified Data by Criterion Case C

In the initial step, all measured data are plotted with solid red as shown in Figure 14. Then the first criterion of Case A is applied in Figure 14, where intensity of the data plot below this criterion is turned from red to blue. Once the colour is changed to blue, that colour still remains in the following steps. Consequently, the colour of all measured data changed blue, therefore all values are confirmed to be within the assumed criteria from Figure 16. Table 4 shows the correlation factors of both

Ship A and Ship B. According to the resulting formulae, no strong correlation is observed among δ_1, δ_2 and δ_3.

Table 4: Correlation Factors obtained from Two Ships

SHIP	CASE	a1	a2	a3
A	A	1.00	0.65	0.25
	B	0.20	1.00	0.60
	C	0.45	0.55	1.00
B	A	1.00	0.60	0.40
	B	0.20	1.00	0.30
	C	0.45	0.55	1.00

4. WHIPPING

Measured time history of the longitudinal bending stress indicates that the whipping vibration was caused in rough sea condition and that this whipping stress superposed the longitudinal stress due to wave bending. Single stress amplitude due to whipping stress observed on Ship A is about 30MPa [1]. This additional whipping stress may increase the fatigue damage factor, therefore the effect of whipping and practical design method for the fatigue strength was investigated.

4.1 MEASURED DATA

Figure 17 shows an example of the measured time history of wave height and longitudinal bending stress. High cycle stress from 0.2Hz to 2.0Hz and low cycle from 0.0Hz to 0.2Hz stress were divided from measured data by filtering process. According to this figure, measured longitudinal bending stress is the combination of the lower cycle stress of the frequency 10 seconds/cycle and the higher cycle of about 1.5 seconds/cycle. Therefore, 2-nodes of whipping vibration induced by slamming causes this high cycle stress.

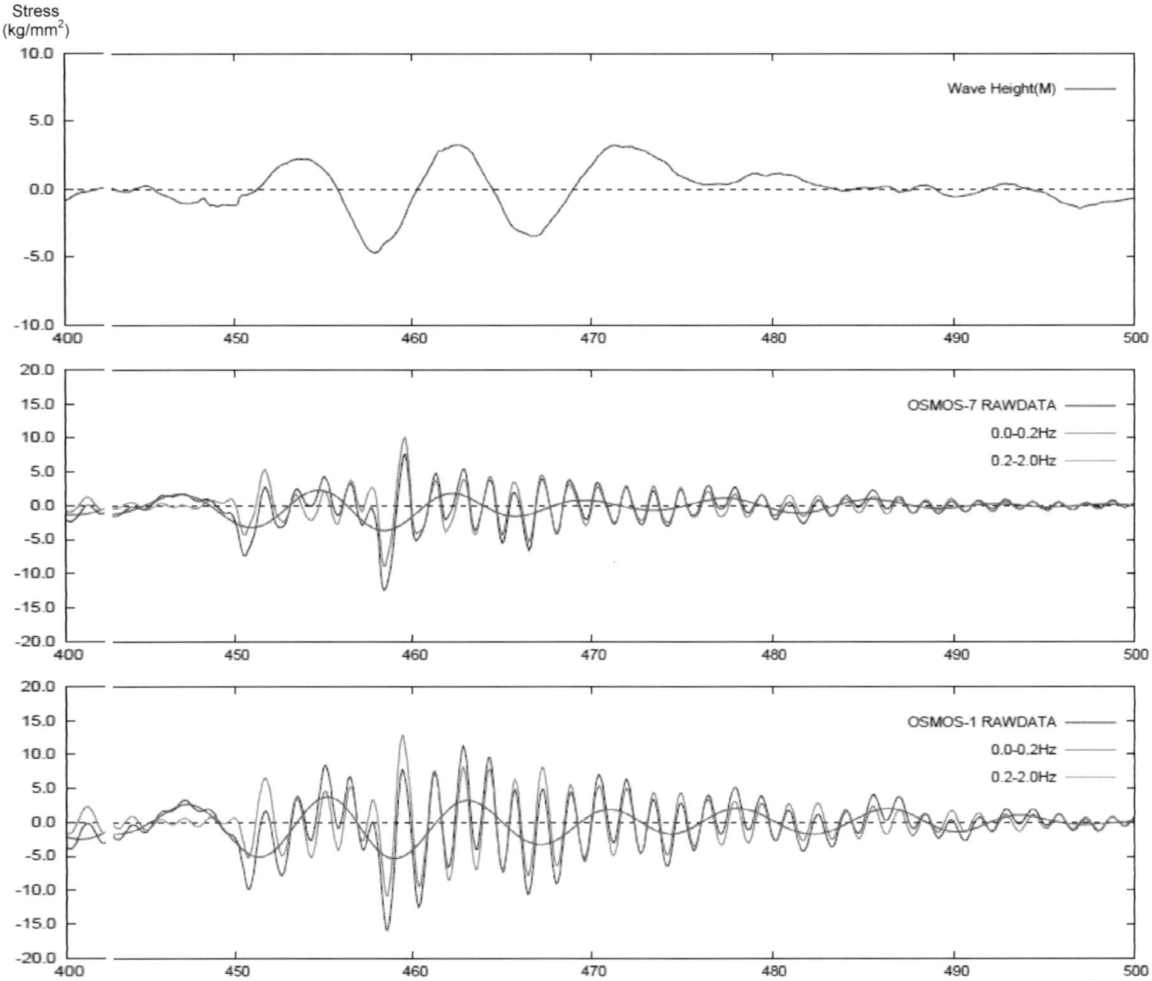

Figure 17: Measured Time History of Longitudinal Bending Stress of Ship B

Whipping vibration was occurred at the 450 second in Figure 17, and then 2nd impact induced large stress range. This stress range is the maximum in all measuring period. Although, design stress range due to wave bending moment is 200MPa, which is calculated from vertical bending moment of IACS UR S11, the maximum stress reached to 215MPa during two and a half years measurement.

Figure 17 shows that there is not large difference between whipping stress of OSMOS-1 and OSMOS-7, therefore, attention for fatigue strength on whole hold region shall be paid.

4.2 LONG-TERM PREDICTION OF BENDING STRESS

Time history of bending stress has two components of stress due to wave bending moment and whipping vibration, i.e., measured stress has two physical phenomena. Therefore, it is difficult to count the number of waves and estimate the extreme value of stresses on the basis of measured long-term distribution.

Table 5 shows the predicted bending stresses, which are the 10^{-8} extreme value based on Weibull long-term distribution. These are predicted from lower frequency stresses where the higher frequencies more than 0.333Hz are cut out by filtering process.

Table 5: Long-term prediction of bending stresses

	Predicted Stress (MPa)
OSMOS-1 (Mid)	84
OSMOS-3 (Aft)	71
OSMOS-7 (Fwd)	90

4.3 EFFECT ON FATIGUE STRENGTH

Although there is no implicit method of counting the number of cycles which takes account of the whipping effect, an effect on fatigue strength is investigated based on measured stresses.

When ship is expected to sail in rough sea, ship's master will avoid rough sea by detouring and ship's master will reduce ship's speed. Therefore, to investigate an effect of whipping, judgement of master shall be included. Regarding this point, the measurement data have the consequents of master's judgement when ship sail in rough sea condition.

10,596 data have been measured during two and half years of measuring period. 73 measured data which has more than 10^{-4} of cumulative damage factors by the fatigue sacrificial specimen [4] of 20 minutes are chosen from all data. And component of whipping vibration is extracted from the data by filtering process. From the resulting time history is shown in Figure 18. Figure 18 (a) and (b) show that several times of whipping measured. And whipping vibrations were heavily-damped in 30 seconds at the longest. From the Figure 18 (c), small range of vibrations are found, however small stress rage will no effect on the cumulative damage factor, since those are smaller than the fatigue limit.

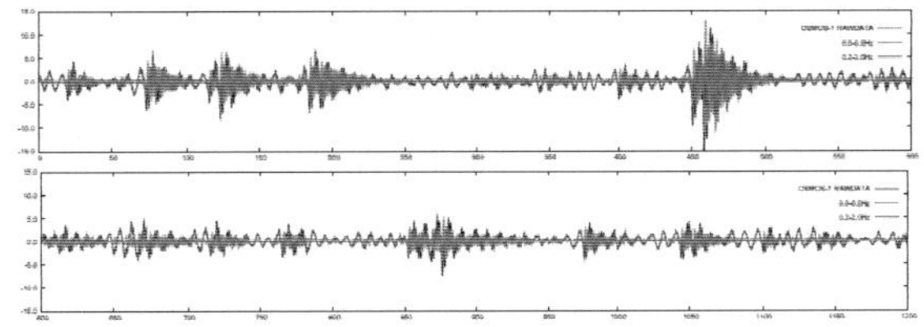

(a) Measured Data of the Maximum Cumulative Damage Factor in 20 Minutes

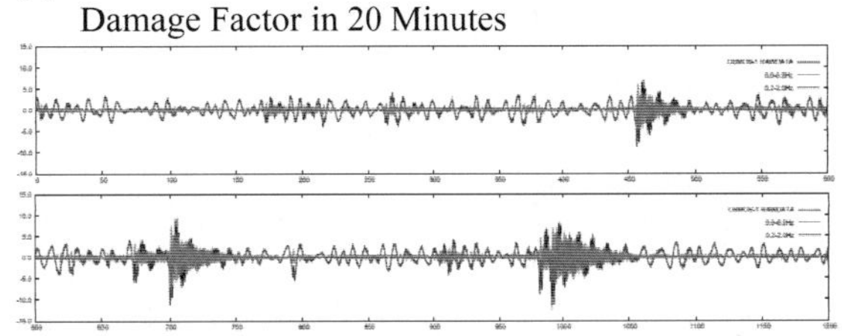

(b) Measured Data of Second Largest Cumulative Damage Factor in 20 Minutes

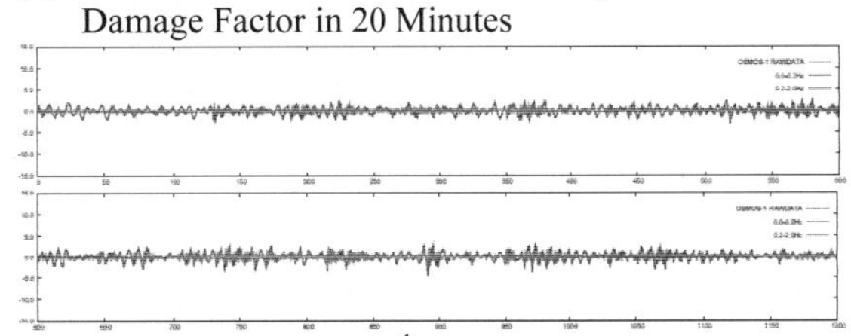

(c) Measured Data of 54th Largest Cumulative Damage Factor in 20 Minutes

Figure 18: Measured Time History of Bending Moment after the Filtering Process

The number of occurrences of whipping vibration was divided to three ranges by stress range, 1st range is more than 100MPa, 2nd range is between 40MPa and 100MPa and 3rd range is smaller than 40MPa. The number of occurrences of whipping vibration of 1st range and 2nd range are 11 times and 64 times, respectively. Those numbers are counted during the measurement period of two and a half years, number of occurrence is extrapolated to ship's life of 20 years. In order to calculate the cumulative damage factors, D curve of HSE for butt joint are used. From Miner rule and Heibach correction for low stress range, the cumulative damage factor of two ranges is totally 0.04, considering 30 seconds dumping. This result shows that there is not large effect on fatigue strength of hull girder. This damage factor is calculated from high cycle stress range, the cumulative damage factor shall be added to cumulative damage factor induced by lower cycle.

4.4 DESIGN APPLICATION INCLUDING WHIPPING EFFECT

As mentioned 4.3, the cumulative damage factor is increased by 0.04 according to the measurement result of ship B, which experienced severer whipping than ship A.

Therefore, some safety space for fatigue strength is required.

As to conventional ships, whipping effect has been covered by implicit safety margin. Enlargement of container ships may turn this effect into a more important factor. Therefore, investigation into the difference of this effect among various designs is important.

5. CONCLUSIONS

The full-scale measurement of stresses and deformation of two post-panamax container ship was carried out, and data were collected. The major found and considerations are follows;

(a) The authors carried out full scale measurement of two post panamax container ships.
(b) Procedure to derive components of stress and deformation was developed and actual values were calculated based on actual measurement.
(c) Long-term prediction of fore-aft deflection of cross deck strip was much smaller than design assumption. One explanation for this fact is that container weight actually loaded is in general quite lower than the design load for deformation.
(d) From the result of measurements, there is no implicit correlation between three deflection modes. And new formulae to combine those components are proposed.
(e) From the measurement, two node vibrations due to whipping was observed, and measured stresses including whipping effect reached design stresses due to wave bending moment.
(f) Since there is not large differences between stresses at fore and middle of hold part, attention for fatigue strength on whole hold region shall be paid.
(g) According to the investigation of whipping effect to fatigue strength, it was found that a small amount of damage factor is added due to this effect. Because conventional ships cover this effect by implicit safety margin, investigation into the difference of this effect among various designs is important.

6. ACKNOWLEDGEMENTS

Full scale measurements had been carried out as a part of the joint research project between Nippon Kaiji Kyokai and IHI Marine United Inc. The authors wish to gratefully acknowledge the effort of all people associated with this work.

7. REFERENCES

1. T.Okada, Y.Takeda, T.Maeda, 'On board measurement of stresses and deflections of a Post-Panamax containership and its feedback to rational design', *Marine Structures 19*, 2006

2. K.Miyahara, R.Miyake, N.Abe, A.Kumano, M.Toyoda, Y.Nakajima, 'Full-scale measurement on hull response of a large-container ship in service', *OMAE 2006*, 2006

3. M.Toyoda, H.Kusumoto, T.Okada, E.Kobayashi, 'Advanced structural design of containership to achieve overall safety and economy', *Design & Operation of Container Ships RINA*, 22-23 November 2006

4. N.Yamamoto, K.Okamoto, A.Kumano, 'Full-Scale Stress Measurements and Observation of Fatigue Sacrificial Specimen', *Conference Proceedings the Japan Society of Naval Architects and Ocean Engineers Volume 2E*, 2006

8. AUTHORS' BIOGRAPHIES

Masanobu Toyoda is PE of Ship & Marine Engineering working with Basic Design Department at IHI Marine United Inc. He is responsible for hull structures basic design of container ship.

Tetsuo Okada Ph. D, holds the current position of manager at Basic Design Department, IHI Marine United Inc. He is responsible for hull structure basic design.

Tomoki Maeda is working in Structural Strength Department, Research Laboratory at IHI Corporation.

Toshiyuki Matsumoto is the manager of the Research Institute at Nippon Kaiji Kyokai.

MEASUREMENTS OF WAVE INDUCED VIBRATIONS AND FATIGUE LOADING ONBOARD TWO CONTAINER VESSELS OPERATING IN HARSH WAVE ENVIRONMENT

G. Storhaug, and **S. Heggelund**, Det Norske Veritas, Norway

SUMMARY

Ship vibrates locally and globally from various sources of excitation. While most literature focus on propeller and engine excited vibrations, which can be handled by proper design, wave induced vibration can not easily be avoided by moderate changes of the hull lines. The waves may cause the whole hull girder to vibrate due to springing (resonance) and whipping (transient response), which increase the fatigue and extreme loading. DNV has done extensive research on this topic the last decade, and recently, it has also become an industry concern. Commercial modern hull monitoring systems (such as HMON systems that satisfy class requirements) in combination with sophisticated model tests are the best tools to answer the key questions: How important is the wave induced vibrations, and does it have to be included in design rules?

Full scale measurements have been carried out for two different container vessels operating in the North Atlantic. Model tests have also been carried out for the larger vessel. This paper focus on the additional contribution to fatigue damage from wave induced vibration. Results are obtained both amidships and in front of the superstructure. The extensive measurements illustrate that the wave induced vibrations are significant, and furthermore, the general fatigue loading can be high for this trade. For some details, the loading including vibration can also be higher than considered in design rules. This is also confirmed by observed fatigue cracks after relatively short time of operation. State-of-the-art design procedures have to be used to avoid such cracks in this harsh trade, but measurement systems can also be used to decrease the fatigue loading by awareness and routing.

1. INTRODUCTION

Recently, DNV documented that wave induced vibrations can contribute considerably to fatigue damage on blunt vessels, [1]. It actually dominated the fatigue loading in deck amidships on a large 220'dwt ore carrier in ballast condition.

The conventional perception is that the importance of springing (and whipping) increases with size of the vessel. This is a simplified statement, which is based on the idea that larger ships have lower natural frequencies, which will be excited more easily by longer waves with more energy. Unfortunately, it is more complicated. This is illustrated in [2], where the smaller 175'dwt ore carrier has a higher contribution to fatigue damage from the wave induced vibrations in the North Atlantic (57% of total fatigue damage), than from the 220'dwt ore carrier (41%). Further more, measurements from one single ballast trip of a small 18' dwt tanker indicated that 87% of the fatigue damage came from wave induced vibrations [3]. The fatigue loading was however low for the small tanker, i.e. 1/5 of the design fatigue loading. From these studies of blunt vessels, it is evident that the bow draft is a very important parameter. Lower draft increases the vibration levels, so it is not only a matter of natural frequency and length.

The size of container vessels has increased significantly recently, and there are a couple of hundred Ultra Large Container Ships (ULCS) ordered, but only a couple of such ship designs are already sailing. The experience is thereby limited. The size increase has raised the concern about the wave induced vibrations also on these vessels. They have however very different excitation mechanisms compared to the blunt vessels. The draft in cargo condition is relatively low, and the speed is high, hence the concern is not necessarily any less than on blunt vessels. Proper documentation is still lacking. Fortunately, the largest vessels are currently sailing on a trade with moderate loading, but some time into the future this may change.

To approach this topic DNV decided to study the effect of wave induced vibrations on a trade that matters, i.e. where high loading can be expected often. Firstly, preliminary assessment was carried out of a 4000TEU vessel operating in the North Pacific. 39% and 46% fatigue damage came from the vibration for two 3 month periods (spring and winter), respectively [4]. Secondly, model tests were carried out in the North Atlantic wave environment for a 4400TEU vessel. In head seas 37% of the fatigue damage in deck amidships came from vibrations [4]. Numerical calculations were confirmed to be somewhat unreliable, but the theory is indicated to be more suitable for container vessels than for blunt vessels [5,6]. Thirdly, the real 4400TEU operating in the North Atlantic was instrumented in 2006 [7]. Further, a smaller vessel, 2800TEU, was also instrumented [3]. This vessel experienced cracks after a short time of operation, and it is operating in the same trade as the 4400TEU vessel. The 2800TEU vessel was not classed originally to DNV, and the details that cracked, can not be found onboard the 4400TEU vessel. These two vessels are of interest in order to study possible size effects, and measurements for these vessels will be presented herein.

It should be noted that in January 2007, MSC Napoli broke in two in the English Channel. DNV did extensive analyses, which confirmed that the total loading in way of the engine bulkhead exceeded the capacity [8].

MSC Napoli was an old design 4400TEU Post-Panamax vessel. Model tests of a similar sized, but newer design 4400TEU Panamax vessel, as presented in [4], indicated that whipping may contribute significantly in way of the engine bulkhead area. In the DNV's report [9], included in appendix of MAIB's final report, it was stated that the whipping contribution may have increased the dynamic loading by 10 to 50%. The uncertainties are considerable because of the uncertainties of sea states, speed, wave event and course etc. It should be noted that percentage contribution from whipping refers to conditions that differs from what is most commonly referred to as an extreme event in an extreme sea state. The latter has also been investigated by model tests of the 4400TEU vessel, and the increase in the dynamic loading was then 20% in head seas [10] at low speed.

In the sypnosis of the MAIB report, [9], it is stated: "Recommendations have been made to the International Association of Classification Societies, which are intended to increase the requirements for container ship design, consolidate current research into whipping effect, and to initiate research into the development and use of technological aids for measuring hull stresses on container ships." The current DNV research campaign on container ships was actually established with this in mind, although prior to the MSC Napoli event. It should however be noted that the HMON systems used onboard these vessels actually fulfils the intention by MAIB. Currently these systems are used on a voluntary basis. For illustration, some 'extreme' whipping events based on the measurements will be shown for motivation of further research.

2. THE SHIPS

Main characteristic data of the two ships considered herein is listed in Table 1. Both are Panamax vessels, and the larger is slightly faster.

For the extreme whipping impacts the flare angle is considered important for container vessels. For the 4400TEU vessel, the flare angle at 0.1Lpp aft of FP is about 35° (while locally at the upper part of the flare 0.05Lpp aft of FP the angle is about 46°). For the 2800TEU vessel the similar angles are 33° and 40°, respectively. For comparison, the same angles for one 8600TEU design were 57° and 61°, respectively, indicating that the flare increases significantly for the modern Post-Panamax vessels. This is one of the concerns regarding extreme whipping effects on the larger vessels. The angles are defined according to DNV rules for flare slamming, which has proven to be a necessity for container vessels, and which is used by several of the main yards [11].

Coming back to MSC Napoli, the flare angle was 48° and locally 53°, respectively. This is more than on the 4400TEU vessel in **Table 1**, but less than the 8600TEU vessel. The body lines in the forward part of MSC Napoli may also differ from modern design.

Table 1: Main characteristic data

Max. TEU	4400	2800
L_{PP} (m)	281	232
L_{OA} (m)	294	245
Beam, moulded (m)	32.26	32.2
Depth, moulded (m)	21.5	19.0
Scantling/design draught (m)	10.78	10.78
Scantling block coefficient (-)	0.69	0.685
Deadweight (tonnes)	47754	40900
Typical transit draught, (m)		9.5
Typical transit stern trim, (m)		0.5-1.0
Speed at design draft (kn)	23.0	21.3
Horsepower MCR at 104.7/93rpm (kW)	37275	25786
Vertical 2-node frequency (Hz)	0.66	0.76
Built	2003	1998

2.1 THE MEASUREMENT SYSTEMS

The measurement systems onboard the vessels differ. The system onboard the 4400TEU vessel is a modern Light Structures hull monitoring system, which satisfies the revised HMON requirements [12]. The main sensors are long-life-time optical sensors. It is however slightly extended for the purpose of research. The system onboard the 2800TEU vessel is a Miros based system, similar to those which was used on the ore carriers [1,2,6]. Conventional strain sensors are used. Both systems determine the additional fatigue damage caused by wave induced vibration. The Light Structures system does also give trends and warnings when exceeding 80 and 100% of the design loading level, and the system is far more flexible.

Many sensors are utilized onboard the vessels, but the main strain sensors on the 4400TEU vessel are located amidships. 8 strain sensors are used in one cross section to separate 4 global stress components, to investigate side shell fatigue and the effect of the wave induced vibrations. On the 2800TEU vessel the 4 main strain sensors are located in deck amidships and in front of the engine bulkhead. They are used to investigate the effect of wave induced vibrations and to separate global vertical bending stress and axial stress from warping stress and horizontal bending stress. Both systems are supplemented with Miros Wavex directional wave radar, bow accelerometer and many other sensors.

From both systems half hour statistics and continuous time series are saved from all sensors. These are also shown on the display on the bridge continuously.

Further explanation of the setup of the measuring systems can be found in the references [3] and [7].

3. RESULTS FOR THE 2800TEU VESSEL

3.1 TRADE AND MEASUREMENT PERIOD

The trade can simply be illustrated by a position plot. The vessel is operating in the North Atlantic between northern Europe and Canada.

The 2800TEU vessel was instrumented in August 2007, and almost continuous measurements exist from 19th of August until current time. The position plot up till 4th of May is presented in Figure 1. This represents 23 crossings including one winter period. During this period 11066 half hours (231 days=0.63 years) have been collected.

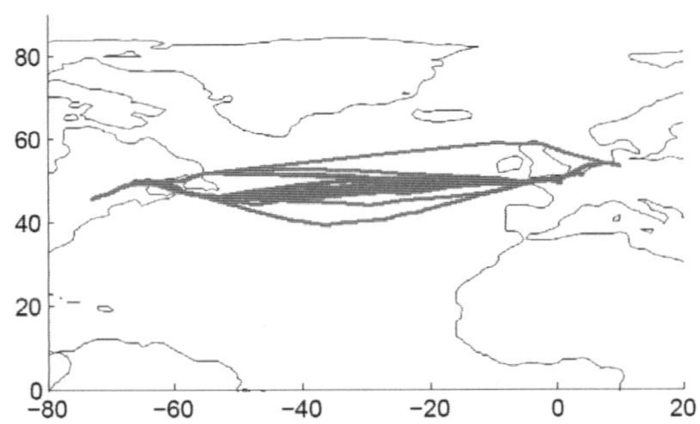

Figure 1: Route from August 2007 to May 2008.

From the speed measurements it is noted that the maximum speed over ground was 25.2kn, significantly above the service speed. If we consider the ship as sailing when the vessel has speed above 10kn, the average speed is then 18.6kn, and the time at sea is 75.6%. If we consider the vessel to be in the North Atlantic between 5 and 60° west, then the average speed is 18.8kn, and the maximum speed is 23.3kn. The time in this region is 49.8%. The speed versus time is presented in **Figure 2**. Significant speed reduction is occasionally observed especially during winter time. This is involuntary speed reduction caused be added resistance in waves, but it can also be voluntary speed reduction by reducing the rpm. Speed as low as 11kn is observed, and occasionally even lower.

There is also some variation in the RPM from the time in this area as seen in **Figure 3**. The first 100 days the RPM sensor was not working properly and a new was then installed on the next service trip onboard. The maximum RPM was 87, and typically above 80.

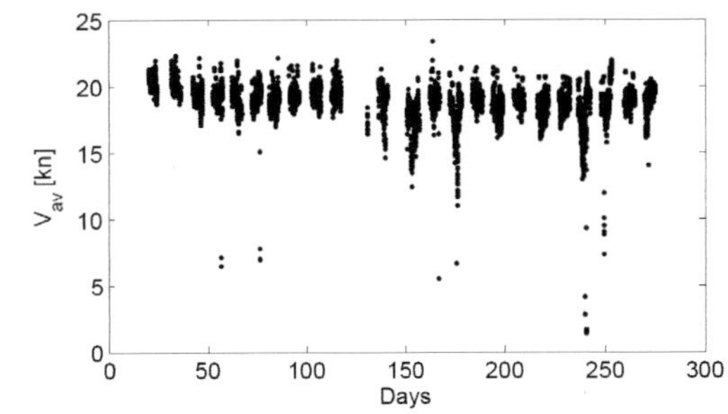

Figure 2: Speed when between 5 and 60° west. 0 corresponds to 1st of August. Each dot represents a half hour average.

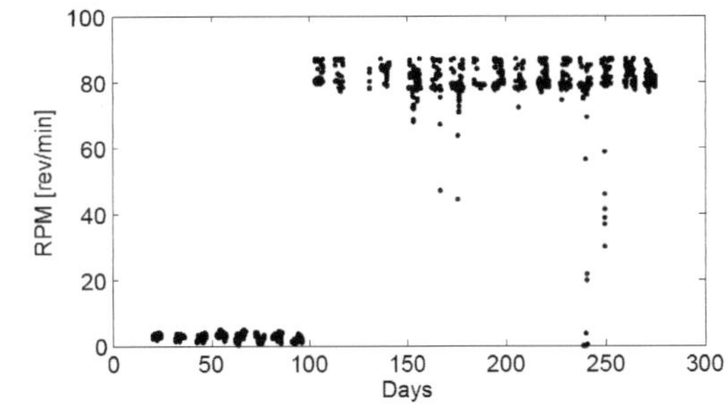

Figure 3: RPM when between 5 and 60° west. 0 corresponds to 1st of August.

If we zoom in to one of the crossing where the vessel encountered some weather, the speed and rpm can be seen in **Figure 4**. It can be observed that the RPM is more or less constant, but that the captain has reduced the RPM slightly when encountering some heavy seas, and then increased it slightly afterwards to win back the lost time. It can also be observed that at one time instant the speed and rpm is reduced dramatically. It is common practise to reduce to safe speed for checking/securing deck. The engine is otherwise controlled on fixed rpm by adjusting the fuel, but if encountering heavy seas the rpm will go down also involuntarily as the fuel limiter becomes activated.

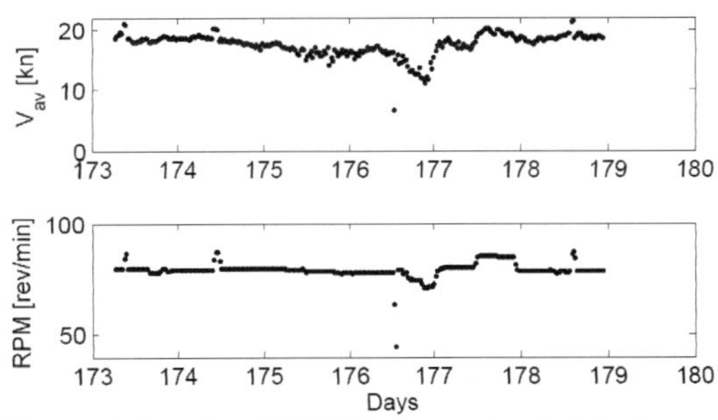

Figure 4: Speed and RPM for a crossing in late January.

3.2 SEA STATES

The directional wave radar has not been calibrated for this particular vessel, so there is a large chance that the wave radar overestimates the wave heights (based on experience). By a spot check using hind cast data and by observations onboard it was stated that the wave height is about 30% lower than measured at 5m significant wave height. Some further spot checks were made, when the vessel was passing NOAA buoys. The results are shown **Figure 5**. It is clear that the wave radar overestimates the waves at higher sea states.

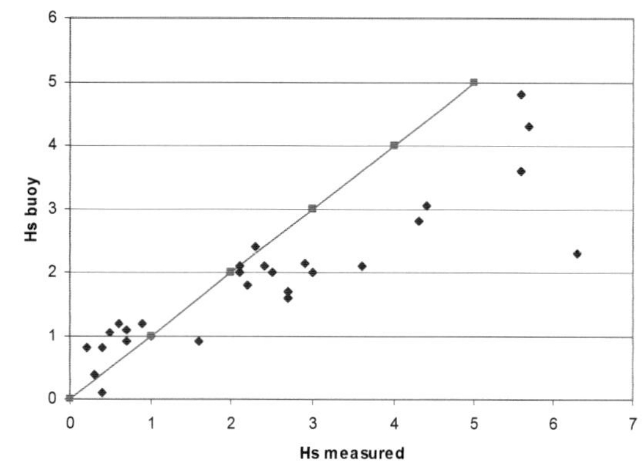

Figure 5: Comparison of wave heights between buoy data and measurements, when the ship is passing buoys.

The corresponding comparison of periods is shown in **Figure 6**. It can not be stated that the period is deviating from the buoy measurements with significant confidence.

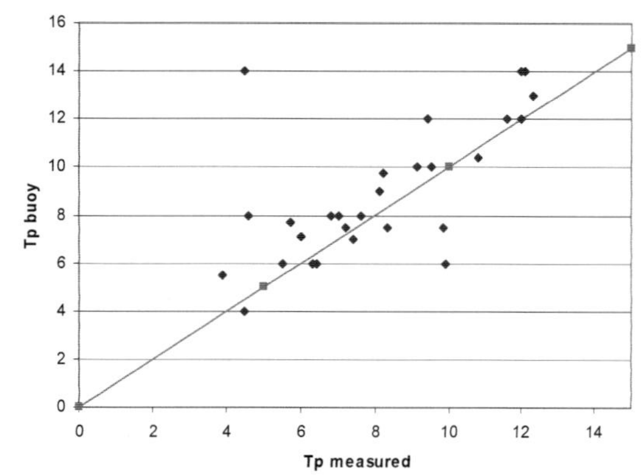

Figure 6: Comparison of peak periods between buoy data and measurements, when the ship is passing buoys.

The measured wave heights are adjusted by multiplying the measured wave heights with a factor of 0.7 (slope of fitted linear line going through zero with minimum error = 0.69). This is not very scientific, but it is necessary since the calibration is not carried out by the supplier.

The probability of the measured wave height with the 0.7 factor is compared to the probability based on DNV North Atlantic (NA) and World Wide (WW) scatter diagram [13] in **Figure 7**. Note that the DNV World-Wide trade includes 16% time in the North Atlantic and 18% in the North Pacific, so totally 34% in harsh area. The measurements have an over-representation of small waves, possibly because the scatter diagram does not refer to close-to-shore data, where the vessel spends significant amount of time. The measurements are below NA except for at 1m Hs, and the measurements are also in better agreement with WW. It is reasonable that the vessel has lower probability of encountering larger waves than from the NA scatter diagram. This indicates that some routing is carried out to avoid the highest waves. For wave heights from 7m Hs, the measurements have also lower probability than the WW data. The effect of routing is illustrated also in Figure 1, when the vessel is not choosing the shortest track.

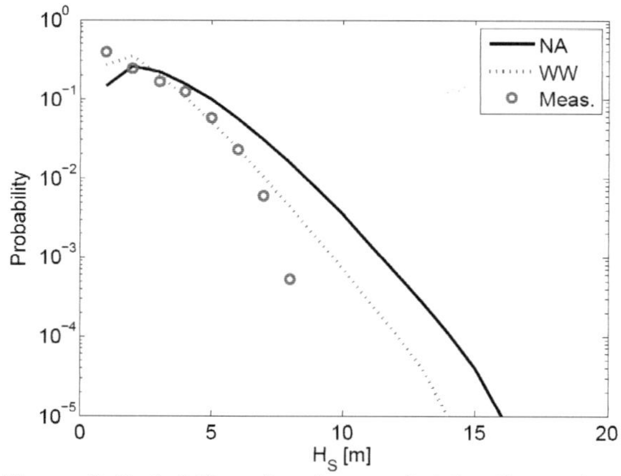

Figure 7: Probability of each wave height. Comparison between DNV North Atlantic (NA) scatter diagram and measurements corrected with 0.7.

The period is also important for the response, and the peak period, T_P, (wave with most energy) is shown versus the measured wave height in **Figure 8**. It appears that as the wave height increases, the period becomes less spread. For the steep waves, this is not unexpected, as the minimum period will increase for higher waves. It is however somewhat surprising that the maximum periods decrease for increasing wave heights. Swell seas may partly explain this behaviour at lower sea states.

The speed was reduced in heavy sea both due to voluntary and involuntary speed reduction. The speed as a function of significant wave height for head sea ±30° relative heading is shown in **Figure 9**. Only the records where the RPM was above 70 was kept, but this means that it accounts to some extent also for voluntary speed reduction. A quadratic fit is made to the records and the equation is stated in the figure. It illustrates for instance that the average speed is 17.8kn in 4m Hs.

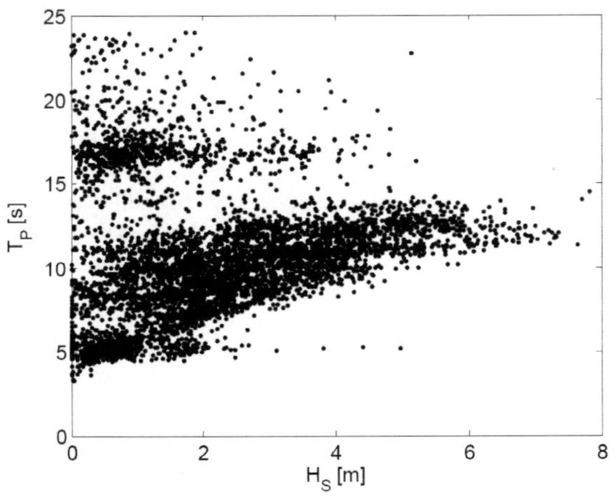

Figure 8: Measured significant wave height (*0.7) versus the peak period.

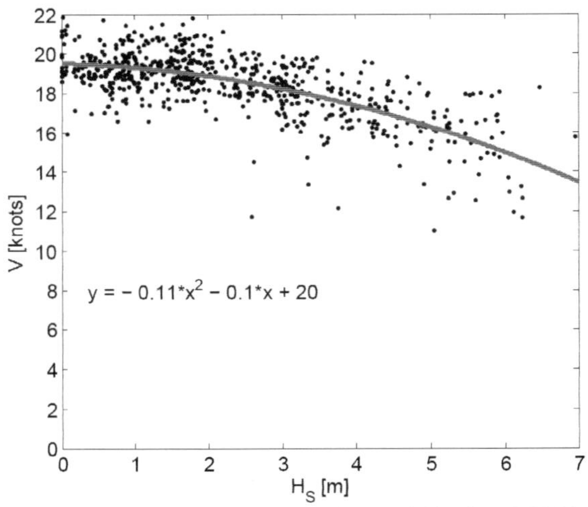

Figure 9: Speed reduction in head sea ±30° when RPM is above 70.

3.3 FATIGUE DAMAGE AND VIBRATION

The main objective of the research is to determine the contribution to fatigue damage from wave induced vibrations. To do this the fatigue damage is calculated by the "raw" time series (just filtered to remove the noise). The results are referred to as "Total fatigue damage". The fatigue is also calculated by the wave frequency part of the stress. In this case, the signal is filtered to remove the vibration effect. The results are referred to as the "WF fatigue". The difference between the Total and WF fatigue is the fatigue damage from the wave induced vibrations, referred to as "HF fatigue" or "Vibration fatigue". If fatigue damage had been calculated based on the high frequency pass filtered stress, with only vibrations included, this estimate would be less than the HF fatigue. To base the fatigue damage on the vibration isolated would be wrong, but when the WF stress approach zero and becomes relatively small, then these two estimates will converge.

The fatigue damage is based on SN-curves for structural details protected from corrosive environment assuming a stress concentration factor (SCF) of 2.0. Further details are given in reference [3] and [7].

The total fatigue damage measured in deck so far is shown in **Figure 10**. It is based on the port sensor on the outer deck longitudinal. The vibration damage (HF) is significant and it is 27% and 35% of the total damage at the midship and after quarter length, respectively. The absolute total damage is 0.0512 and 0.0039 at the two cross sections. The average fatigue damage rate compared to design is 163% and 13% for the two cross sections. Hence, the fatigue damage at the aft quarter length is not a concern in deck, even though the vibration contribution is higher.

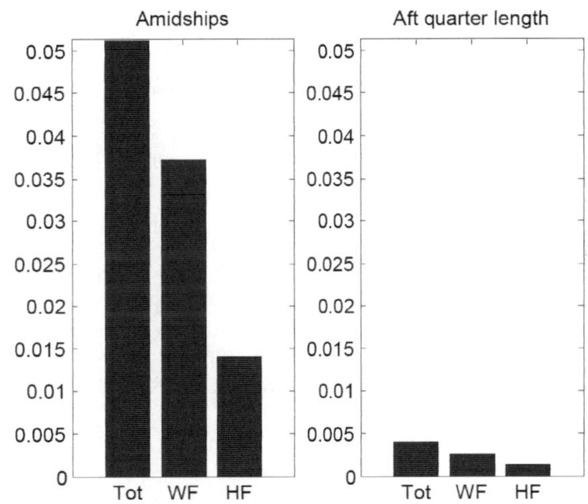

Figure 10: Total measured fatigue damage for midship and aft port sensor.

The fatigue rate needs a clarification. A fatigue rate of 100% is defined as average according to design for a service life of 20 years for the 2800TEU vessel. The fatigue rate is calculated based on statistics every half hour, and the average fatigue rate is the average for all measured half hours. The value of 163% indicates that the fatigue loading is higher than according to design, and the fatigue life (in terms of the definition of the SN-curve) can then be estimated as 20years/1.63=12 years. The value of 163% is slightly biased by the fact that the measurement period cover more from the winter period than the summer period, hence during the coming summer months, this value will be reduced. Still, it is clear that the vessel is operating in a harsh trade, and fatigue cracks can be expected for poorly designed details before the end of the service life. Actually many cracks have already been identified on this vessel and the sister vessel operating in the same trade after 8 years of service [3].

The fatigue rate as a function of time is shown in **Figure 11**. It is observed that the high fatigue rate is starting from December (1st of January is after 153 days). The high fatigue rate is however only typically observed 1-2 days for each crossing. The highest fatigue rate of about 6000% means that 1 day in such a storm is equivalent with 2 months of average fatigue loading according to design (assumed SCF=2.0).

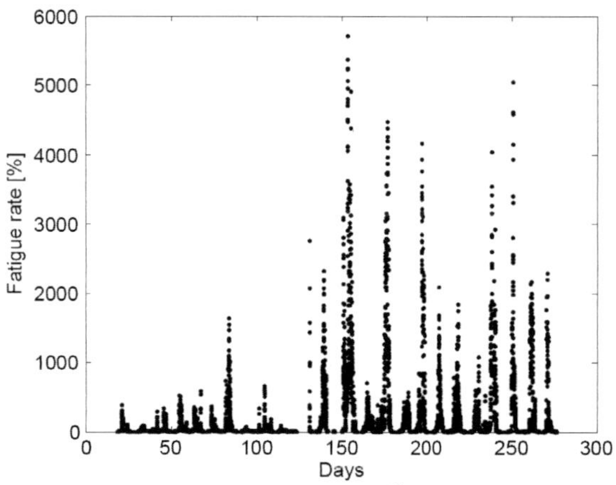

Figure 11: Total fatigue rate from 1st of August 2007 for the port midship sensor.

A zoom in on the worst storm so far around New Year is shown in **Figure 12**. Actually, the vessel appears to encounter 3 significant low pressures. Fatigue rates of up to 4000-6000% are typically encountered on winter crossings, but it is also here the effect of good seamanship will be most effective wrt. fatigue loading.

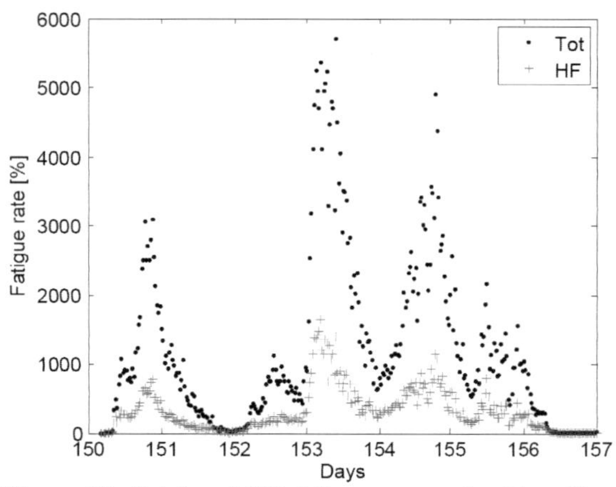

Figure 12: Total and HF fatigue rate at the New Year crossing. Port midship sensor.

The fatigue damage of the port deck sensor amidships is plotted as a function of the measured significant wave height in **Figure 13**. An increasing trend is observed, but something happens for 6-7m significant wave height. It should be noted that as observed in **Figure 8**, there are not that many recordings for these wave heights, and **Figure 13** is independent on direction. Hence, it may happen that these records refer to, for instance, beam seas.

The total fatigue damage as a function of significant wave height is displayed in **Figure 14**. It is observed that 4-5m significant wave height dominates the contribution to the fatigue damage. This is in fair agreement with [4], where the dominating wave height for fatigue damage in head seas is 5m significant wave height, Hs. Also in [1], 5m Hs is the dominating sea state. It is also indicated that the relative importance of the HF damage is relatively constant for the different Hs (varying in between 19 and 29%). Note that the total fatigue damage of 0.0415 in **Figure 14** does not totally add up to number of 0.0512 in **Table 2**. This is because only the records (totally 5557) when the wave radar is turned on is counted in **Figure 14**. Close to shore the radar is used for navigation. Even though this is 50% of the number of half hour stress records it accounts for 81% of the fatigue damage.

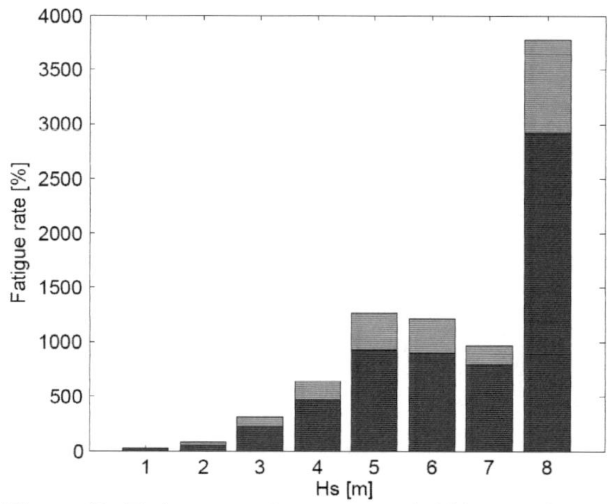

Figure 13: Fatigue rate (average per half hour) of deck port sensor amidships as a function of Hs. Upper part is HF and lower is WF damage.

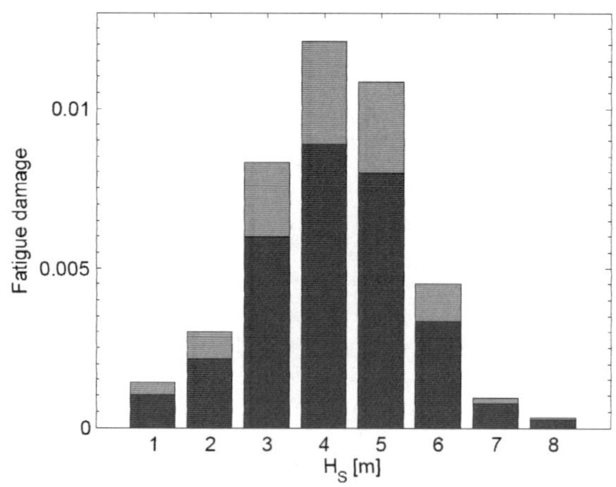

Figure 14: Fatigue damage of deck port sensor amidships as a function of Hs. Upper part is HF and lower is WF damage.

3.4 FATIGUE AND GLOBAL LOAD EFFECTS

Two sensors in deck can be used to separate the global vertical bending and axial stress from the global horizontal bending and warping stress. The sensors are located on the longitudinal stiffener in deck close to the side shell on port and starboard side. The fatigue damage has been assessed based on the port and starboard sensors separately, and with the time series of stresses from these two sensors added and then divided by two. The resulting stress represents then the stress from the vertical bending and axial stress only. The difference between this artificial sensor and either of the port or starboard sensor indicates the importance of the horizontal bending and warping stress component. For the midship section this is illustrated in **Figure 15**. The aft section in front of the engine bulkhead is displayed in **Figure 16**.

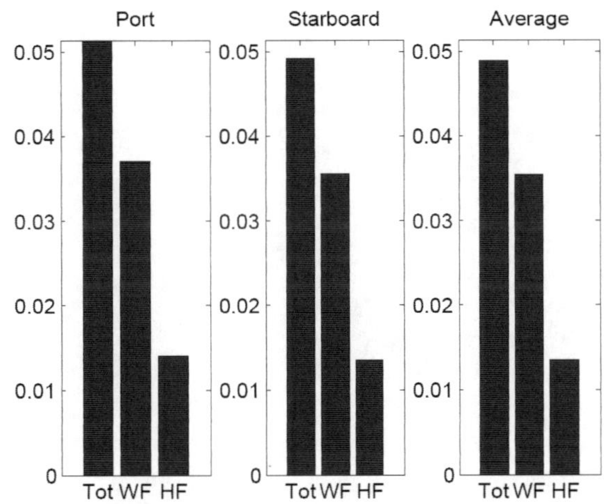

Figure 15: Midship fatigue for two sensors and for the artificial sensor.

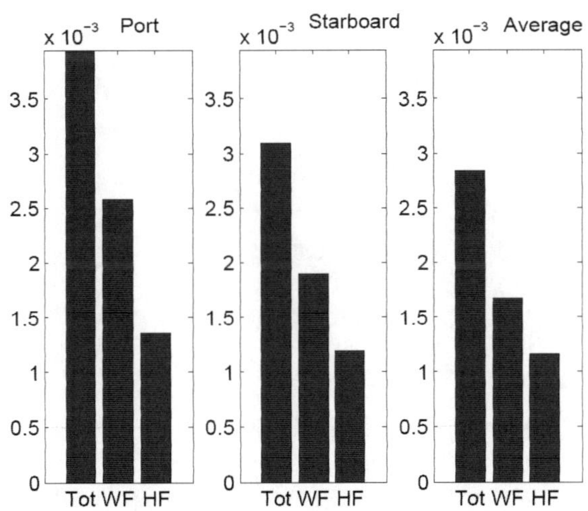

Figure 16: Aft fatigue for two sensors and for the artificial sensor.

The numerical values from **Figure 15** and **Figure 16** are listed in **Table 2**. As observed from the figures, the port sensor gives the highest values, and starboard sensors give less. This may come from slightly asymmetric heading profile, which is not unusual. Further it is seen that the artificial "average" sensor gives even less damage, hence the horizontal bending and warping stress increases the fatigue damage slightly. The increase due to the addition of horizontal bending and warping stress for the WF damage is about 5% amidships and 54% at aft ship. The warping stress was originally believed to reduce the fatigue damage at the deck level, while increase it at the bilge area. It is however also a matter of horizontal bending.

Table 2: Fatigue damage for deck sensors at midship and aft including artificial average sensors

	Port	Starboard	Average
Midship Tot	0.0512	0.0492	0.0489
%	100	96.1	95.5
Midship WF	0.0372	0.0355	0.0354
%	100	95.4	95.2
Midship HF	0.0141	0.0136	0.0135
%	100	96.5	95.7
Aft Tot	0.0039	0.0031	0.0028
%	100	79	72
Aft WF	0.0026	0.0019	0.0017
%	100	73	65
Aft HF	0.0014	0.0012	0.0012
%	100	86	86

3.5 EXTREME LOADING

As indicated in **Figure 9**, the vessel has encountered sea states of up to 6m Hs in head seas. In the following the "extreme" loading will be investigated.

The total versus WF maximum or minimum stress per half hour is presented in **Figure 17**. This figure refers to the midship section. If there had been no vibration (HF) stress, the total would be equal to the WF amplitudes and all the points would stay on the diagonal. Hence, the deviation from the straight line shows the importance of high frequency vibrations. It is clearly seen that the vibration increases the stress level both in sagging and in hogging (positive stress). The stress is nominal, so it does not include any global or local stress concentration. At higher stress levels it also appears that the vibration effect becomes more pronounced, although the relative difference can be significant at low levels. The former may be the effect of whipping and the latter of springing. This is to some extent also illustrated in **Figure 18**, where the HF stress is shown versus the total stress. A cubic polynomial has been fitted through the data illustrating that the HF stress is about 25% of the Total stress at lower levels, and the importance increase for higher levels. For the aft quarter length, just in front of the engine bulkhead, the total stress versus the WF stress is shown in **Figure 19**. Also in this figure the port sensor is used. The resemblance with **Figure 17** is good, but the relative importance of the vibration appears slightly higher at the aft quarter length in hogging.

In **Figure 20**, the WF stress (without vibration) is shown versus the WF stress for only vertical bending (and axial) stress. If there was no warping and horizontal bending stress everything would fall on the diagonal. As observed there are some effect of the horizontal bending and warping stress present. **Figure 21** shows the same for midship section. It is clear that the maximum stress is dominated by the vertical bending moment and not horizontal bending or warping stress in this area.

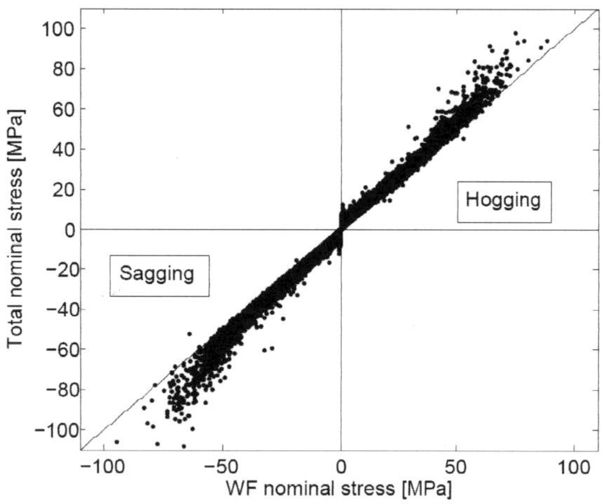

Figure 17: Total versus WF maximum (minimum) stress amplitude per half hour for midship section.

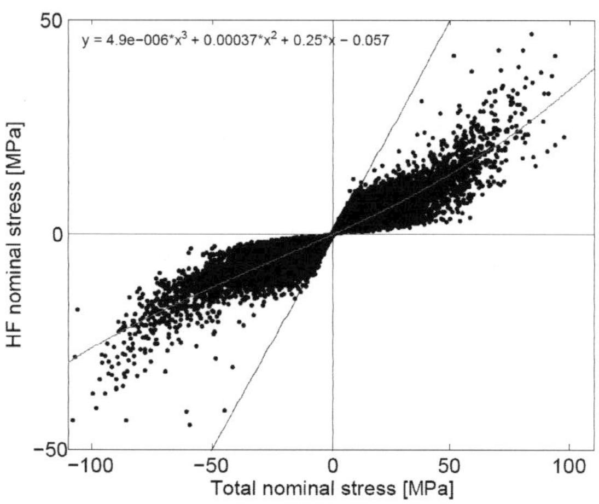

Figure 18: HF versus Total maximum (minimum) stress amplitude per half hour for midship section.

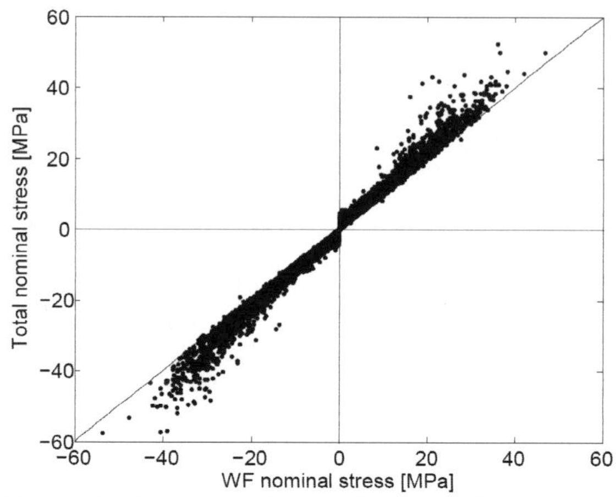

Figure 19: Total versus WF maximum (minimum) stress amplitude per half hour for aft quarter length.

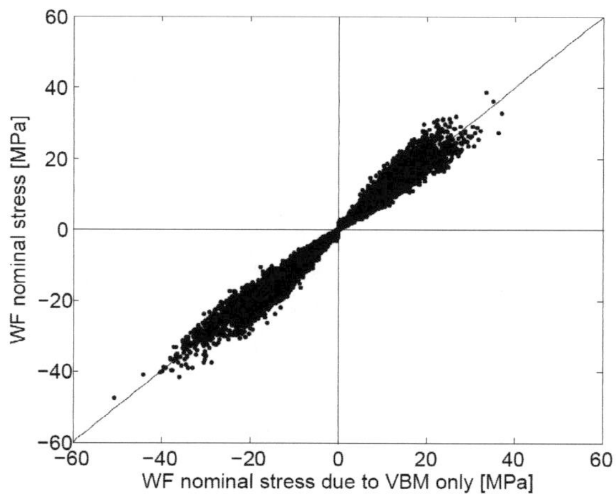

Figure 20: WF stress versus WF stress with only vertical bending (and axial) stress for aft quarter.

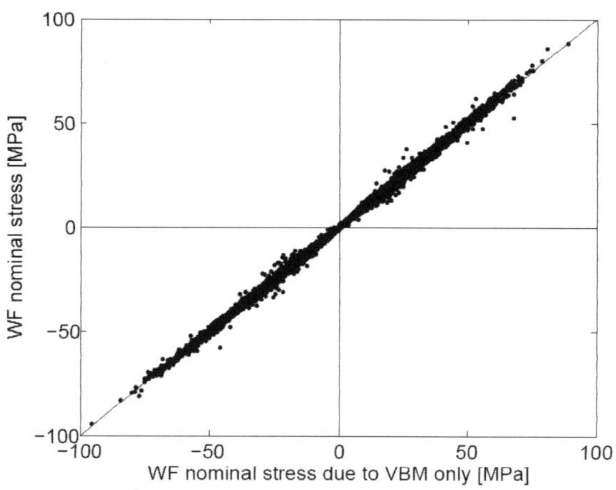

Figure 21: WF stress versus WF stress with only vertical bending (and axial) stress for midship.

The standard deviation of the stress can also be shown to be slightly affected by the vibration, in particular at lower stress levels. The total divided by the WF standard deviation stress is displayed in **Figure 22**.

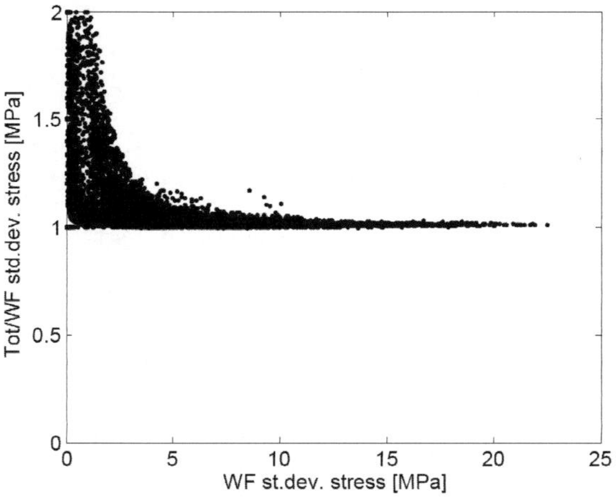

Figure 22: Total/WF standard deviation stress as a function of WF starndard deviation stress for midship section, port sensor.

3.6 EXAMPLE OF AN "EXTREME" EVENT

A half hour on the evening of 24th of January 2008 is considered. The significant wave height without correction read 7.5m, hence it was probably between 5 and 6m significant wave height. The peak period was 11.4 seconds. The waves were coming from 226° (true) with relative heading of 316°, i.e. bow quartering seas coming in on port side. The position was 45°N 40°W. This is certainly not an extreme event, and the masters have in general managed to avoid the worst weather during the measurement period. This illustrates however one simple way to visualise the consequence and some physics based on a significant event.

The time series with significant vibration is shown in **Figure 23**. Both the total and the WF filtered signal are shown with the vibration being the difference. The static stress is not included in this figure. The vibration is increasing the dynamic stress from about 60 to 100MPa in hogging, i.e. 67% increase. It does not appear to be a clear whipping, and the first large vibration cycles are actually observed on the hogging side. It could be more springing type of vibration. The vibration frequency is about 5.1 rad/s, and the encounter frequency of the rigid response is about 0.7rad/s (9s). There were also some natural frequencies at 10 and 15rad/s, but these were insignificant.

The corresponding plot for the after quarter length is displayed in **Figure 24**. Here the vibration appear to be even more dominating, increasing the dynamic stress from 30 to above 60MPa, i.e. more than a doubling of the stress level in hogging. It should be remembered that this is not an extreme sea state.

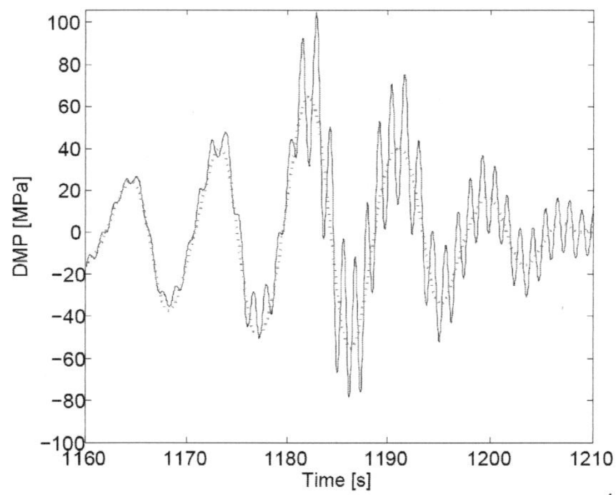

Figure 23: Port sensor midship, DMP, time series at 24th of January, 23.00UTC, 2008. Tension is positive.

The difference between the port sensor stress and the artificial sensor, which represents the vertical bending and axial stress, is displayed in **Figure 25**. The dashed line then represents the combined effect of the warping and horizontal bending stress, and it can be seen that this stress is somewhat out of the phase with the total stress which includes significant vibrations. The small vibrations can be caused by imperfect calibration of the port and starboard sensor.

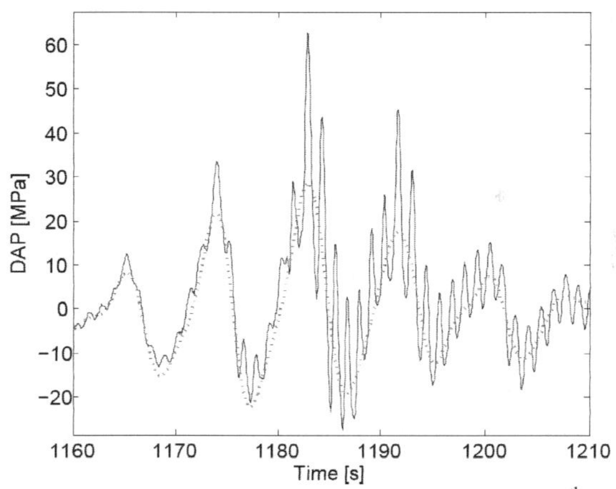

Figure 24: Port sensor aft, DAP, time series at 24th of January, 23.00UTC, 2008.

The corresponding time series of bow acceleration is shown in **Figure 26**. Positive acceleration is downwards, meaning that an upwards impact gives positive acceleration felt by the sensor. It is observed that there is apparently a small impact at about 1181 seconds, inducing some higher order vibrations. This occurs when the WF acceleration is positive, which may be the case when the bow is going into a wave. The phase relations between the acceleration and stress may not be as clear as this occurs in irregular sea. Apart from that, the vibration response is building up, indicating springing rather than the impulse like whipping type. The vertical acceleration amplitude due to vibration is up to 2m/s^2, while the total dynamic maximum amplitude is about 6m/s^2.

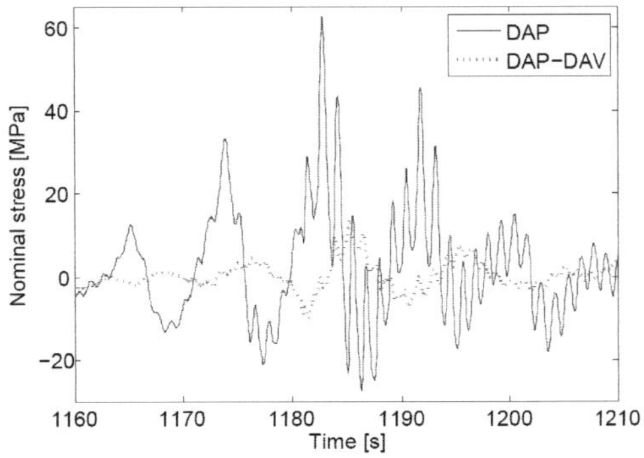

Figure 25: Port sensors and the combined effect of warping and horizontal bending stress. Time series is taken from 24th of January, 23.00UTC, 2008.

The mean RPM during this half hour is 71.2, but significant variations are present as illustrated in **Figure 27**. The RPM is measured with 5Hz, and the maximum is 87.6 at about 1190 seconds. The encounter period of the main waves is reflected in the period of the rpm fluctuations. Fixed RPM control is not considered to be the best way to control the engine in heavy seas, still it is commonly used.

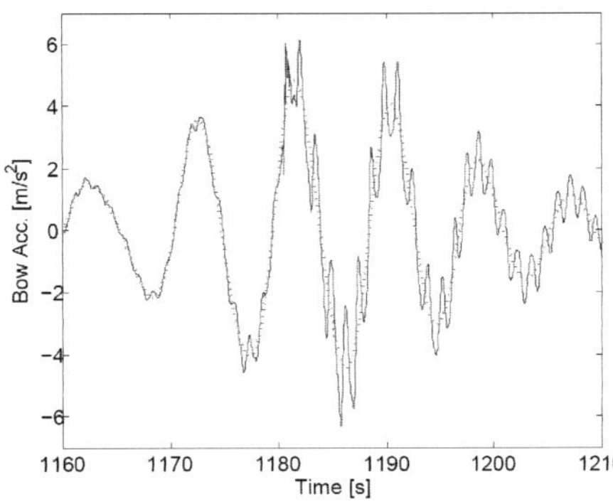

Figure 26: Vertical bow acceleration time series at 24th of January, 23.00UTC, 2008.

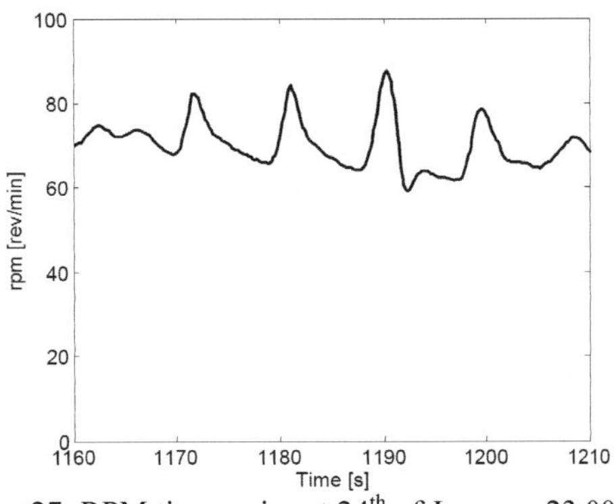

Figure 27: RPM time series at 24th of January, 23.00UTC, 2008.

The mean rudder angle (measured at 5Hz) is 2.2° during the half hour, which match well with waves coming in on port side. **Figure 28** shows that the rudder is working hard and affected by both ship motions (10s periods) and slowly varying motions at (50s periods).

During a half hour the vessel speed fluctuates between 11 and 15kn (measured at 1Hz). The mean is 13.3kn. The corresponding time series of the speed is shown in **Figure 29**. The speed variation due to the encounter waves is observed, but when encountering a few larger waves, the speed drops as much as 2 knots within 25 seconds. This is due to additional added resistance in waves.

The mean course during this half hour is 269.5°, but varying between 261 and 279°. The corresponding course plot is shown in **Figure 30**. Significant variations in the course are observed based on single encounter waves. The rudder angle also to some extent reflects the course changes. In this particular sea state, the rudder is controlled by active steering, and not an autopilot. The large amplitudes in the rudder angle are not considered optimum for the fuel consumption, but in heave seas there are more important concerns.

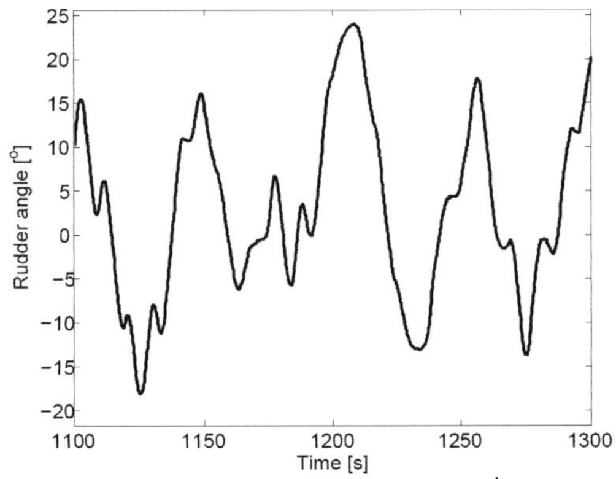

Figure 28: Rudder angle time series at 24th of January, 23.00UTC, 2008. Positive angle turns ship to port.

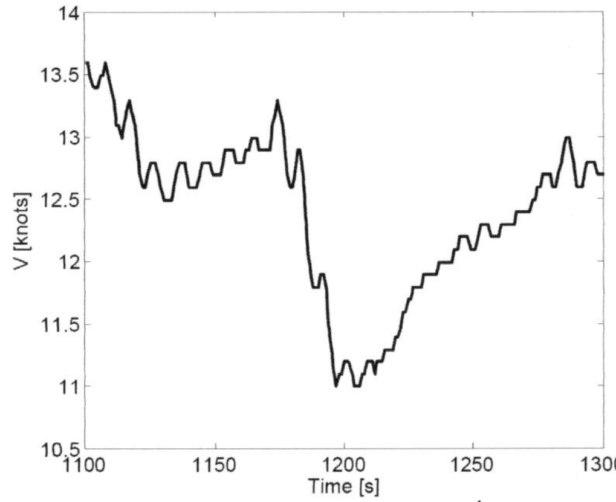

Figure 29: Speed time series at 24th of January, 23.00UTC, 2008.

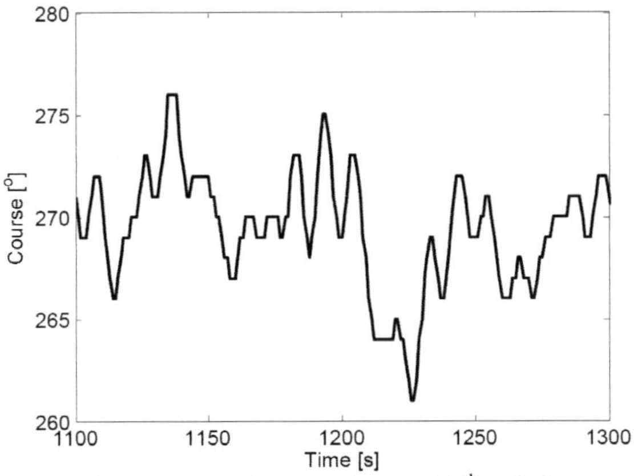

Figure 30: Course time series at 24th of January, 23.00UTC, 2008.

The heading time series (measured at about 0.6Hz), may differ from the course. This is possible, since the heading shows the direction of the bow, while the course shows how the vessel moves over ground. For instance in zero speed, the vessel may drift side ways in beam seas, so that there is a 90° difference. In forward speed and with current it may not be as easy to know how the result becomes. The heading is shown in **Figure 31**. It can be observed that this has less variation (varying between 266 and 275° during the half hour). The difference is also affected by sway, which is included in the course, but not in the heading. The mean heading is 270.0°, hence 0.5° different from the course in this case.

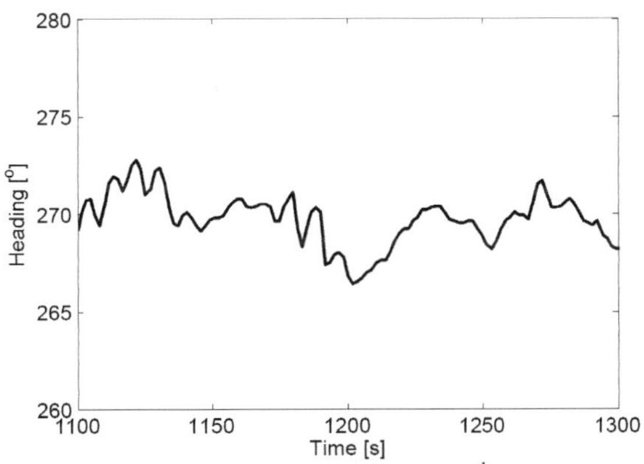

Figure 31: Heading time series at 24th of January, 23.00UTC, 2008.

The motions are also measured. The roll is shown in **Figure 32**. The dominating roll period is about 15 seconds, and it is within ±5° within the half hour. The mean angle is -0.13°. The pitch is shown in **Figure 33**, and the mean is 0.09°, and it is within ±4° during the half hour. The peak encounter period is about 9.5s. At the time when the vibration increase (about 1180s), the pitch is relatively large. The bow is at maximum negative angle at 1181 seconds, i.e. the bow is "submerged". The heave is within ±3.0m (the MRU reference point is close to the vessel's COG). At time 1181s, the vessel is moving down, but at close to zero (mean).

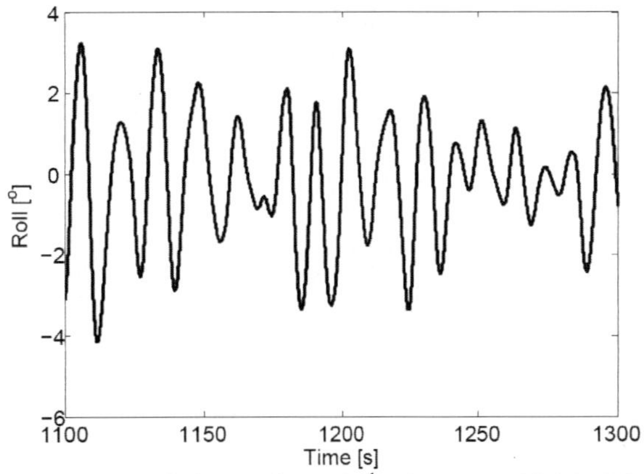

Figure 32: Roll time series at 24th of January, 23.00UTC, 2008. Roll to port is positive.

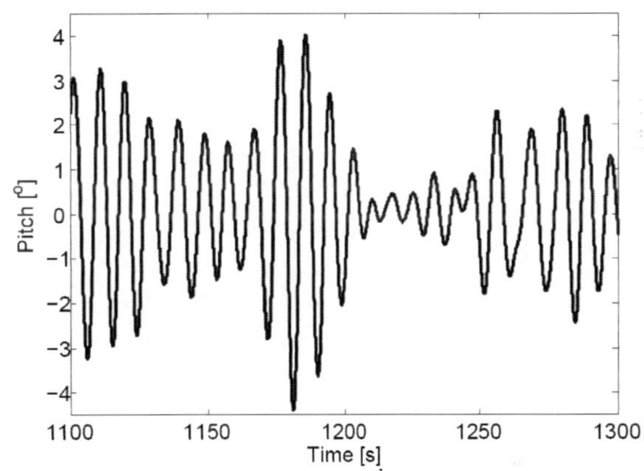

Figure 33: Pitch time series at 24th of January, 23.00UTC, 2008. Bow up is positive.

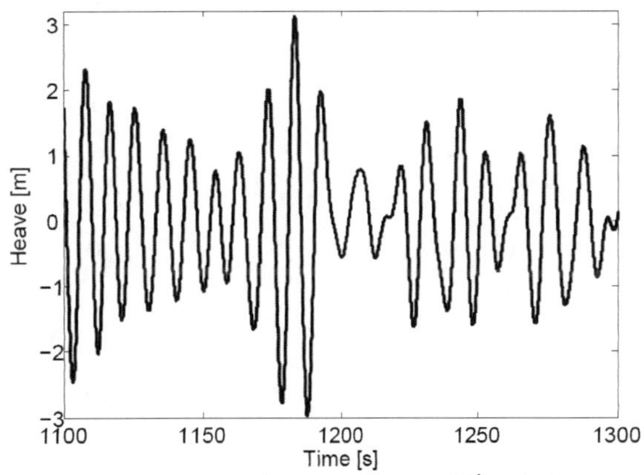

Figure 34: Heave time series at 24th of January, 23.00UTC, 2008. Ship going down is positive.

The sway and yaw are shown in **Figure 35** and **Figure 36**, respectively. The sway is within ±4.5m during the half hour. The sway has a main period close to the rudder angle period of 50 seconds. At 1180s the sway amplitude is towards starboard. The yaw angle is ±4° within the half hour. The mean yaw angle is somewhat meaningless, since it depends on when the zero setting of the MRU was taken. The mean is therefore subtracted. The yaw is affected by the rudder angle period (or visa versa) as well as the ordinary ship motions. At 1181s the yaw angle is close to zero (mean).

The surge motion is not measured, but the surge velocity and acceleration are included. The surge acceleration is maximum 0.23 m/s^2 and minimum -0.33 within the half hour. The time series is shown in **Figure 37**. It is noted that the longitudinal acceleration from vibration is pronounced relative to the ship surge motion, even though the MRU is only 2.6m above the neutral axis (but almost 38m aft of midship). Relative to the vertical vibration acceleration in the bow, the longitudinal vibration acceleration is however only about 5% in this case. Further towards the ends and high above the neutral axis, e.g. at the bridge, this longitudinal vibration should be easily felt by the officers on watch.

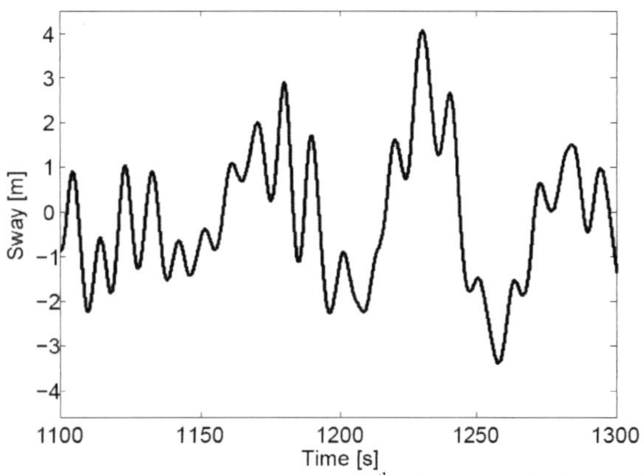

Figure 35: Sway time series at 24th of January, 23.00UTC, 2008. Ship going to starboard is positive.

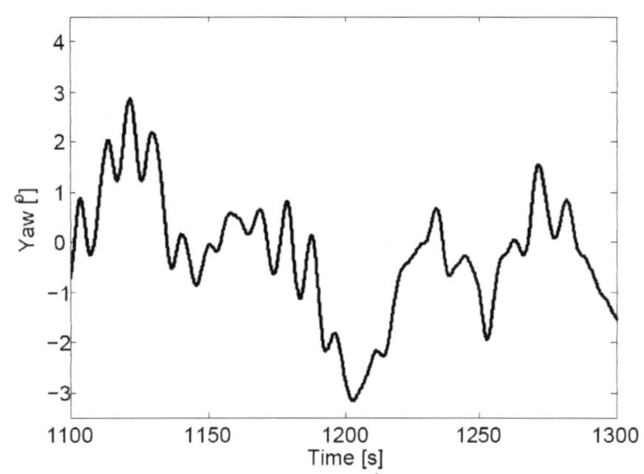

Figure 36: Yaw time series at 24th of January, 23.00UTC, 2008. Ship turning to starboard is positive.

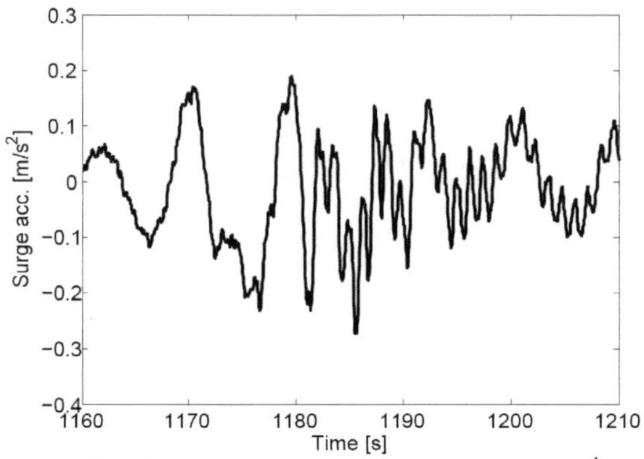

Figure 37: Surge acceleration time series at 24th of January, 23.00UTC, 2008.

4. RESULTS FOR THE 4400TEU VESSEL

4.1 TRADE AND MEASUREMENT PERIOD

The 4400 TEU vessel operates in the same trade between northern Europe and Canada. Almost continuous measurements exist from 28th of October until present. The position plot for the period 2007-11-01 to 2008-05-13 is given in **Figure 38**. During this period, 18 crossings have been performed. Data for 8853 half hours (184 days) have been collected.

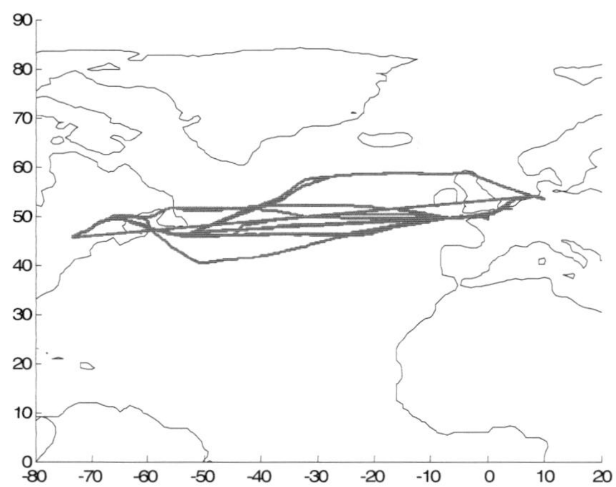

Figure 38: Route from November 2007 to May 2008.

The vessel is between 5 and 60° west 43.2% of the time. In this region, the average speed is 19.7kn. The maximum speed is 22.7kn. The speed versus time is shown in **Figure 39**.

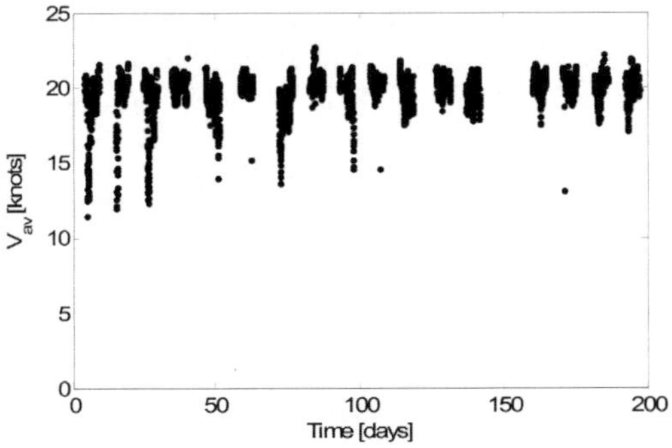

Figure 39: Speed when between 5 and 60° west. 0 corresponds to 28th of November. Each dot represents a half hour average.

The engine RPM in **Figure 40** is typically between 80 and 90. The maximum value is 95.

The worst storm during the actual time period was encountered on the crossing between 1st and 6th of November. The speed and rpm for this period is shown in **Figure 41**. The most extreme response occurred on the 2nd of November. This corresponds to Time = 5 days on the plots presented in this subsection. It is seen that at this occasion, the RPM is reduced slightly, whereas the ship speed is halved. This means that the added resistance in waves is large.

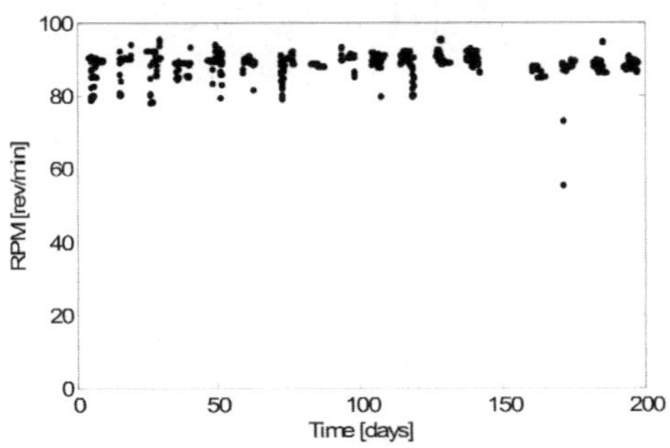

Figure 40: RPM when between 5 and 60° west. 0 corresponds to 28th October.

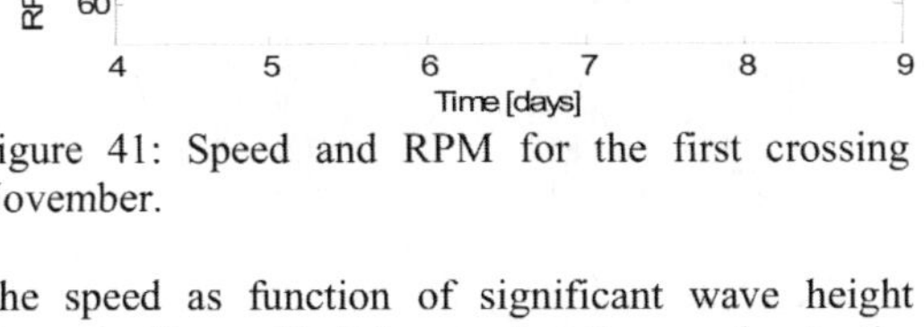

Figure 41: Speed and RPM for the first crossing in November.

The speed as function of significant wave height is shown in Figure 42. It is seen that the speed reduction is large for the westbound crossings. In this case, head seas are dominating. For the eastbound crossings, the dominating wave headings are close to following seas. In addition, the encounter frequency is lower and slams occur more rarely. Hence, the speed reduction is much less on the eastbound crossings. A quadratic fit is made to the data from the westbound crossings. The result is shown together with the corresponding curve for the 2800TEU vessel. It is seen that the curve for the 2800TEU vessel actually gives a better fit for the highest waves.

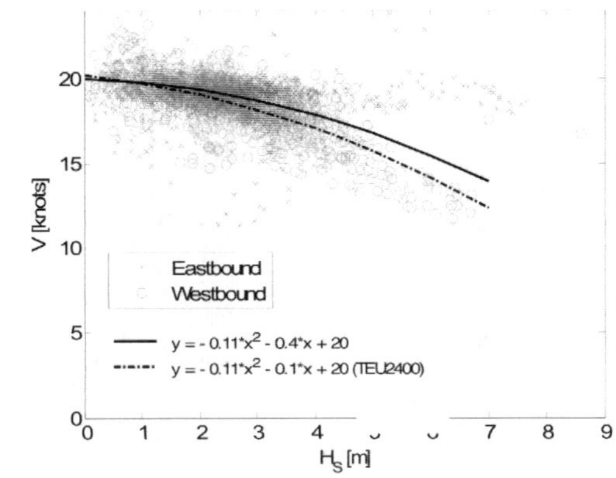

Figure 42: Speed reduction when RPM is above 70 (all wave directions included).

4.2 SEA STATES

The same type of wave radar was used on both vessels. Also for the 4400TEU vessel, the measured significant wave heights have been compared to buoy measurements. The results are shown in Figure 43. This resembles the results for the 2800TEU vessel, but the overestimation at higher waves appears to be less. Minimizing the error for a linear line going through zero gives a slope of 0.86. However, very high waves are not included or measured often in the comparison. The captain has also not reported significant deviation at high sea states. It has been decided not to correct the measurements for this vessel herein. The periods from measurements and buoys were in good agreement (as for the 2800TEU vessel).

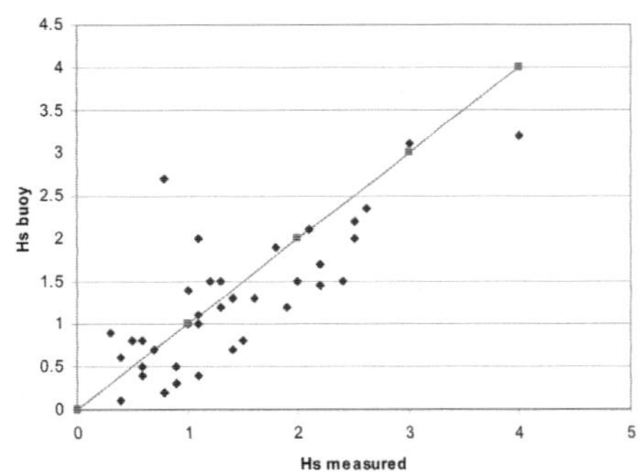

Figure 43: Comparison of wave heights between buoy data and measurements, when the ship is passing buoys.

4.3 FATIGUE DAMAGE AND VIBRATION

The fatigue damage measured in the period 2007-11-28 – 2008-05-13 (i.e. mainly in the winter period) is shown in Figure 45. The damages for the port and starboard deck longitudinals are given. The fatigue damage caused by vibration constitutes about 30% of the total fatigue damage. The average fatigue rate compared to design is 222% for this winter period (assuming SCF=2.0).

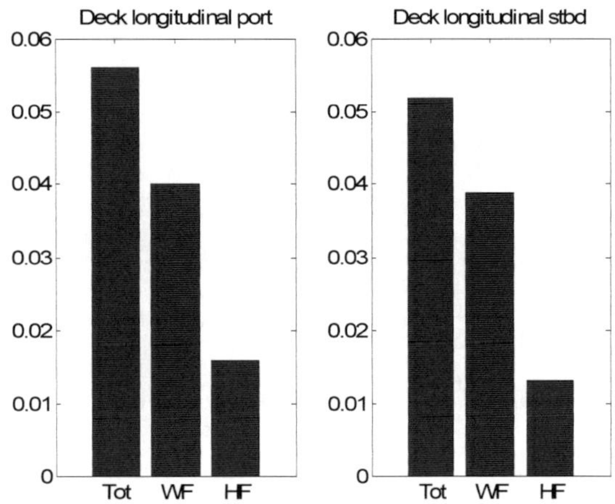

Figure 44: Fatigue damage measured for the 4400TEU vessel in the period 2007-11-28 – 2008-05-13.

If the whole measurement period since November 2006 is considered, the average fatigue rate is 215% based on 16407 records (i.e. 0.94years effective duration). The fatigue damage from wave induced vibration accounts for 27%. If the SCF is set to 2.0, the fatigue life is 9 years. The measurements are still, however, biased by more time during the winter periods. For a SCF of 1.8, which is more typical for the stiffener through frame transitions, the fatigue life would increase to 13 years. Even with overrepresentation by the winter period, some cracks may be expected to occur before the end of the service life. This will also depend on the annual variations, which is not available yet.

The fatigue damage for the side shell longitudinals are shown in **Figure 45**.The WF fatigue is shown with the lower parts of the bars, while the upper part shows the HF contribution. The importance of high frequency vibrations is still significant, since the sensor positions are about 3.23m below the neutral axis. Also the encounter response with above 0.3Hz (i.e. 3.33s) is regarded as vibration, and it may get contribution from smaller waves (for the global sensors, the ship does not display significant bending response for such high frequencies).

The total fatigue damage in **Figure 45** is very high for the side shell longitudinals, and much higher than for the deck longitudinals. It should be noted that the hot spot stress is estimated by a simplified combination of two sensors as explained in [7]. Basically, one sensor is used to establish the global stress (located at moment zero point for local bending, as well as the neutral axis for the stiffener plate configuration). In addition, the local bending stress is established as the difference between the two sensors. In this case the tripping of the L-profile, which is causing additional stress, is neglected in the simplified method. The SCF for local bending is set to 2.4 instead of 2.0, so the tripping effect becomes more important. This results in an overestimation of the fatigue damage. It was also checked for errors in the set-up and software, but all turned out to be OK. If for instance the cables were switched, the fatigue damage would drop to about 40% of the original, but it is still high. In any case, the side shell is pointed out as the critical area, and this side shell longitudinal was also selected because calculations identified this one as most prone to fatigue damage. Time will show how far off the measurements are from the truth, but it can be pointed out that no cracks have been identified yet after 5 years of operation.

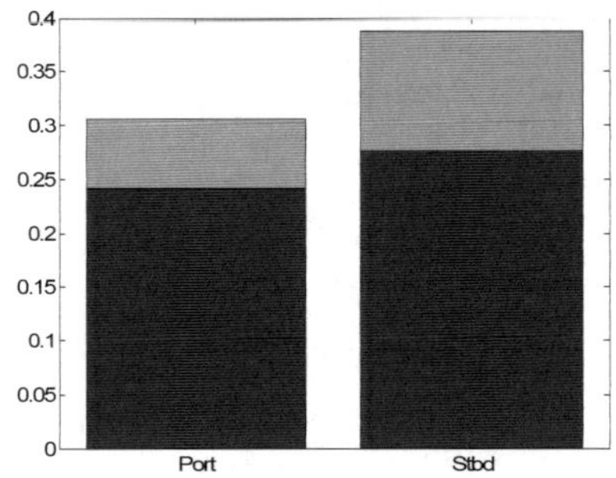

Figure 45: Side shell fatigue damage measured in the period 2007-11-28 – 2008-05-13.

The warping stress is identified and taken away from the time series as explained in [7]. The fatigue damage is then recalculated. Unfortunately, a SCF=1.0, was given to this artificial sensor. In Figure 46, the fatigue damage without warping has been corrected to a SCF=2.0, raised to the power of 4.0 for the WF part and 3.7 for the HF part. This is taken from [2]. The figure then shows that the warping reduces the fatigue damage in deck longitudinals, as expected, with about 25%.

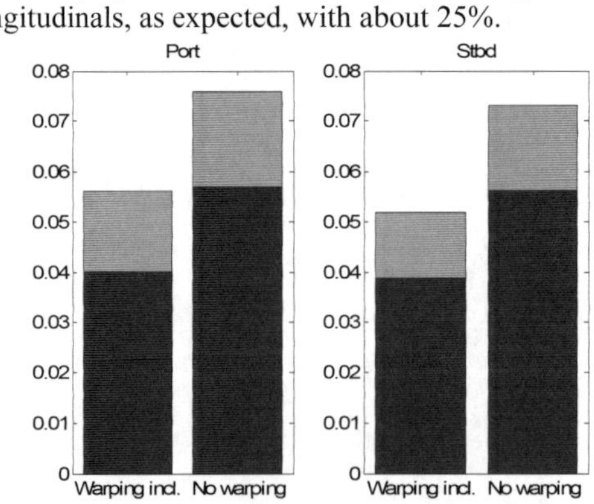

Figure 46 Fatigue damage in deck with and without warping from period 2007-11-28 – 2008-05-13

Time plot of the fatigue rate is shown in **Figure 47**. The largest fatigue damage occurs on the crossing between 1st and 6th of November, i.e. the first crossing shown in the figure. The highest fatigue rate is close to 10 000%. In this storm, one half hour is equivalent to 2 days of average fatigue loading according to design.

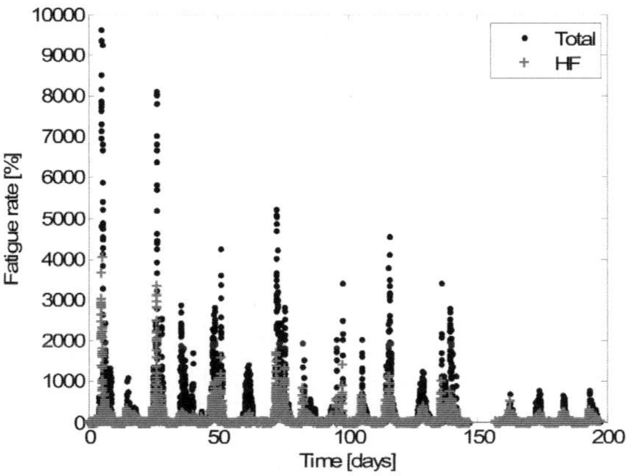

Figure 47: Total fatigue rate from 28th of October 2008 (deck longitudinal, port).

The fatigue damage for the first crossing is shown in Figure 48. The HF contribution is significant, particularly for the first (and worst) part of the storm.

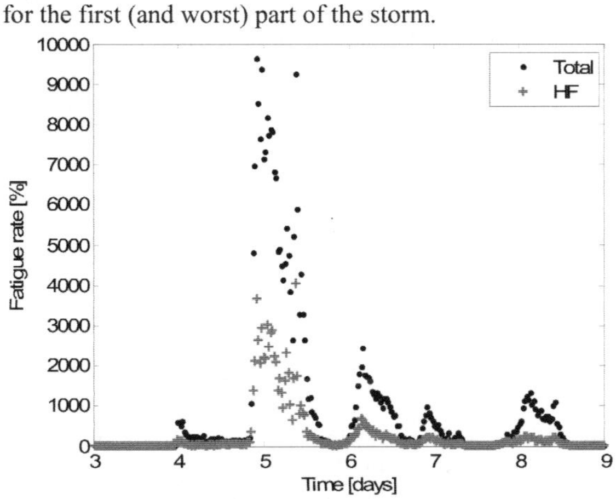

Figure 48: Total and HF fatigue rate for first crossing in November 2007 (deck longitudinal, port).

A corresponding figure for a later crossing is shown in **Figure 49**. The difference between these two figures is the direction the vessel is going. The first crossing is from Europe to Canada, i.e. from east to west. **Figure 47** shows that the fatigue rate is highest for these crossings. The fatigue rate, and also the amount of vibration damage, is much less for the eastbound crossings. Overall, for the Westbound crossings, the mean fatigue rate is 673%. Vibrations constitute 30% of the total damage. For the Eastbound crossings, the mean fatigue rate is 308%. Vibrations constitute 20% of the total damage. For the eastbound crossings, the dominating wave headings are close to following seas. In this case, the encounter frequency is lower and slams occur more rarely. This explains the lower importance of HF vibrations for crossings in this direction.

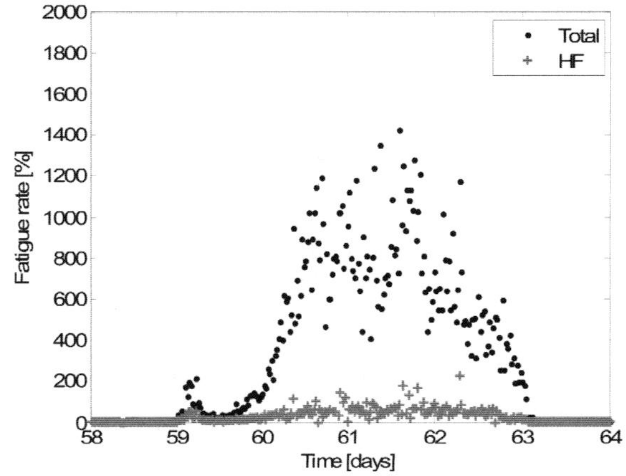

Figure 49: Total and HF fatigue rate for crossing in New Year crossing.

The fatigue damage of the port deck sensor amidships is plotted as a function of the measured significant wave height in **Figure 50**. The fatigue damage is increasing with increasing significant wave height. The relative importance of the HF damage is largest for the lowest wave heights. This effect is relatively small, however. For H_S = 1 m, the HF damage constitutes 32% of the total damage. For H_S = 6m, the corresponding number is 25%.

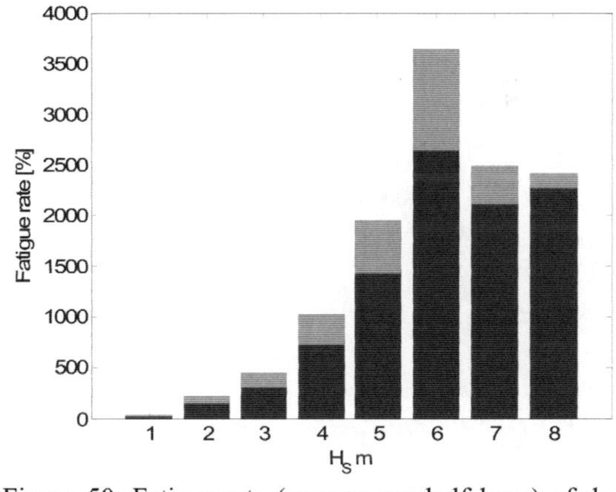

Figure 50: Fatigue rate (average per half hour) of deck port sensor amidships as function of H_S. Upper part is HF and lower part is WF damage.

The total fatigue damage as function of significant wave height is shown in Figure 51. The dominating wave height is $H_S \sim$ 3-4m. This is slightly less than for the TEU2800 vessel, but it may be affected by the short measurement period and the encountered sea states during this period. It may also be affected by the wave radar overestimating smaller sea states.

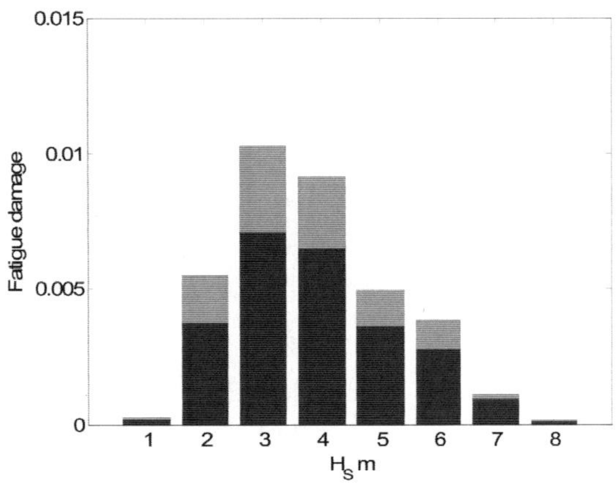

Figure 51: Fatigue damage of deck port sensor amidships as function of H_S. Upper part is HF and lower part is WF damage.

4.4 EXTREME LOADING

The total versus WF maximum or minimum stress per half hour is shown in **Figure 52**. As for the 2800TEU vessel, the vibration clearly increases the stress level both in sagging and hogging. Again, it is seen that the vibration effect is more pronounced at the highest stress levels.

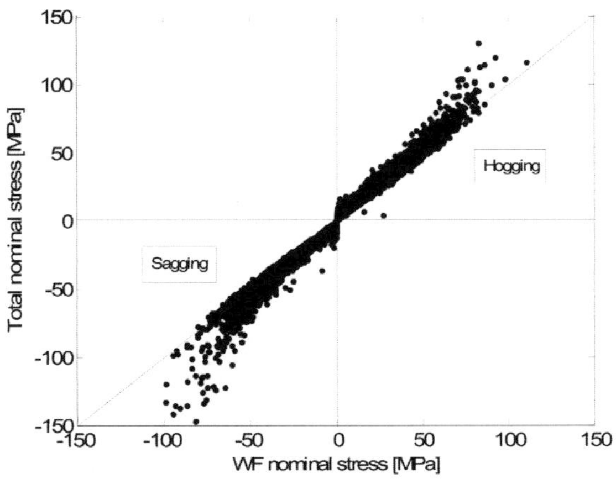

Figure 52: Total vs WF maximum (minimum) stress amplitude per half hour for midship section.

Figure 53 shows the HF stress versus the total stress. The cubic polynomial for the 2800TEU vessel is also shown. This figure also shows that the effect of HF stress is increasing for increasing values of WF nominal stress.

Also the standard deviation of stress is affected by vibration. This is particularly the case at the lowest stress levels. This is seen in **Figure 54** showing the total divided by the WF standard deviation of the stress at the deck longitudinal (port).

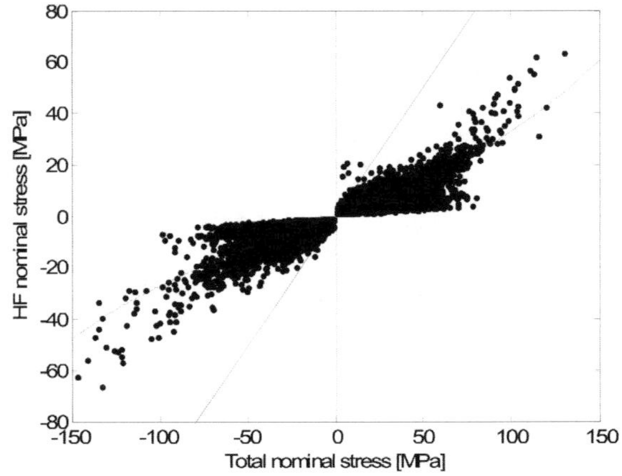

Figure 53: HF vs total maximum (minimum) stress amplitude per half hour for midship section.

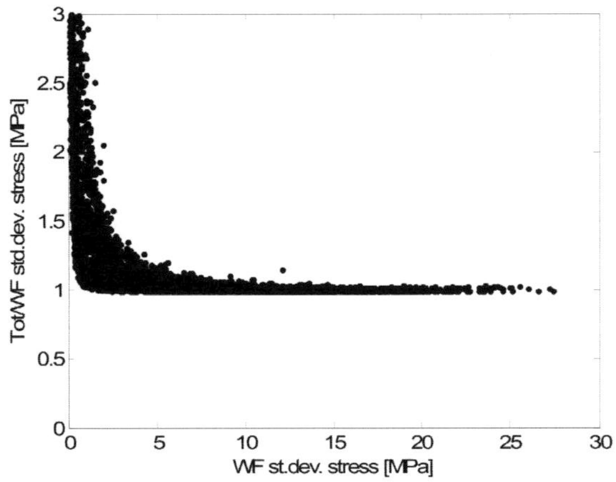

Figure 54: Total/WF standard deviation stress as function of WF standard deviation stress for midship section, port sensor.

4.5 TIME SERIES FOR EXTREME EVENT

The largest fatigue rate is measured at 2nd November 2007 around 16:00 hours. The time series of the dynamic stress (i.e. mean stress is subtracted) at the port deck longitudinal is shown in **Figure 55**. Both the total and the WF filtered signals are shown. In this case, the large dynamic response starts when the vessel is in sagging condition. This is typical for a whipping event. The maximum stress is about 80% higher in hogging and 95% higher in sagging due to whipping.

The fatigue rate was very high also on 23rd November 2007 around 14:00 hours. The significant wave height was 6 m and the peak period was 12.4s. The relative heading is about 45°, with waves coming in on starboard side. The time series of the stress at the port deck longitudinal is shown in **Figure 56**. In this case, the maximum stress is increased by 70% in hogging and 80% in sagging due to whipping.

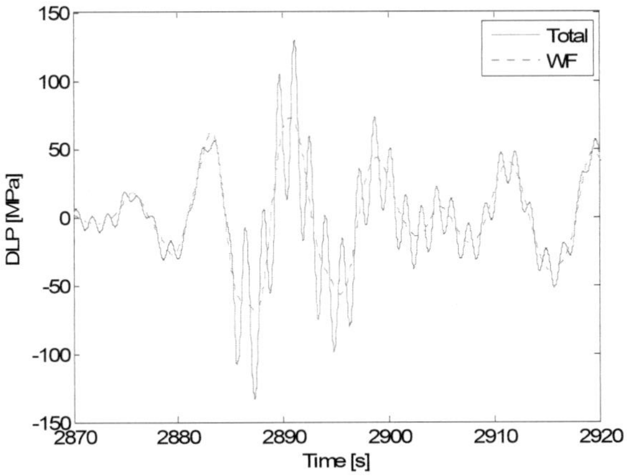

Figure 55: Time series for dynamic stress at port deck longitudinal (DLP) at 2nd November 2007, 16:00 UTC. Tension is positive.

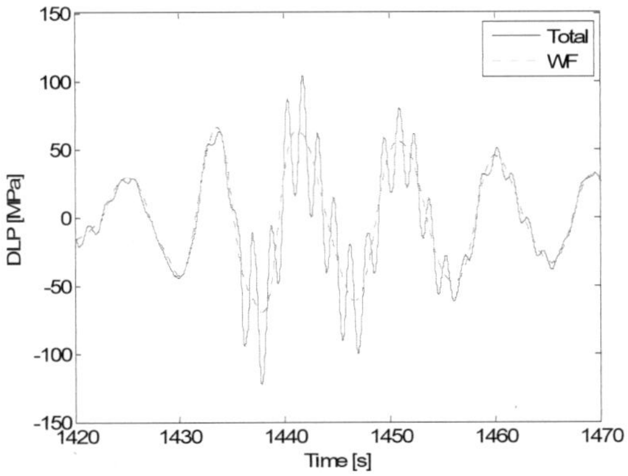

Figure 56: Time series for dynamic stress at port deck longitudinal (DLP) at 23rd of November 2007, 14.00 UTC. Tension is positive.

The RPM time series is shown in **Figure 57**. This is similar to the time series for the 2800TEU vessel, and fluctuations due to encountered waves are observed.

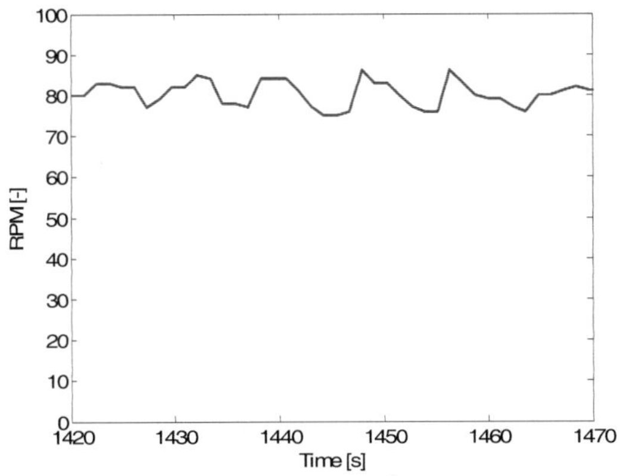

Figure 57: RPM time series at 23rd of November 2007, 14.00 UTC.

The rudder angle is shown in **Figure 58**. It is not confirmed if autopilot was used in this case. The speed in **Figure 59** shows a drop of 3knots within a minute, while **Figure 60** confirms small heading changes.

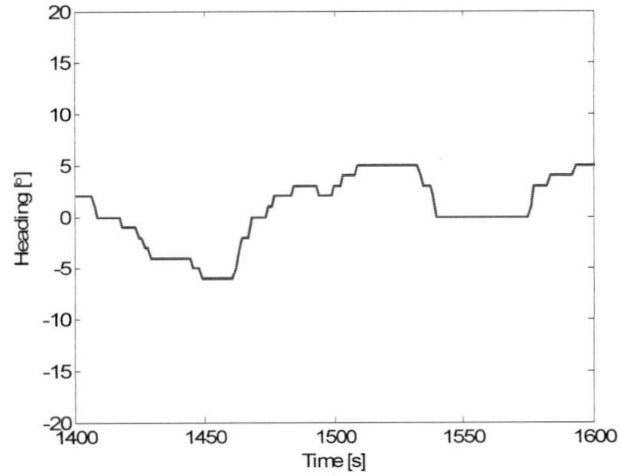

Figure 58: Rudder angle at 23rd of November 2007, 14.00 UTC. Positive angle turns ship to port.

Figure 59: Speed time series angle at 23rd of November 2007, 14.00 UTC.

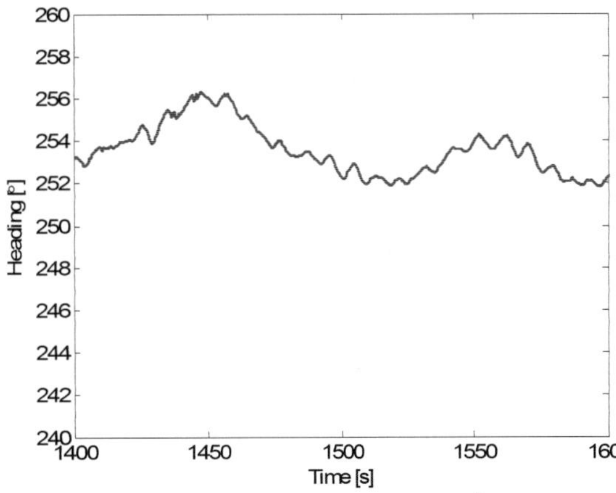

Figure 60: Heading time series angle at 23rd of November 2007, 14.00 UTC.

In **Figure 61** to **Figure 66** the 6 motions are shown. The roll is small. The sway and surge is about 1m. The yaw is affected by the rudder angle, and displaying small effect of the encounter waves. The heave is significant, but is measuring heave at the super structure, so it is affected by pitch.

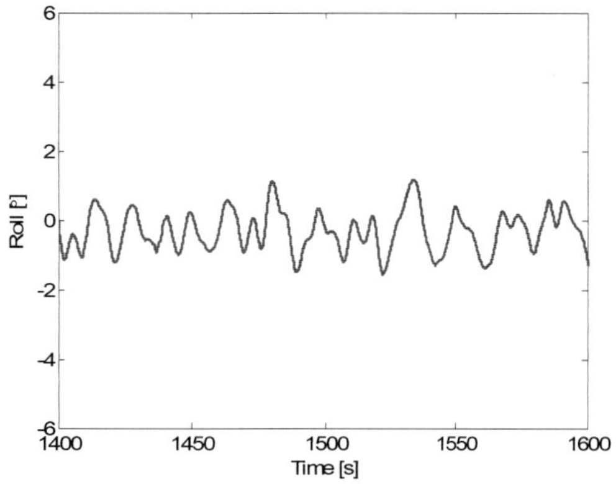

Figure 61: Roll time series at 23rd of November, 14:00 UTC. Roll to port side is positive.

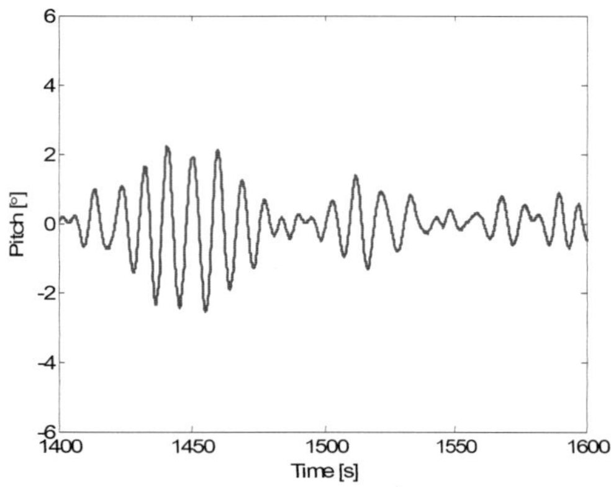

Figure 62: Pitch time series at 23rd of November, 14:00 UTC. Bow up is positive.

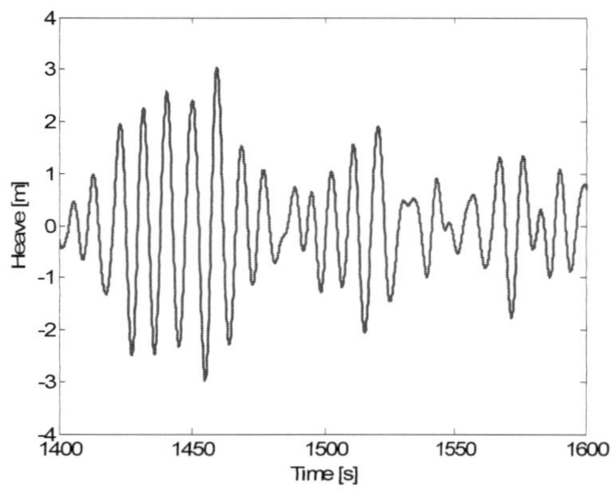

Figure 63: Heave time series at 23rd of November, 14:00 UTC. Downwards motion is positive.

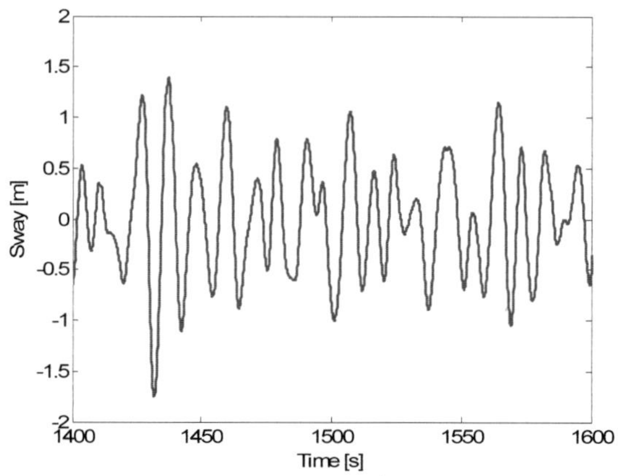

Figure 64: Sway time series at 23rd of November, 14:00 UTC. Motion to starboard is positive.

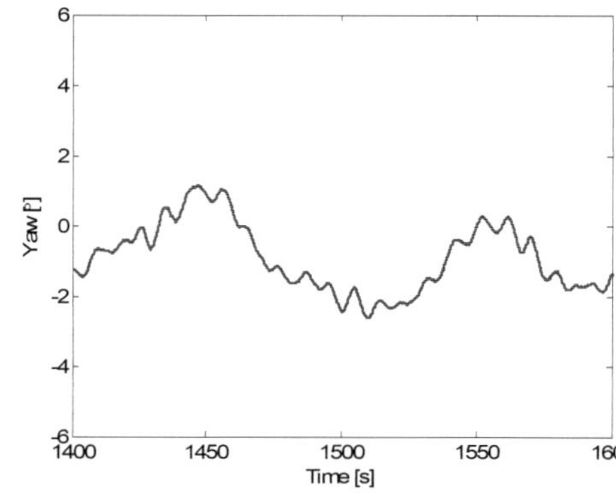

Figure 65: Yaw time series at 23rd of November, 14:00 UTC. Turning to starboard is positive.

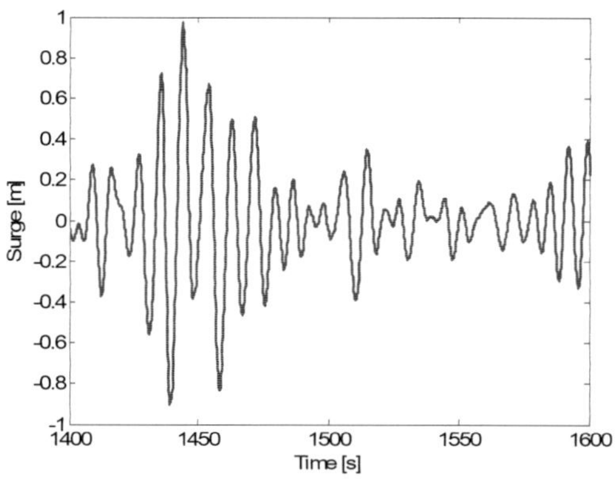

Figure 66: Surge time series at 23rd of November, 14:00 UTC.

The surge acceleration is shown in **Figure 67**. It is seen that the acceleration is large at the same time instant as the large vibration stress. Also in this case, longitudinal accelerations from vibration can be clearly seen. This supports the conclusion that a severe bow slam event has occurred in this case. The acceleration is much larger than for the 2800TEU vessel, but in this case the MRU is located higher up in the super structure (deck below the bridge).

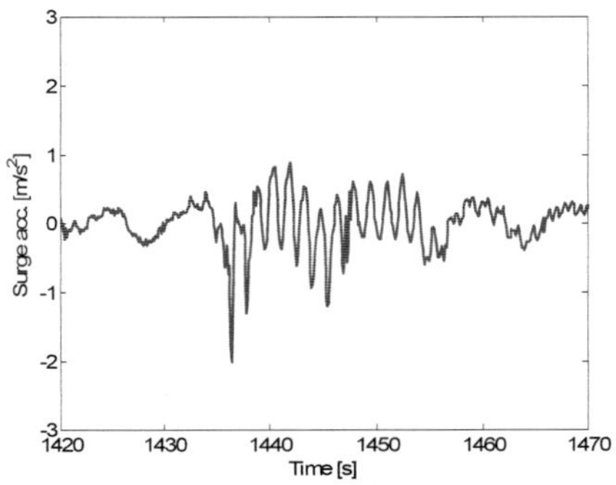

Figure 67: Surge acceleration time series at 23rd of November, 14:00 UTC.

5. CONCLUSIONS

Full scale measurements from two container vessels, operating in the North Atlantic during winter season 2007/2008, have been reported.

The measurements show that the vessel speed is reduced considerably in higher sea states. A large part of the speed reduction is caused by added resistance in waves. This effect is largest on westbound crossings where head seas are dominating.

For the 2800TEU vessel, the wave radar overestimates the waves in higher sea states. This is adjusted by multiplying the measured wave heights with 0.7. For the TEU4400 vessel, the difference between buoy and radar measurements is less, but with some overestimation in particular at lower sea states. The measured wave heights indicate that routing is performed, and that it appears to be effective for sea states above 6m Hs.

The measurements of the fatigue loading in deck shows that the wave induced vibrations contribute with 27% of the total fatigue damage on both ships. The effect of size is then not evident. The importance of vibration damage relative to the conventional WF damage is almost constant with the increase in significant wave height. The maximum vibration amplitudes are however more pronounced at the highest stress levels.

The fatigue rate is high in periods, typically of about one to two days on each crossing. However, also in average the fatigue rate in deck is high, and about 163% and 215% for the smaller and larger vessel, respectively. This is based on a SCF of 2.0, and it is biased by more measurements collected during the winter half year. Still, it indicates that the loading is high, and that fatigue cracks may be anticipated before the end of the service life. Cracks have already been observed at stringer level 1 on the older smaller vessel after 8 years of operation. Further, the measurements confirm that the side shell is sensitive to fatigue damage, although the fatigue estimates by the measurements are conservative. These details are suggested for follow up during inspections.

The fatigue damage per half hour increases with increasing H_S, but at higher sea states the trend becomes unclear due to few recordings and possible off head sea headings (beam/stern quartering/following seas). The largest contribution to the overall fatigue damage comes from 3 to 5m Hs.

Examples of single extreme events have been shown. For the 4400TEU vessel, the time series show a typical bow slam induced whipping incident. For the 2800TEU vessel, the stress time series indicate a more springing type of phenomenon with large vibration in hogging. For both vessels, high frequency vibrations increase the extreme dynamic stress with up to 80% in hogging amidships. At the engine bulkhead at the aft quarter length, the relative importance appears to be even higher and double the dynamic hogging moment. It should be noted that these events refer to sea states of about 6m Hs, and the vessels have significant forward speed of about 13-14knots. These values can then not be directly used in extreme sea states.

The combined effect of horizontal bending stress and warping stress has been confirmed to be more important at the engine bulkhead than amidships wrt. fatigue loading based on the 2800TEU vessel. These two loading components increased the fatigue damage at deck level. It was significantly at the aft cross section, where the increase was 50% without the vibration effect. It should however be noted that at the engine bulkhead the fatigue damage in deck was insignificant compared to midship section. Amidships in deck the combined effect gave an increase of 5%. For the 4400TEU vessel, the warping stress reduced the fatigue damage in the longitudinals with 25%. Hence to neglect it, which is common, is conservative in deck.

It has also been noted that the surge motions can be highly influenced by whipping and vibration events, and a whipping surge acceleration of 0.2g was measured.

To summarize, the wave induced vibrations can have a significant effect on the fatigue damage (but less than on similar sized blunt ships), but an even higher effect on the extreme loading. The latter then refers to what is measured as extreme events, and does not refer to extreme design sea states. The warping stress of deck longitudinals amidships is significant, while the combined effect of warping stress and horizontal bending is not. The method for estimation of side shell longitudinals is not straight forward and it can be improved. The measurements onboard can be used to increase the awareness for such special events, to identify critical areas, to reduce the risk of overloading and to reduce fatigue damage. Routing appears to be an essential part of this.

6. ACKNOWLEDGEMENTS

The authors would like to thank the owner, manager, superintendents, masters and crew onboard the container vessels for their kind assistance in the measurement project. Thanks also to the owner and DNV for funding these R&D projects. The opinions expressed in the paper by the authors do not necessarily represent the views of the Det Norske Veritas.

7. REFERENCES

1. Moe, E., Holtsmark, G. and Storhaug, G., 'Full scale measurements of the wave induced hull girder vibrations of an ore carrier trading in the North Atlantic', *Transactions of RINA, Design & Operation of Bulk Carriers*, 2005.

2. Storhaug, G., Moe, E. and Holtsmark, G., 'Measurements of wave induced hull girder vibrations of an ore carrier in different trades', *Journal of Offshore Mechanics and Arctic Engineering*, Vol. 129, Issue 4, Nov. 2007.

3. Storhaug, G., Moe, E., and Lopes, T.A.P., 'Whipping measurements onboard a midsize container vessel operating in the North Atlantic', *Proceedings of RINA, CMP & Shanghai SNAME, International Symposium on Ship Design & Construction*, 28-29th of November, 2007.

4. Drummen, I., Storhaug, G., Moe, E. and Moan, T., 'Experimental and full scale investigation of the importance of fatigue damage due to wave-induced vibration stress in a container vessel', *Transactions of RINA, Design & Operation of Container Vessels*, 2006.

5. Drummen, I., Storhaug, G. and Moan, T., 'Experimental and numerical investigation of fatigue damage due to wave-induced vibrations in a containership in head seas', *Journal of Marine Science and Technology*, Vol. 13, no. (to appear), 2008.

6. Storhaug, G., Vidic-Perunovic, J., Rüdinger, F., Holtsmark, G., Helmers, J.B. and Gu, X., 'Springing/whipping response of a large ocean going vessel – A comparison between numerical simulations and full-scale measurements', *Proc. of 3rd Int. conf. on Hydroelasticity in Marine Technology*, UK, 2003.

7. Storhaug, G. and Moe. E, 'Measurements of wave induced vibrations onboard a large container vessel operating in harsh environment', *Proc. of PRADS*, Houston, USA, 2007.

8. Steen, E., Storhaug, G., Kippenes, J., Sado, O., Austefjord, H. and Dugstad, G., 'Ultimate strength assessment of the engine room area of the container vessel MSC Napoli - January 2007 Incident', *Transaction RINA, Design and operation of container vessels*, UK, July, 2008.

9. 'Report on the investigation of the structural failure of MSC Napoli, English Channel, on 18 January 2007', MAIB report no. 9/2008, April 2008, Annex D Part 1.

10. Drummen, I. and Moan, T., 'Experimental investigation of the application of response conditioned waves for long-term nonlinear analysis', *Proc. of PRADS*, Houston, USA, 2007.

11. 'Strengthening against bow impact', *DNV Rules for classification of Ships*, Pt.3 Ch. 1 Sec. 7, E100, January 2008.

12. 'Hull monitoring systems', *DNV Rules for classification of Ships*, Pt.6 Ch. 11, January 2005.

13. 'Fatigue Assessment of ship structures', DNV *Classification Note 30.7*, July, 2005.

8. AUTHORS' BIOGRAPHIES

Gaute Storhaug is a Principle Engineer at the Section for Ship Structures in DNV. His daily work involves trouble shooting, research, consultancy services, and responsibility for full scale measurements and the voluntary class notation HMON (Hull MONitoring). His previous experience includes general hydrodynamic and structural analyses, full scale measurements, model tests and responsibility for fatigue calculations within the Condition Assessment Programme (CAP). Both the MSc and the PhD thesis dealt with the topic of Hydroelasticity. The former was written at MIT and the latter defended at the Centre of Excellence for Ships and Ocean Structures (CeSOS) at the Norwegian University of Science and Technology (NTNU). Email: Gaute.Storhaug@dnv.com

Svein Erling Heggelund currently holds the position of Senior Project Engineer at the Section for Ship Structures in DNV. He is currently working with structural analyses and full scale measurements. He was formerly employed by MARINTEK working with a wide range of problems related to sea loads and load effects on marine structures. In 2001 he defended his PhD on the topic of loading response analysis of catamarans at the Department for Marine Technology at NTNU. Email: Svein.Erling.Heggelund@dnv.com

NEW RULES FOR BREAKWATERS ON CONTAINERSHIPS

A. Kahl, **H. Rathje**, and **T. Schellin**, Germanischer Lloyd, Germany

SUMMARY

Cargo safety demands and recurrent breakwater damages initiated activities to assess current design practice and to establish new design rules for breakwaters on containerships. Linear seakeeping computations yielded extreme vertical relative displacements (motions) between wave contour and breakwater base. These displacements served as the critical ship response parameter to identify wave conditions that led to maximum green water loads on breakwaters. Spectral analysis methods resulted in long-term extreme values for ships advancing at constant forward speed in irregular waves. Based on these extreme values, equivalent regular design waves were selected to model extreme load scenarios. A validated procedure based on user-coded RANS routines calculated time dependent nonlinear pressure distributions of green water loads on breakwaters. New design rules for breakwaters were based on these design pressures. Application of loads according to the new rules led to the reported damages of the breakwaters.

1. INTRODUCTION

Cargo safety demands and recurring breakwater damage as shown, for instance, in Fig. 1 already led to breakwater scantlings that, for many newly built ships, significantly exceed old rule requirements. Also, discrepancies exist between experimental model test measurements of loads on breakwaters and the old classification society rules [1]. Old rule based pressure loads and scantlings of breakwaters correspond to the IACS UR S3 strength requirements of end bulkheads of superstructures and deckhouses [2]. These requirements do not include requirements specifically designated for breakwaters. Consequently, the need arose to establish new design rules specifically for breakwaters.

Fig. 1: Typical damaged breakwater

First, we examined model test measurements of green water loads on breakwaters. These tests were performed with a 195 m long modern containership, the 175 m long S175 containership, and a 127 m long general purpose cargo ship [3 – 7]. All the documented pressure measurements exceeded design pressures based on the old rules, some by as much as a factor of four.

To assess current design practice and to establish new design rules for breakwaters on containerships, we performed theoretical investigations for three representative containerships. Table 1 lists main particulars of the investigated ships. Ratio $(x/L)_{breakwater}$ specifies the dimensionless longitudinal position of the breakwaters. Here L stands for the ship's scantling length, and x is the longitudinal coordinate measured from the aft perpendicular.

Table 1: Main particulars of investigated containerships

	Feeder	Panmax	Post-panmax
Length	126 m	219 m	317 m
Breadth	21.3 m	32.3 m	45.6 m
Depth	12.3 m	19.5 m	27.2 m
Design draft	8.5 m	12.2 m	15.0 m
Design speed	18.2 knots	22.0 knots	26.9 knots
Block coeff.	0.689	0.673	0.660
$(x/L)_{breakwater}$	0.984	0.954	0.963

A linear seakeeping panel code [8] yielded vertical relative displacements (motions) at location x/L as the critical parameter of ship response to waves. Based on long-term values of these relative vertical displacements in irregular seas, we identified equivalent regular design waves that caused maximum relative displacements. Next, a validated procedure based on user-coded Reynolds-averaged Navier-Stokes (RANS) routines [9] calculated time dependent pressure distributions that caused maximum horizontal forces and base bending moments acting on the breakwaters. Finally, we established new rules that complied with loads recalculated from selected damage cases and compared as-built scantlings with those based on old and new rules.

2. SEAKEEPING COMPUTATIONS

We performed seakeeping computations for the ships advancing at constant one-half design speed in regular unit amplitude waves. The three-dimensional numerical procedure, performed in the frequency domain, relied on the use of the zero-speed green function and a forward speed correction based on the encounter frequency approach. Wave periods ranged from five to 34 s; wave headings, at 15 deg. intervals, from zero to 180 deg. (180 deg. wave heading denotes head waves). Over all phases of each wave, we obtained maximum vertical displacements between the wave contour and a point on the sip's still waterline directly below the breakwater. Plotted against wave period, theses displacements are shown as response amplitude operators (RAOs) in Fig. 2.

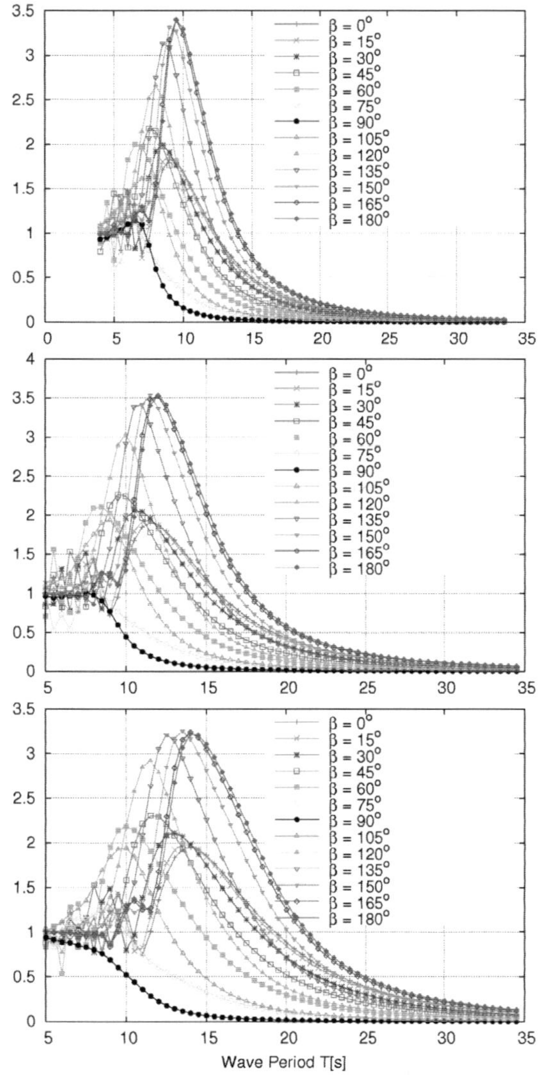

Fig. 2: RAOs of maximum vertical relative displacements for feedership (top), panmax containership (center), and post-panmax containership (bottom)

Then, taking the computed heave and pitch motions into account, we determined RAOs of maximum vertical displacements measured from the breakwater base to the wave contour. Based on these latter RAOs, long-term statistical analyses in irregular waves resulted in extreme vertical relative displacements. For the three sample ships operating for 25 years in seas represented by the standard IACS wave climate [10], Fig. 3 presents these extreme displacements as functions of their cumulative frequencies of occurrence.

Next, we selected equivalent regular design waves that yielded these long-term values of relative vertical displacements. We chose only design waves with steepness ratios of $H/\lambda < 0.10$ (H is wave height and λ wave length). As expected, the breakwaters were subject to maximum loads under head wave conditions, with design wave amplitudes of 7.0, 8.2, and 9.3 m for the feedership, the panmax ship, and the post-panmax ship, respectively. Sea states of significant wave height equal to these deign wave heights occur with a probability of about $3.2 \cdot 10^{-4}$ for the feedership and $1.6 \cdot 10^{-3}$ for the panmax ship and the post-panmax ship. For the operating time of 25 years, these probabilities correspond to about 70 hours at sea for the feedership and 350 hours at sea for the panmax ship and the post-panmax ship.

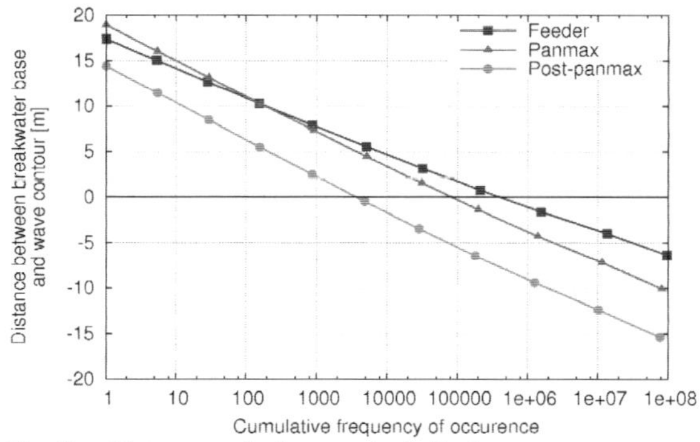

Fig. 3: Extreme relative vertical displacements as functions of the cumulative frequency of occurrence

3. RANS COMPUTATIONS

RANS computations of the three ships in head waves yielded maximum local pressures and total forces acting on the breakwaters as well as the corresponding base bending moments. These computations were performed for the ships under conditions as summarized in Table 2.

Table 2: Design wave condition for RANS computations

	Feeder	Panmax	Post-panmax
Wave amp.	7.0 m	8.2 m	9.3 m
Wave length	141 m	225 m	306 m
Heading	180 deg.	180 deg.	180 deg.
Ship draft	8.3 m	10.9 m	13.0 m
Ship speed	9.1 knots	11.0 knots	13.5 knots

The numerical procedure consisted of dividing the fluid domain into a finite number of control volumes, so-called cells. For each cell the RANS equations are solved in their integral form at successive time intervals. An additional transport equation solves for the volume fraction of water and air. The numerical procedure, specifically developed to predict slamming loads, was validated against model test measurements [11]. As a sample, Fig. 4 shows computed time histories of the resulting longitudinal force and the corresponding base moment acting on the breakwater of the panmax containership. Figure 5 shows a screen shot of this ship under design wave conditions at the time instant of maximum loading on the breakwater.

Table 3 lists RANS-computed results and comparable results according to classification society rules [12]. Symbol p_{rule} stands for rule based pressure either at the base or the top of the breakwater, M_{rule} is the rule based moment about the breakwater base, M_{RANS} is the RANS computed moment about the breakwater base, and M_{RANS} / M_{rule} is the ratio of the computed and the rule based moment. Rule based moments turned out to be significantly less than computed bending moments, especially for the two larger ships.

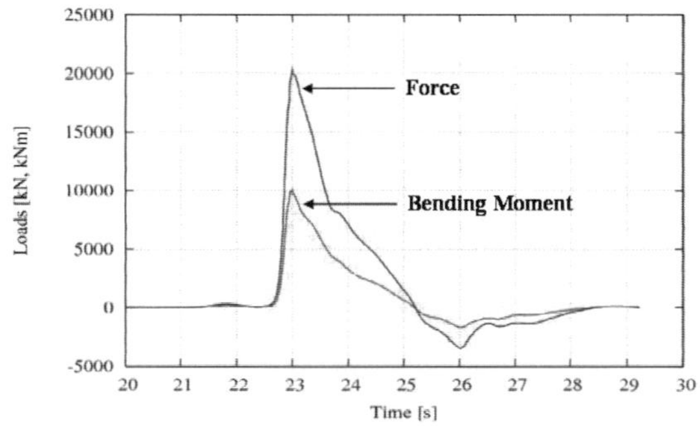

Fig. 4: Time histories of breakwater loads for the panmax containership

Fig. 5: Screen shot of the panmax containership under design wave conditions

Table 3: Rule based and RANS computed loads on breakwaters

	Feeder	Panmax	Post-panmax
p_{rule} at base	136 kN/m²	83 kN/m²	28 kN/m²
p_{rule} at top	55 kN/m²	23 kN/m²	25 kN/m²
M_{rule}	11 243 kNm	5 383 kNm	4 683 kNm
M_{RANS}	17 062 kNm	20 382 kNm	18 320 kNm
M_{RANS} / M_{rule}	1.52	3.78	3.95

4. FE DAMAGE ANALYSES

For two additional panmax containerships, we performed structural finite element (FE) analyses of damaged breakwaters to assess loads that caused this damage. Table 4 lists main particulars of these two ships. Figure 1 shows the damaged breakwater of one of these ships. (Ship 2 is a sister ship of the panmax containership investigated above.)

Table 4: Main particulars of panmax containerships that sustained damage to breakwaters

	Ship 1	Ship 2
Length	264 m	219 m
Breadth	32.2 m	32.3 m
Depth	21.5 m	19.5 m
Draft	13.0 m	12.2 m
Design speed	24.0 knots	22.0 knots
Block coeff.	0.678	0.673
$(x/L)_{breakwater}$	0.960	0.954

Generated FE models of the breakwaters of both ships, seen in Fig. 6, included structural components of their foundations beneath the forecastle deck.

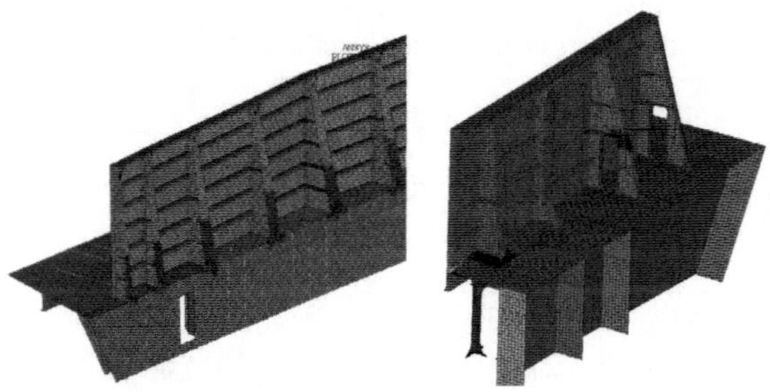

Fig. 6: FE models of breakwater structural configurations for ship 1 (left) and ship 2 (right)

We performed implicit nonlinear analyses that accounted for large structural deformations. The multilinear material law relied on a 15 percent increase in yield stress to account for the additional strength that generally characterizes shipbuilding steels. To ensure the occurrence of structural instability characterizing the observed damage, we subjected the FE models to scaled buckling eigenforms in accordance with allowable structural imperfections. A continuous rule-based trapezoidal lateral load acted on the breakwaters. Its magnitude was increased until the FE computation diverged and consequently stopped. This divergence, caused by geometrically strong nonlinear effects of structural components, characterized the observed instabilities of damaged breakwaters.

Damage predictions of FE computations agreed favorably with observed damage. Table 5 lists bending moments about the base of the breakwaters resulting from rule based loads and FE predicted loads. Symbol M_{rule} stands for rule based bending moment, M_{FE} is the FE predicted bending moment, and M_{FE} / M_{rule} is their ratio. Here again, rule based bending moments were significantly less than FE predicted bending moments.

Table 5: Rule based and FE predicted base bending moments on breakwaters of ships 1 and 2

	Ship 1	Ship 2
M_{rule}	7 187 kNm	5 383 kNm
M_{FE}	26 087 kNm	18 895 kNm
M_{FE} / M_{rule}	3.62	3.51

For the failed primary supporting members, we assessed buckling strength against current rule based loads. They had adequate safety margins against instability. For Ship 2, at some locations of primary supporting members allowable stresses were already exceeded under current rule based loads. Nevertheless, the FE computations diverged and stopped only when loads reached the level that caused M_{FE} given in Table 5.

5. THE NEW RULES

Old classification society rules treat breakwaters as unprotected superstructure and deckhouse end bulkheads. According to old rules, the design pressure, p_A, to determine breakwater scantlings is

$$p_A = n \cdot c\left(b \cdot c_L \cdot c_0 - z\right) \qquad (1)$$

with

$$b = 1.0 + 1.5\left[(x/L - 0.45)/(C_B + 0.2)\right]^2 \qquad (2)$$

where n is a tier height dependent coefficient that increases with scantling length L, c is a factor that accounts for the width of the breakwater relative to the ship's breadth, b is a factor that accounts for the ship's motion, c_L is a length coefficient that equals 1.0 for ships longer than 90 m, c_0 is the so-called wave coefficient that depends on the ship's scantling length, and z is the height from the summer load line to the load center of the structural member where p_A acts. Term ($b \cdot c_L \cdot c_0 - z$) in (1) represents the pressure head for a load center at z, and coefficient c_0 is comparable to the amplitude of an equivalent design wave from our RANS computations. Plots in Fig. 6 show that, on average, our chosen design wave amplitudes turned out to be about 18 percent less than the corresponding wave coefficients c_0 according to GL class rules [11] and UR S3 IACS requirements [7] although their functional relationships correlated favorably.

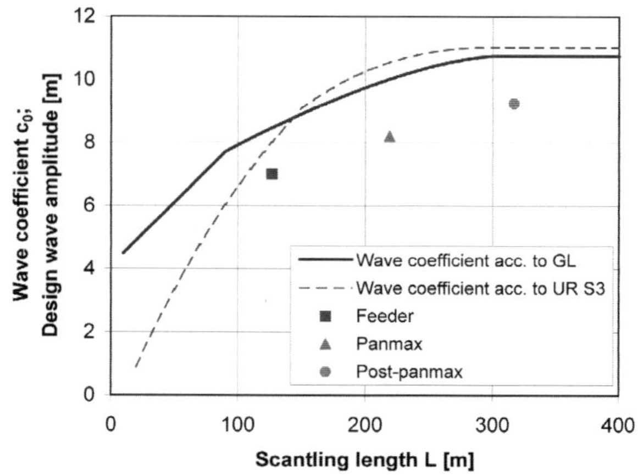

Fig. 7: RANS-computed equivalent design wave amplitudes and comparable wave coefficients

Figure 7 displays, here for the panmax containership under equivalent design wave conditions, representative samples of three computed time series of vertical motions z, namely, the wave elevation at the ship's longitudinal position of the breakwater, the vertical motion of the breakwater base, and the corresponding relative vertical displacement between wave contour and breakwater base. Adding to the largest (positive) values of relative vertical displacement the height extending from the ship's summer load line to the breakwater base resulted in maximum relative vertical displacements of the breakwater base. These values can be compared to the term ($b \cdot c_L \cdot c_0$) in (1). This comparison, given in Fig. 8, demonstrates that the relationship based on the old rule resulted in maximum values of relative vertical displacements that, on average, are about 19 percent less than the corresponding RANS-computed values.

The new rule defines coefficient $n = 10 + L/20$ in (1), sets factor $c = 1.0$ in (1), and increases factor 1.5 to 2.75 in (2) to read

$$b = 1.0 + 2.75\left[(x/L - 0.45)/(C_B + 0.2)\right]^2 \qquad (3)$$

The resulting plots of maximum vertical displacements based on these changes, also shown in Fig. 8, now agree favorably with the RANS computed values.

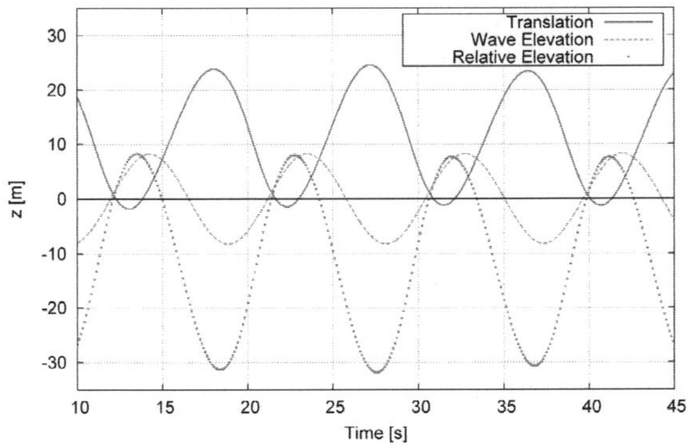

Fig. 7: Time series of vertical motion (Translation), wave elevation, and relative vertical displacement (Relative Elevation) at breakwater base of panmax containership

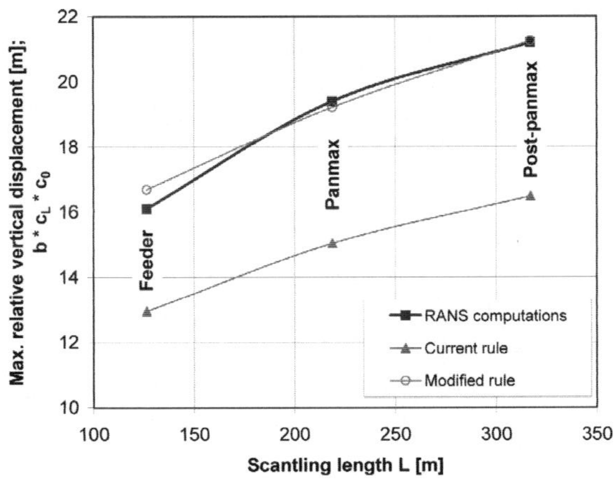

Fig. 8: Maximum relative vertical displacements

The new rules yielded pressure loads that led to forces and bending moments that closely agree with RANS computed loads. Table 6 lists the comparative loads. Symbols p_{rule}, F_{rule}, and M_{rule} refer to values according to the new rules; symbols F_{RANS} and M_{RANS}, to values based on RANS computations.

Table 6: Loads on breakwaters based on new rules and RANS computations

	Feeder	Panmax	Post-panmax
p_{rule} at base	169 kN/m²	191 kN/m²	152 kN/m²
p_{rule} at top	105 kN/m²	107 kN/m²	62 kN/m²
F_{rule}	9 582 kN	10 500 kN	11 309 kN
M_{rule}	17 315 kNm	19 031 kNm	16 988 kNm
F_{RANS}	9 026 kN	10 080 kN	11 652 kN
M_{RANS}	17 062 kNm	20 382 kNm	18 320 kNm
F_{RANS} / F_{rule}	0.94	0.96	1.03
M_{RANS} / M_{rule}	0.99	1.07	1.08

6. RAMIFICATIONS

We assessed the damaged breakwater of Ship 1, shown in Fig. 1, by subjecting it to design pressures according to the new rule formulas (1) and (3). This resulted in pressures of 206 and 95 kN/m² at the breakwater's base and top, respectively. Predicted damage occurred when 81 percent of the load was applied. Presumably, no damage would have occurred had the breakwater been built according to design pressures specified by the new rule.

For the feedership, the panmax containership, and the post-panmax containership, Table 7 summarizes comparative plate thickness, t, of the breakwater plating and section modulus, W, of the stiffeners. Listed are values of t and W for the breakwater as built, for the breakwater according to the old rules, and for the breakwater according to the new rules. Index "old" refers to values according to the old rules; index "new", to values according to the new rules. Also given are their respective ratios t_{old} / t_{new} and W_{old} / W_{new}. They indicate that plate thicknesses as well as section moduli were inadequate. They must be increased to comply with the new rules.

Table 7: Plate thickness and section modulus of breakwaters

	Feeder	Panmax	Post-panmax
t as built	8.0 mm	12.0 mm	15.0 mm
W as built	115 cm³	223 cm³	220 cm³
t_{old}	8.2 mm	7.8 mm	7.2 mm
W_{old}	133 cm³	157cm³	42 cm³
t_{new}	9.1 mm	11.2 mm	9.6 mm
W_{new}	178 cm³	393 cm³	222 cm³
t_{new} / t_{old}	1.11	1.44	1.33
W_{new} / W_{old}	1.34	2.50	5.30

7. FURTHER REQUIREMENTS

Based on our investigation, a breakwater or an equivalent protective facility, such as a "whaleback" or "turtle deck" arrangement, must always be provided if cargo on deck is to be carried forward of $x/L \geq 0.85$. Scantlings of these breakwaters are to be determined according to design pressures specified by the new rules.

The height of breakwaters can be deduced from relative vertical displacements between wave contour and breakwater base. The term $(b \cdot c_L \cdot c_0 - z)$ in (1) represents the maximum value of this displacement, and the breakwater height should preferably be 80 percent of this value, with 60 percent as a minimum. This also applies to the mean height of "whalebacks" or "turtle decks". The breakwater must be at least as broad as the width of the area behind the breakwater intended for carrying deck cargo.

We also investigated effects of cutouts of moderate size on breakwater loads and found that they did not significantly reduce total loads. However, if cutouts are envisioned, their individual areas should be less than 0.2 m^2 and their combined area less than 3.0 percent of the total area of the breakwater. Cutouts in webs of primary supporting members should be reduced to a minimum, and stiffeners are to be installed to support free edges of these cutouts.

The new rules specify that the plate thickness of breakwaters must not be less than the plate thickness of the lowest tier of an unprotected superstructure end bulkhead according to the old rules. An equivalent stress level of 230/k in primary supporting members must not be exceeded, where k is the material factor. The new rules also require that an adequate structural foundation be installed below the breakwater. Furthermore, all structural components must be designed for adequate buckling strength.

8. CONCLUSIONS

Table 8 summarizes old and new rule formulas for the design pressure on breakwaters. Symbols b' and B' refer to the width of the breakwater and the local breadth of the ship, respectively.

Table 8: Summary of old and new rule formulas

	Old rule	New rule
b	$1.0+1.5\left[\frac{x/L-0.45}{C_B+0.2}\right]^2$	$1.0+2.75\left[\frac{x/L-0.45}{C_B+0.2}\right]^2$
c	0.3 + 0.7(b'/ B')	1.0
n	Depends on height of tier	10 + L / 20
$p_{A\,min}$	Depends on height of tier	For lowest tier
t_{min}	Depends on height of tier	For lowest tier

The new rules for breakwaters yielded design pressure loads that closely agreed with RANS computed loads. Predicted damage of one of the investigated breakwaters, based on design pressures according to the new rules, was in close compliance with observed damage.

The new rules also recommend minimum heights of breakwaters or other protective facilities, specify minimum values for plate thickness of breakwaters and equivalent stresses levels in primary supporting members, and require adequate structural foundations.

9. REFERENCES

1. Kahl, A., el Moctar, O., and Schiff, C., 'Sea Loads for Breakwaters on Containerships', Report NRA 2007.003, Germanischer Lloyd, Hamburg, 2008

2. International Association of Classification Societies Ltd., Requirements Concerning Strength of Ships, IACS UR S3, London, 1973

3. Varyani, K.S., Pham, X.P., and Crossland, P., 'Green Water Investigation for a Containership', *Ship Techn. Res.*, Vol. 51, 2004, pp. 151-161

4. Hamoudi, B. and Varyani, K.S., 'A New Approach to Wave Loading on Deck Mounted Equipments of Offshore Structures/Vessel', *Int. Shipbuilding. Progr.*, Vol. 44, No. 440, 1997, pp. 321-339

5. Hamoudi, B., 'Experimental Investigation of Deck Wetness for Container Ship', *Proc. Int. Cong. of the Int. Maritime Ass. of the Mediterranean*, Vol. 2, Varna, Bulgaria, 1983

6. Fonseca, N. and Guedes Soares, C., 'Experimental Investigation of the Shipping of Water on the Bow of a Containership', *J. Offshore Mech. and Arctic Engg.*, Vol. 127, 2005, pp. 322-330

7. Kapsenberg, G.K. and de Kat, J.O., 'Effects of Freeboard on Green Water Loads for a General Cargo Ship', *Proc. Osaka Seakeeping Colloq.*, Osaka, 2000

8. Papanikolaou, A.D. and Schellin, T.E., 'A Three-Dimensional Panel Method for Motions and Loads of Ships with Forward Speed', *Ship Tech. Res.*, Vol. 39, No. 4, 1992

9. Schellin, T.E., and el Moctar, O., 'Numerical Prediction of Impact-Related Wave Loads on Ships', *J. Offshore Mech. and Arctic Engg.*, Vol 129, No. 1, 2007, pp. 39-47

10. IACS 2001, 'Recommendation No. 34, Standard Wave Data', International Association of Classification Societies Ltd., London, 2001

11. El Moctar, O., Brehm, A., and Schellin, T.E., 'Prediction of Slamming Loads for Ship Structural Design Using Potential Flow and RANSE Codes', *Proc. 25th Symp. on Naval Hydrodynamics,* St. John's, National Academy Press, Washington, DC, 2004

12. Germanischer Lloyd, 'Rules for Classification and Construction, I – Ship Technology, Part 1 – Sea-Going Ships, Chapter 1– Hull Structures', Hamburg, 2007

10. AUTHORS' BIOGRAPHIES

Adrian Kahl graduated as a naval architect in 2002 and is project engineer in the Department Analysis of Hull Structures and Damages of Germanischer Lloyd. Before joining GL, he worked as a ship repair manager at a German shipyard and as a research engineer at Hamburg University of Technology. His technical background primarily comprises structural strength analysis with focus on fatigue strength. Besides damage assessment, he is involved in various rule development projects at GL and within the IACS Hull Panel.

Helge Rathje is head of the Department Analysis of Hull Structures and Damages at GL. Before joining GL in 1994, he worked for MAERSK Shipping Line as a freight coordinator in Hamburg. His technical background primarily comprises seakeeping analysis and statistical evaluation of wave loads and ship motions. He is responsible for rule development of ship loads and structural strength and the assessment of rule related ship damages. He was GL's representative in the former IACS AHG on Wave Data and Sea Loads and chaired the group from June 2002 until its termination in December 2004.

Thomas E. Schellin is a naval architect in the Department Analysis of Hull Structures and Damages of Germanischer Lloyd, working on a consultancy basis after his retirement in 2004 as deputy head of GL's Hydromechanics Department, where he was responsible for the development of hydrodynamic analysis methods for ships and offshore structures. He coordinated nationally and internationally sponsored research projects and contributed to the development of class rules for offshore structures and fast ships. He was GL's representative in the IACS AHG on Mooring and Anchoring and served on the 2001 and 2003 ISSC Committee I.2 Loads.

SEAKEEPING ISSUES IN THE DESIGN OF CONTAINERSHIPS

R. Dallinga, F. van Walree, R. Grin and **J. Koning,** MARIN, The Netherlands

SUMMARY

While the advanced art of minimising the installed power for a given speed plays a clear role in the design of container ships, the place of seakeeping is less well established. The incidental character of seakeeping problems at sea in combination with the fact that building for good seakeeping may decrease the container capacity and increase the building costs seems a major reason for this, in addition to the sheer complexity of the issues. The present paper addresses the last point with an effort to give a complete review of the seakeeping issues in containership design. Based on recent experience from model tests the work explores the physical nature of the involuntary speed loss in waves and reasons for a voluntary speed reduction (green water loads, whipping accelerations due to bow and stern slamming, engine racing) or change in course (exposure of containers to wave crests, rolling). In addition a review is given of the extreme behaviour that the master would like to avoid altogether, like excessive heel due to loss of stability in following seas and parametric roll. Based on the results tentative design guidance is formulated.

1. INTRODUCTION

Several aspects of container ship behaviour in waves can be understood quite well in terms of the linear superposition of harmonic motion components. This computationally very convenient "linear" theory neglects the effects of variations in the immersed hull geometry on the motion induced reaction forces as well the effects of interactions between the motions and the incident waves on the excitation forces.

In general the interaction between variations in immersed hull volume and the motions and the incident wave introduces in the mathematical description products of the amplitudes of the motion and wave components, leading to a non-linear relation between the wave height and the response.

As will be shown the character of the behaviour (linear vs. non-linear) affects the character of the statistics of the behaviour. Linear motions show typical extreme amplitudes that are some 2.5 - 4 times the mean amplitude. Non-linear motions show a considerably larger dynamic range; the extreme values are some 5 to 11 times the mean amplitude.

2. SEAKEEPING HYDRODYNAMICS OF CONTAINER SHIPS

2.1 LINEAR MOTIONS

2.1 (a) Vertical plane motions

Heave and pitch

The fact that transfer functions obtained from tests in irregular waves from different wave conditions (at the same speed and heading) show mutually good agreement and the fact that the distributions of the amplitudes follow the character predicted by linear superposition demonstrate that heave and pitch are fairly linear in character.

The response per metre wave height is mainly a function of the ship heading and, to a lesser extent, of ship speed. Figure 1 shows the typical character of the pitch response for a 230 m ship. A point that is not always appreciated is the fact that the pitch response in oblique waves is relatively high over a rather broad heading sector, implying that a change in course is not always remedial in solving problems related to pitch.

For ships longer than 250 m the peak in the pitch response in head seas migrates to longer periods that do not occur in normal operational wave conditions (the grey area in the graph with wave peak periods between say 7 and 15 s). For these large ships, waves from oblique directions are the main source of pitch.

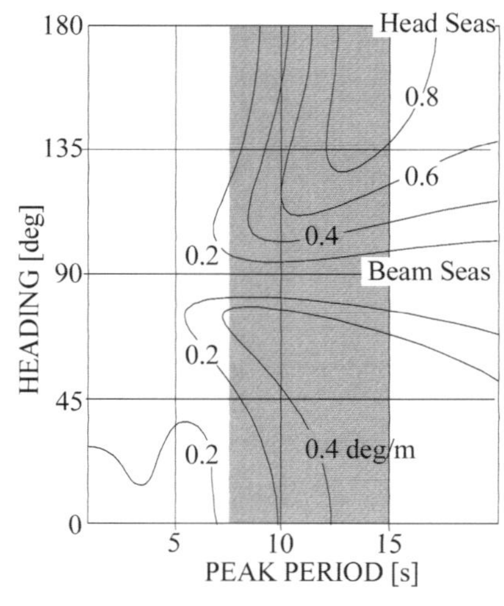

Figure 1: RMS pitch in irregular seas per m RMS wave elevation as a function of heading and wave peak period

Vertical accelerations

The combined (rigid-body) heave and pitch motions determine the local vertical motions, which govern, in combination with the square of the wave encounter frequency, the vertical accelerations.

Because the frequency of wave encounter is low in waves from the stern and from the stern quarter the vertical accelerations are highest in waves from forward directions. Figure 2 illustrates this character for a point in the forward part of the ship.

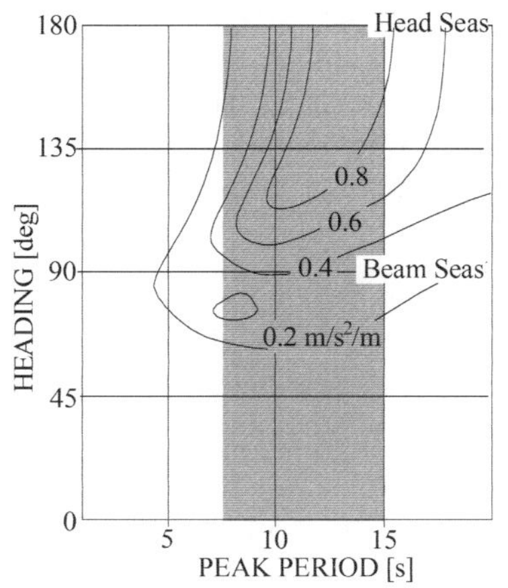

Figure 2: RMS vertical acceleration in irregular seas per m RMS wave elevation as a function of heading and wave peak period

The phasing of the heave and pitch motions governs the effect of the longitudinal position on the vertical accelerations. Figure 3 shows the general character; the lowest accelerations are obtained around one-third ship length from the stern. The accelerations near the bow are nearly six times higher than those at the best position.

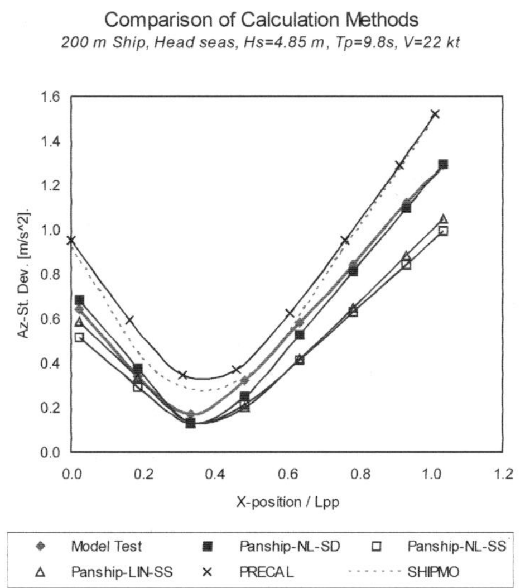

Figure 3: RMS vertical acceleration in irregular seas over the length of the ship

The fact that the character of the vertical motions and accelerations follows linear superposition theory makes it tempting to assume that the traditional implementations of linear potential theory (like 2D and 3D source-sink panel codes) will offer a reliable basis for predictions. Although this is often a reasonable assumption for heave and pitch motions, this is not necessarily the case for the local vertical accelerations in the stern area of modern hull forms with rather flat (fuel efficient) sterns with low submergence.

Figure 3 illustrates the problem with the results of model tests in head seas and three types of calculations. A commonly used 3D source-sink frequency domain code with zero-speed Greens functions (denoted by "PRECAL") clearly over-predicts the response.

An alternative calculation method (PANSHIP, [1]) based on time domain simulations was developed to account for the steady wave system, forward speed effects in the propagation of radiated and reflected waves and non-linear hull geometry effects in the restoring and excitation forces. The version with source-sink description of the waves (PANSHIP-LIN-SS) under-predicts the response. Also when accounting for the non-linear effects in the buoyancy (PANSHIP-NL-SS). With additional account for lift effects below the stern (by adding doublets) a very satisfactory result is obtained (PANSHIP-NL-SD).

A traditional strip theory (SHIPMO) calculation performs slightly better than the 3D panel code.

2.1 (b) Relative wave elevations

Similar to the vertical accelerations, the combined heave and pitch motions and the incident wave govern the water surface elevation with respect to the ship (the "relative motions"). In waves from oblique directions the reflected and radiated wave components magnify the relative wave elevations on the weather side of the hull. In the bow area the variations in local draft lead to changes in the steady wave system (due to forward speed). For ships with a full bow and considerable flare this "dynamic swell-up" can lead to considerable magnification of the larger crest heights.

In many calculation methods the reflected and radiated waves as well as the magnification of the steady wave system are neglected. Not entirely surprising this approach still yields a fair approximation of the relative wave elevation at the bow. Figure 4 shows the character of the response in irregular waves. Note the resemblance with the pitch response in Figure 1. The highest response is typically 2 - 3 times the incident wave. It increases with finer fore bodies, hull forms with a low beam-draft ratio and high speed.

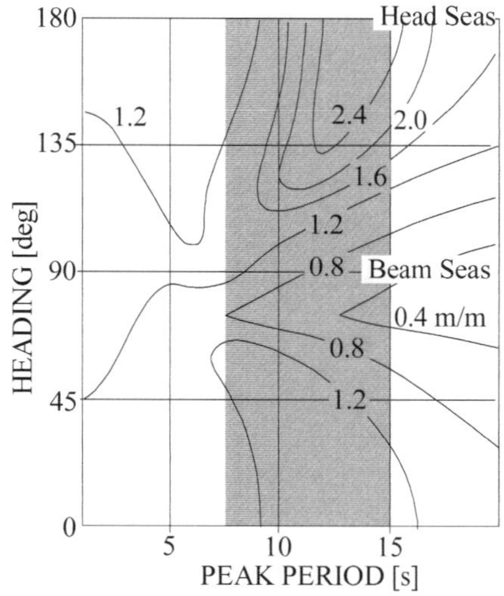

Figure 4: Rms relative wave elevation at the bow per m RMS wave elevation in irregular seas

Wave elevation in the sides

Direct exposure of containers to wave crests is governed by the crest elevation along the weather side. The reproduction of test results with a 3D source-sink panel code learned that, because of the account for the reflected and radiated waves, the prediction of the relative wave elevation in the sides is rather accurate. Figure 5 shows a typical result. Note the relatively sharp increase in the response in the very forward part of the ship in longer waves. In short waves the results approach the level corresponding with full reflection (a value of 2).

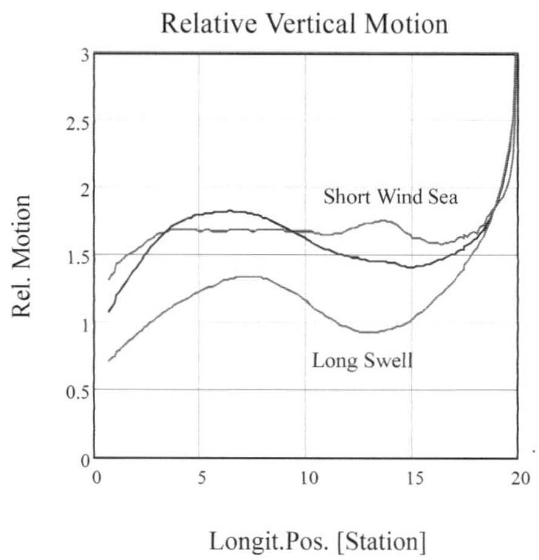

Figure 5: RMS relative wave elevation along the weather side per m RMS wave elevation in irregular seas from the bow quarter

Stern emergence

Stern emergence drives stern slamming and ventilation of the propeller. Considering the relatively large errors in the prediction of the vertical motions in the aft body discussed in the previous section it is not surprising that conventional means do not offer a good prediction of stern emergence.

2.1 (c) Propeller load variations

Often propeller ventilation is the only consideration in the evaluation of propeller load variations. This neglects the important effects of wave and ship motion-induced variations in the propeller inflow. Because (at constant rpm) the thrust and torque variations are nearly proportional to variations in the angle of attack on the propeller blades linear theory offers a reasonable prediction. Figure 6 shows a transfer function of the thrust variation in head seas. It compares the results of measurements with those of calculations with a 3D source-sink method.

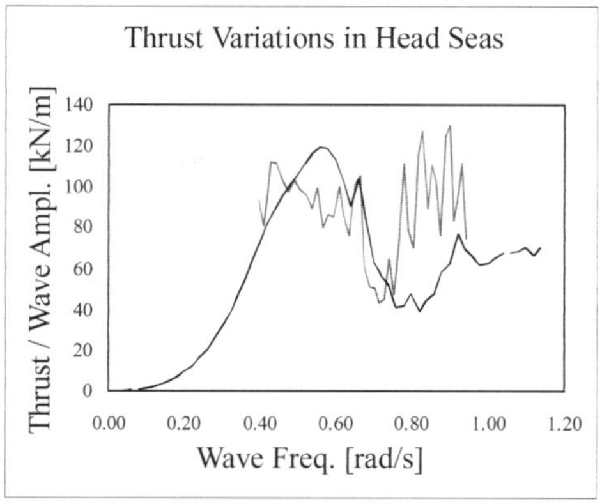

Figure 6: Transfer function of propeller thrust variations in head seas

A missing element in the interpretation of torque variations is the dynamic response of the engine-propeller system. Work at the Delft University of Technology [2] is underway to complete the picture.

2.1 (d) Roll

The "linear" roll response of container ships can be understood largely in terms of the wave-induced roll excitation and the high response at the combinations of heading and wave frequency that yield resonant tuning. Because of the relatively low stability in full load conditions, which often yield natural roll periods beyond 20 s, rolling is hardly perceived as an issue in the design stage.

In practice ships are not always fully loaded. In these conditions the corresponding shorter natural period of roll makes unfavourable tuning with the incident wave much more likely.

Figures 7 and 8 indicate the magnitude of the roll response and related transverse accelerations at the top of the container stacks as a function of heading and wave period for a full load and a partly loaded condition. The most unfavourable heading migrates from the stern-quarter (for full load) to almost abeam. Note (because of the increasing roll inertia component) the increase in the transverse accelerations in partly loaded condition.

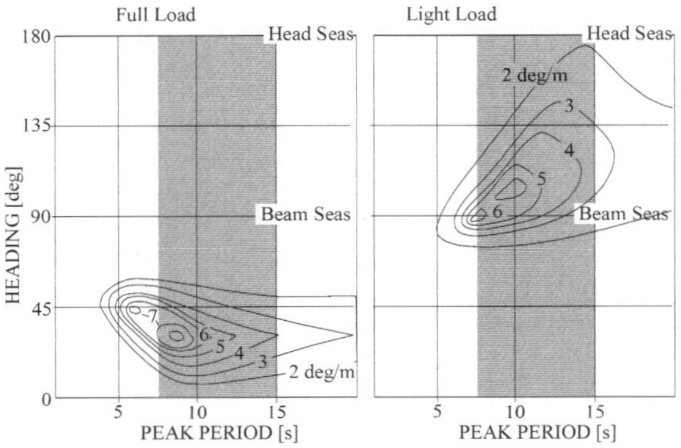

Figure 7: Roll response in full load and light load condition (RMS roll per mRMS wave elevation in irregular waves)

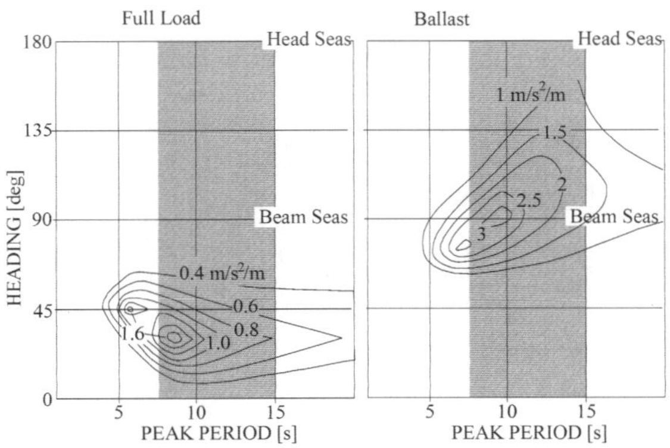

Figure 8: Transverse accelerations in full load and light load conditions (RMS per m RMS wave elevation)

Roll stabilisation
Bilge keels play an important role in the control over the roll motions because they contribute significantly to the roll damping at reduced speed.

Fin stabilizers are particularly effective at moderate and higher speeds. Because rolling is mostly an issue in lightly loaded conditions it is important to account for these off-design conditions explicitly in the fin sizing and control.

2.2 NON-LINEAR MOTIONS

2.2 (a) Bow and stern slamming and related hull girder vibrations

Relatively high velocities and rapidly changing immersed volumes lead to rather local, fast moving high pressure areas. The related short duration impulsive loads lead to higher harmonics in the rigid body motions and transient (whipping) and resonant (springing) vibrations in the flexural modes.

Except for very light load conditions, where traditional slamming below the fore foot is conceivable, container ships experience impulsive loads mostly in the bow flare and below a flat stern.

The loads in the bow flare are related to the rather high vertical entry velocities that can occur in waves from forward directions or the very rapid changes in exposed area when encountering steep waves from the bow quarter. See Figure 9.

Stern slamming is related to the very large and rapid changes in immersed stern area during re-entry of the flat sections.

The impulsive loads that drive the flexural response are the product of a pressure and an area. In some cases, like stern slamming, the exciting pressures may not be very large (say 30 m water column) while they still evoke a substantial flexural response.

Although the prediction of slamming loads has been the subject of a considerable volume of research, the numerical evaluation of the very rapid highly local phenomena in a large fluid domain is still quite difficult.

Figure 9: Bow flare impact in steep irregular waves from the bow quarter

Flexural response
Measurements in the laboratory and at sea suggest that the damping of a ship in the flexural modes is rather low. Numerical experiments with a homogeneous slender beam suggest that a consequence of the low damping is that the slamming-induced vibrations travel many times through the structure before they lose their energy (or are dampened by a successive impact with the "right" phasing). In practice this means that, when considering the magnitude of the local flexural accelerations, the location of the impulsive excitation hardly plays a role;

the maximum accelerations closely follow the mode shapes of the structure.

The degree as to which higher order mode shapes play a role is determined by the "duration" of the excitation in relation to the natural periods. In practice, and despite the potentially high contribution of the higher order modes in the accelerations (because of the higher eigen frequencies), the two node response is a major feature in the response of flexible models and of ships at sea.

Figure 10 illustrates the above points. The fact that the test data with a segmented model show the same trend in quite different test conditions supports the notion that the effect of the location of the external loads hardly matters in the locally experienced vibrations. Of course the location of the excitation does matter in the magnitude of the vibrations.

In the two cases considered in Figure 10 both the model results and measurements at sea show a similar mode shape with relatively high response at the vessel's extremities. The fact that the trend does not follow the response of a slender homogeneous beam (in the example the 2-node with a small contribution of the 3-node response) emphasises the need to model the actual mode shapes in calculations and model tests.

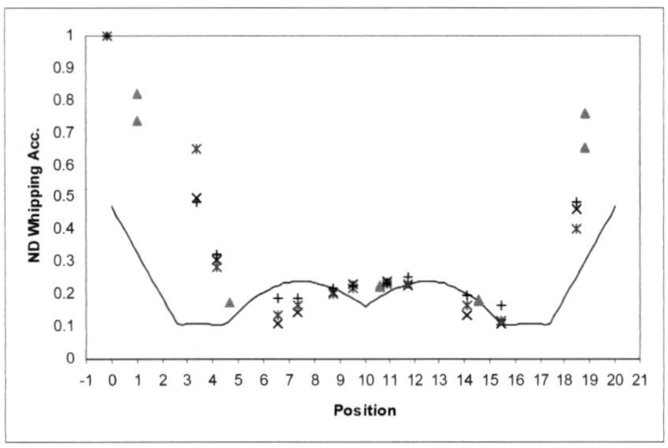

Figure 10: Non-dimensional maximum flexural vertical accelerations over the length of a ship (red triangles from sea trials, the line represents a homogeneous slender beam)

The amplitude of the excursion in the various vibration modes depends on the bow flare, stern submergence, speed, heading and wave height and steepness [3]. Various model test campaigns have shown that in normal operational waves, oblique wave directions cause the largest flexural response. They also showed that, in case the bending stiffness in the horizontal plane is similar to that in the vertical plane, the horizontal bending becomes an issue similar to the vertical bending.

In the introduction it was noted that the linear and non-linear motion component would show a difference in the character of the statistics. This is illustrated in Figure 11, which shows the joint statistics of the flexural and rigid-body accelerations of a container ship. The rigid body response resembles the expected Rayleigh distribution; the flexural response resembles a negative exponential distribution. Note that with increasing exposure duration (or lower accepted risk of exceedance) the slamming induced component becomes more important.

Wave loads and fatigue damage
The slamming induced flexural response discussed in the foregoing contributes in the magnitude of extreme bending loads as well as in fatigue damage [4,5].

Because of the non-linear character the contribution is relatively large in higher waves. This makes it rather sensitive to weather avoiding and routing.

The contribution of springing in the flexural response is of course rather sensitive to the damping of the hull structure. There are indications that for container ships the cargo plays a role in this respect.

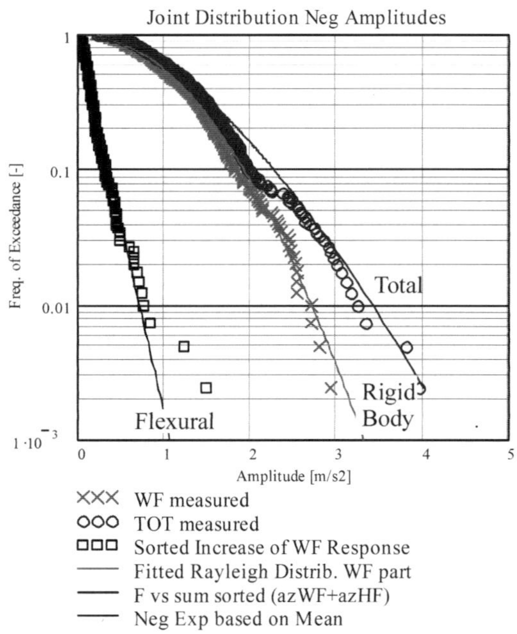

Figure 11: Joint statistics of rigid-body and flexural accelerations (from sea trials)

2.2 (b) Green seas loading

When considering green water on the fore deck there are two mechanisms that play a role. In the first mechanism the bow is submerging at some vertical velocity. Depending on vertical velocity and the pressure from surrounding waves a vertical wall of water rises along the bulwark before this volume of water collapses on the foredeck. In head seas the symmetry of the situation creates a jet of water that travels aft at a very high velocity.

Figure 12: Breakwater impact

In the second mechanism the crest of a steep incident wave sweeps more or less undisturbed over the foc'sle (as in Figure 12 [6]). Because of the wave steepness the vertical velocity of the bow is small and does not play a large role.

Figure 13 shows results of measurements on the loads experienced by a breakwater on a multi-purpose ship in head seas. It suggests that the load statistics follow a negative exponential distribution.

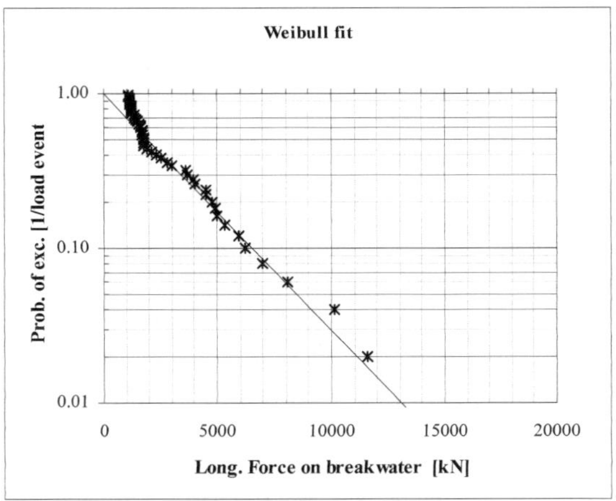

Figure 13: Breakwater load statistics

2.2 (c) Effects of stability variations in waves

The incident wave and ship motions cause variations in the transverse stability. The variations can lead to large roll amplitudes in two quite different ways.

Temporary loss of stability

If the stability variations are of the same order as the stability in calm water and if the variations are sufficiently long in duration (like when travelling in following seas at moderate or high speed) the ship will experience occasional large heel angles due to temporary loss of stability. This phenomenon is observed in higher following seas in full load conditions with a relatively low stability. Experience from model tests suggests that in these conditions the minimum stability rules do not prevent rather large heel angles to occur.

Parametric roll

A second mechanism through which the natural stability variations lead to excessive heel angles is parametric resonance or parametric roll. It develops if three conditions are met simultaneously. First of all the stability variations should be of sufficient magnitude. Secondly, the dominant period of the stability variations should be half the natural period of roll. And thirdly, the roll damping should be relatively small. [7]

In practice parametric roll is observed mostly at low speed in head or following seas in wave conditions where the pitch response is largest.

A reason may be that, because the relative wave elevation is symmetric on both sides, the stability variations are relatively large.

The relatively low roll damping in this speed range causes the sensitivity at low speed. Although fin stabilizers lose part of their effect at reduced speed there is reason to believe that they are still quite effective at modest speeds. An anti-roll tank is known to be very effective as well.

The importance maintaining speed in conditions with a threat of parametric roll was shown in the APL China investigation. In these tests the natural variations in speed caused by natural wave grouping triggered parametric roll.

Because normal severe storm conditions are associated with peak periods of around 12 s ships with a roll period above 20 s are particularly vulnerable. In practice ships with a relatively low stability appear to be relatively sensitive.

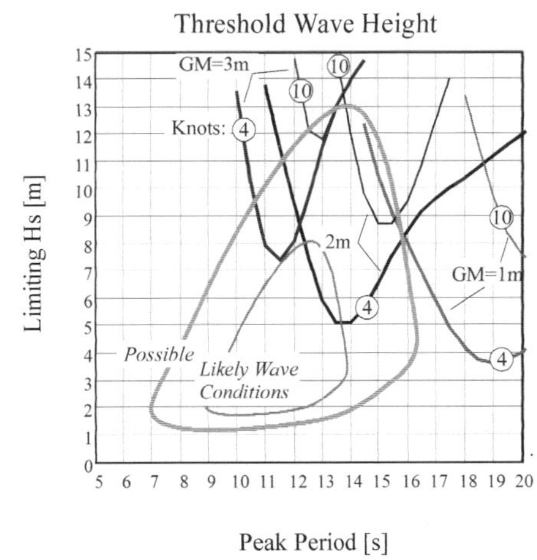

Figure 14: Threshold wave height for parametric roll

Figure 14 shows a very crude estimate of the threshold wave height above which parametric roll may be expected for three different stability values and speeds. The trends suggest that for a given ship and speed there is a stability level for which critical wave conditions are in the normal range.

The depth of the "bucket" depends strongly on the available roll damping. The position of the bucket is governed by forward speed and transverse stability.

2.3 ADDED RESISTANCE

The interaction of the ship motions and the incident waves underlies numerical methods to predict added resistance in waves. It yields a peak in the added resistance in conditions with relatively high pitch response. Effects that are commonly neglected are the drift angle due to the transverse drift forces in oblique waves and the changes in the steady bow wave system due to non-optimum bulb submergence or the immersion of relatively blunt parts of the bow.

The drift angle may contribute to the fact that the added resistance in head waves is usually lower than in waves from oblique directions. In fact, very similar to the character of the pitch response discussed in the first section and the flexural response discussed in Section 2.2, the added resistance remains relatively high over the entire range of forward directions. Figure 15 illustrates this point for a 180 m ferry.

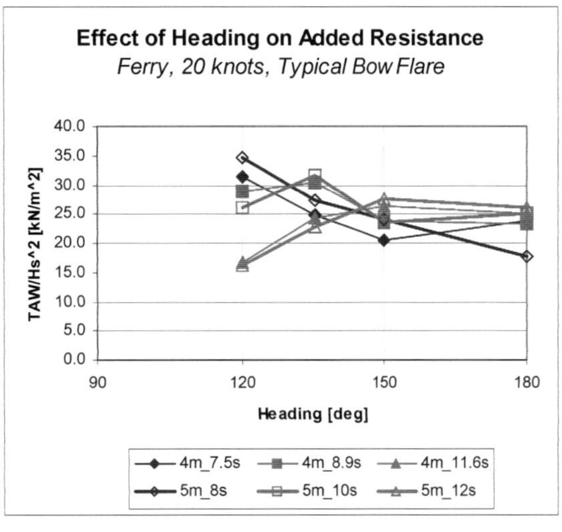

Figure 15: Effect of heading on added resistance in irregular waves

In practice numerical methods to estimate the added resistance in waves are notoriously unreliable. The relatively small quantities are rather sensitive to details in the schematisations that are used.

The added resistance from the wind is usually calculated from the relative longitudinal wind velocity and the drag coefficient at the relative wind direction. As in the case of the waves, the effects of a steady drift angle due to the transverse wind force component are usually neglected.

3. CONTAINERSHIP SEAKEEPING

The issues discussed in the foregoing affect container ship operations and economy.

In moderate wave heights the added resistance from wind and waves causes a loss in speed and a related reduction in propulsive efficiency.

The product of the additional trip duration and the power actually used governs the increase in fuel consumption.

In increasing wave heights the "linear" motions (roll, pitch and related accelerations) will grow more or less proportional with the wave height. The non-linear aspects of the behaviour (slamming-induced accelerations, added resistance) will grow with the wave height squared, leading in increasing number of "incidents" with noticeable slamming vibrations and green water related spray and a rapidly increasing speed loss. Although the acceleration levels are still not very extreme they can lead to damage to containers that are in poor shape, containers with poorly stowed contents or poorly lashed stacks.

In higher waves there will be a point where the master intervenes with a reactive reduction in speed or a change in course. A reduction in speed eases the vertical accelerations, the violence of any green water on the foredeck and slamming-induced whipping accelerations. A change in course is effective in reducing the roll motions.

The perception of the risk of damage, governed by what they see, hear and feel, drives the reactive measures of the crew. Visual clues are the impression of the waves (like white capping), ship motions (like roll) and the appearance of massive spray above the container stacks. Auditory clues relate to the wind, engine rpm variations and mechanical contact between the container stacks. Ship motions and slamming-induced vibrations can of course be felt directly.

The "sensatory" input is masked by the size of the ship (height above water) and the crew's on-board position (which affects the magnitude of local accelerations and vibrations as well as the visual impression of the behaviour of container stacks). Experience with monitoring campaigns indicates that the crews on large ships find it hard to appreciate safety and risk of damage on large ships.

In case of reactive measures the ship faces a delay. The product of the adopted engine power and the delay as well as the subsequent recovery of the delay at a relatively high speed increases the fuel consumption.

If the ship is caught in extreme weather the natural reaction of the master is to sail at reduced speed in head seas. In these cases it is important to be able to maintain a minimum speed to avoid parametric roll and stern slamming

In some cases (in particular for diesel-direct drives) the engine loading becomes an issue. In these cases the speed loss in bad weather can reduce the engine rpm's down to a level at which the turbo chargers are unable to deliver sufficient air for complete combustion. The subsequent loss of control of the ship is something that the crew will avoid at all cost.

As discussed the non-linear contributions to the behaviour and loads cause extreme values that are much larger than the typical event. This may explain why many incidents involving loss of containers and damage are perceived as "freak" events.

This aspect of non-linear behaviour may be an important reason why an experienced master avoids very bad weather all together whenever possible.

If the master decides for pro-active measures the additional route length and the required trip duration govern the additional fuel consumption. The reliability of the weather forecast is off course an important factor in realising the "benefits" of this investment in "safety". Details of the expected weather, like wave period and steepness and the likelihood of freak waves, have a large effect on the actual risk of damage while they are not part of standard weather forecasts.

4. CONTAINER SHIP DESIGN

4.1 GENERAL

Good operational performance requires investments in good seakeeping. The normal investments are adequate structural capacity (ultimate loads as well as fatigue), freeboard at the bow and a break water, bilge keels and perhaps active roll stabilisation. The issues that these measures do not resolve are dealt with by the ship master with prudent or even very cautious seamanship.

One question in container ship acquisition is if the "standard" measures are optimal. A finer bow, with less slamming and added resistance and a somewhat higher transverse stability with less risk of excessive heel due to loss of stability or parametric roll, will lead to less pro-active deviations from the shortest route and less damage and delays. But off course the reduced deck area implies a smaller container capacity.

Another question in acquisition and design is if it is worthwhile to invest in seakeeping issues related to "off-design" operations. Not only the drag associated with non-optimum bulb or stern immergence but also the problems caused by excessive stern slamming and rolling accelerations in partly loaded conditions lead to inefficient shipping. Active roll stabilisation and ballasting options to control the stern draft and limit the stability may be investments that make a return.

4.2 SPECIFIC ISSUES

Power

The power available to overcome the added resistance from wind and waves determines the sustained speed in adverse weather. Sufficient power can have a large influence on the risk of parametric roll.

The current trend towards larger ships with a reduced calm water speed requires a considerably larger service margin.

Motion control

Bilge keels have an important effect on the roll response at reduced speed and the roll in light load conditions. If parametric roll occurs they greatly affect the maximum roll angles.

Stabilizers and anti-roll tanks are also quite effective in suppressing the risk of parametric roll.

Bow flare

The bow flare plays an important role in the slamming-induced flexural response which contributes to fatigue damage and the loss of containers. There are also reasons to believe it contributes to the added resistance in higher waves and the stability variations that underlie parametric roll.

Because excessive slamming is an import reason to reduce speed, a modest bow flare also reduces the risk of parametric roll.

Stern submergence

Although stern slamming is not necessarily associated with very high pressures it can contribute significantly to the flexural response. Because stern emergence decreases with increasing speed, sufficient power is one way to avoid problems. Adequate ballasting options in partly loaded conditions seem another solution.

Freeboard

The height of the foc'sle determines the risk of shipping green water. Apart from the associated risk of damage on the foredeck or the forward containers the natural reaction of the master (to reduce speed) increases the risk of parametric roll.

An important detail seems the arrangement of the foredeck. Measures to absorb the kinetic energy of the green water that reaches the foc'sle deck (without launching it in the direction of or on top of the containers) might greatly reduce the risk of green water damage.

Stability
The transverse stability in operational conditions is an important parameter in the roll response. If it is too low in the full load condition excessive heel due to temporary loss of stability may be expected in high following seas. If it is too high (for instance in partly loaded conditions) the roll angles and related transverse accelerations become very high for headings around beam seas.

Between these two extremes there is a range of stability levels where parametric roll may develop at reduced speed.

4.3 TOOLS

As discussed the prediction of the hydrodynamic aspects of seakeeping is a major technical challenge. Although considerable efforts have been made and although the results of this work have contributed very much to the available know-how, the numerical techniques to predict non-linear aspects like parametric roll and slamming-induced impulsive loads and even also the linear motion components have unfortunately not reached the high level of maturity where their resolution is sufficient to play a role in the detail optimisation (stern shape, bow flare) of a specific design.

Design verification by means of seakeeping tests with segmented model reduces the uncertainties in the operational performance of a newly built vessel. Combining the test results with voyage simulations makes it possible to translate the measured added resistance, the slamming-induced vibrations and fatigue damage and the parametric roll boundaries in the fuel consumption and operational reliability for specific routes. The obtained insight also offers possibilities to develop operational guidance on these complex issues and insight in to-the-point on-board advisory systems.

5. ACKNOWLEDGEMENTS

The present review would not have been possible without the stimulating interaction with our customers on the seakeeping of container ships. We gratefully acknowledge their confidence.

6. REFERENCES

1. Walree, F. van (2002). "Development, validation and application of a time domain seakeeping method for high speed craft with a ride control system". In *Proceedings of the 24th Symposium on Naval Hydrodynamics*, pp. 475-490.

2. Grimmelius, H.T., Mesbahi, E., Schulten, P.J.M., Stapersma, D., (2007) "The use of diesel engine simulation models in ship propulsion plant design and operation.", In *CIMAC Conference, Vienna*,

3. Dallinga R.P.(2002). "Bow flare slamming of container ships and it's impact on operational reliability". In *RINA Symposium on the Design and Operation of Container Ships, 2006.*

4. Aalberts, P.J. and Nieuwenhuijs, M.W., (2006). "Full scale wave and whipping hull girder loads. In *Ohydroelasticity in Marine Technology-2006, Wuxi.*

5. Drummen, I., Storaug, G., Moe, E. and Moan, T. (2006), "Experimental and full scale investigation of the importance of fatigue damage due to wave-induced vibration stresses in a container vessel" In *RINA Syposium on the Design and Operation of Container Ships, 2006.*

6. Kapsenberg G.K. and de Kat, J.O., (2000). "Effects of freeboard and bow height on green water loads for a general purpose cargo ship. In *Offshore Colloquium-2000, Osaka.*

7. France, W.N., Levadou, M.", Treakle, T.W., Paulling, J.R., Michel R.K. and Moore, C. (2001). "An investigation of head-sea parametric rolling and its influence on container lashing systems, *SNAME Annual Meeting, 2001.*I

7. AUTHORS' BIOGRAPHIES

Reint Dallinga holds the current position of Sr. Project Manager in the seakeeping group at MARIN. In this position he is engaged in contract research and the related development of know-how and tools to translate the hydrodynamic characteristics of ships in performance.

Frans van Walree holds the current position of Sr. Project Manager in the seakeeping group of MARIN. The development of tools for the prediction of intact stability of ships and the seakeeping of high speed ships are his main responsibilities.

Rob Grin holds the current position of Project Manager in the seakeeping group of MARIN. Besides contract research he is responsible for developments in the area of scenario simulations and the prediction of added resistance in waves.

Jos Koning holds the current position of Project Manager in the Trials and Monitoring department at MARIN. Among other activities he is responsible for the Lashing @ Sea project.

GUIDELINES FOR APPLICATION OF YP47 STEEL PLATES TO STRENGTH DECK CONSTRUCTION OF ULTRA LARGE CONTAINER SHIPS

Y. Yamaguchi, T. Shigemi, T. Matsumoto and **A. Miura,** Nippon Kaiji Kyokai (ClassNK), Japan

SUMMARY

The application of high tensile steel plates with far higher strength to hull structure has been attracting attention, from the safety point of ships, as an effective solution to the recent trend of using extremely thick steel plates in the hull structures of large container ships caused by the ever-increasing size of container ships.

A large container ship, to which higher tensile steel plates of yield point 47 kgf/mm^2 (460 N/mm^2) (YP47 steel plates) was applied for the first time in the world, was delivered in June 2007 under the classification of ClassNK.

ClassNK has established the Guidelines for Application of YP47 Steel Plates to Strength Deck Construction of Ultra Large Container Ships based on the findings of recent researches and the experiences accumulated in the application of these plates to the actual ship. These are comprehensive guidelines that encompass the technical issues related to the hull structure strength such as hull girder strength, fatigue strength etc., as well as the steel grades of hull structural members, the specifications of steel plates and welding procedures and so on. Furthermore the guidelines specify the brittle crack arrestability as a backup safety measure in connection with the use of material having far higher strength.

These guidelines are expected to contribute toward ensuring the safety of container ships under the recent trend of rapid increasing in their size. An overview of the guidelines and the technical background are introduced in this paper.

1. INTRODUCTION

Amid internationalization of markets and the continued increase in the volume of goods transported by sea globally, there is a growing demand for larger, higher performance ships.

Figure 1 shows the trend in the increase in the size of container ships. The size of container ships has increased rapidly in the last few years, and the first ultra large container ship in the world with a capacity exceeding 10,000 TEU was built in 2006.

A container ship is provided with a large hatch opening in the strength deck to enable cargo handling operations to be carried out efficiently. To ensure adequate hull girder longitudinal strength, comparatively thick steel plates have been used in the strength deck structures (hatch side coaming, upper deck plating, sheer strake, longitudinal bulkhead) of large container ships since the past, but with the trend of increase in the size of container ships, much thicker steel plates are now being used.

Figure 2 shows the trend of plate thickness of hatch side coaming with the increase in carrying capacity of container ships. Higher strength and thicker steel plates have now been used with the increase in size of the ships. There are cases among 8,000TEU class ultra large container ships where extremely thick steel plates with thickness exceeding 70 mm have been used.

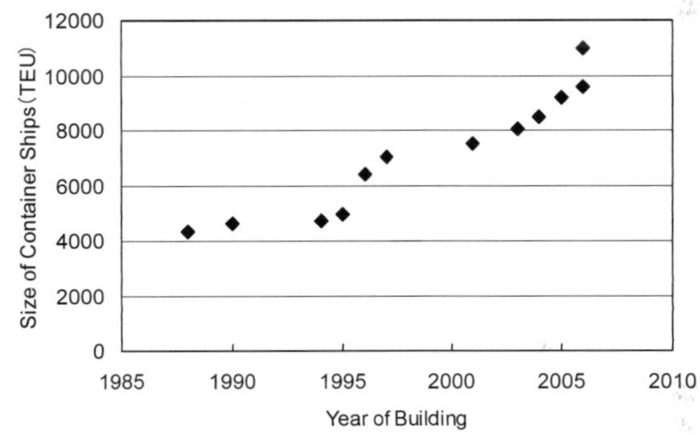

Figure 1: Trend of size of container ships

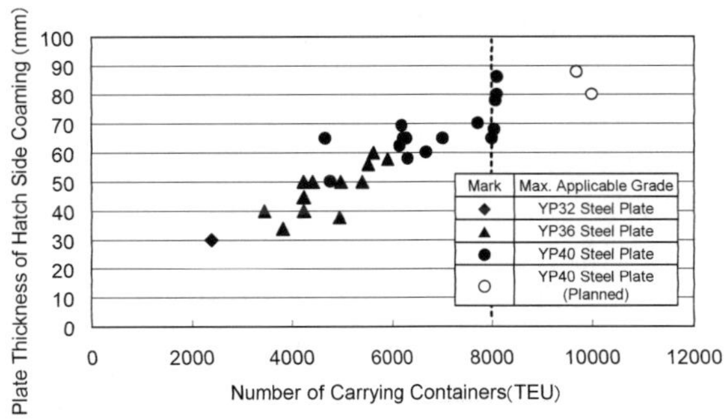

Figure 2: Variation of hatch side coaming plate thickness with increase in size of container ships

Application of such extremely thick steel plates in the hull structure have raised concerns on the following potential issues, and measures to alleviate the issues are being researched.

- Strength to resist brittle fracture in base metal and welded joint (preventing brittle fracture occurrences and arresting the propagation of brittle cracks)
- Issues in assembling, fitting and welding works of hull structures due to increase in weight
- Issues in quality control (welding work, non-destructive inspection, etc.) of welded joints on extremely thick plates

ClassNK started practical research on new higher tensile steel plates for the hull having a specified yield point of 47 kgf/mm^2 (460 N/mm^2) (YP47 steel plates) from 2001 jointly with a shipyard and a steel mill in response to the trend of using extremely thick plates in the hull structure of ultra large container ships [1].

A large container ship with YP47 steel plates to her hull structures has been delivered in June 2006 for the first time in the world under the classification of ClassNK based on the results of the above research.

The developed YP47 steel plate has excellent strength, toughness, workability and weldability as steel plates for hull structures, which makes it possible to inhibit the increase in thickness of hull structural members of ultra large container ships. Currently, the YP47 steel plate has been attracting attention as a solution to the issue of using extremely thick steel plates in ultra large container ships.

Under the circumstances mentioned above, ClassNK has established practical technical guidelines related to the use of YP47 steel plates for the hull in large container ships, called the Guidelines for Application of YP47 Steel Plates to Strength Deck Construction of Ultra Large Container Ships (hereafter referred to as "Guidelines") so as to ensure further safety of the ships in response to the noticeable increase in the size of container ships, based on the findings of recent researches and the experiences accumulated in the application of these plates to the actual ship.

The overviews of the Guidelines are introduced in this paper.

2. CONTENTS OF THE GUIDELINES

Various issues exist to be resolved for practical application of higher strength steel materials to hull structures. The main issues are as listed below.

- Setting appropriate higher tensile steel factor
- Ensuring the necessary strength, fracture toughness, workability and weldability
- Ensuring quality of welded joints including fracture toughness, and establishing welding work procedures to satisfy the quality
- Ensuring adequate fatigue strength

The Guidelines offer the necessary technical guidance for such issues mentioned above, and contain mainly the following:

- Applicability
- Hull structure
  - Hull girder strength and Higher tensile steel factor
  - Fatigue strength
  - Application of steel grades to hull structural members
  - Brittle crack arrest design
- Steel material requirements and approval of manufacturing process (approval of steel materials)
- Welding
  - Welding work
  - Welding procedures (welding joint requirements)
  - Welding materials
- Inspections (non-destructive inspections)

The description of some of the contents mentioned above is given hereafter.

3. SALIENT POINTS OF THE GUIDELINES

3.1 HULL GIRDER STRENGTH (HIGHER TENSILE STEEL FACTOR)

Thick steel plates are used in the strength deck structures of large container ships to have adequate hull girder strength (longitudinal bending strength and torsional strength). Consequently, the Guidelines are defined to be applicable to container ships in which YP47 steel plates contribute to the hull girder strength.

The higher tensile steel factor (reduction factor for strength corresponding to mild steel requirements; hereafter referred to as "HT factor") for the hull girder strength is defined as 0.62 in the Guidelines.

When setting the HT factor, the safety margin for tensile strength of steel plates needs to be considered together with the effect of the increase of yield strength. As observed in Figure 3, the set HT factor of 0.62 has a safety margin for tensile strength equivalent to that of the HT factor of high tensile steels presently being used (YP32, YP36, YP40).

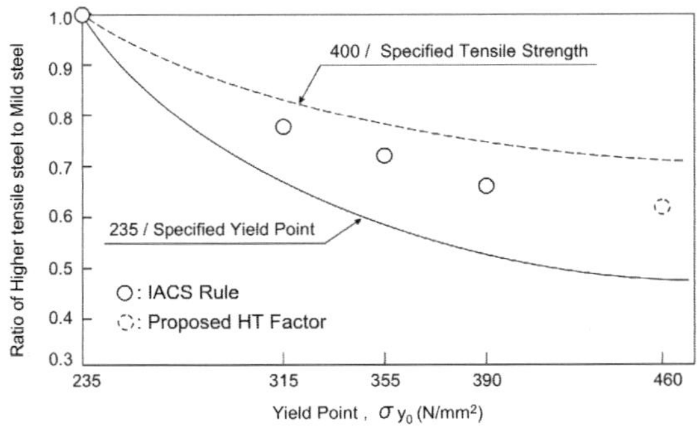

Figure 3: Relation between HT factor and yield point

Figure 4 shows an example of the application of YP47 steel plates to the hull girder strength in the 8,000TEU class and the 12,000TEU class ultra large container ships. The effect of YP47 steel plates in inhibiting increase in thickness of the hull structural members is evident from the Figure. It is also verified that the set HT factor is reasonable in connection with the application to the fatigue strength, bottom side buckling strength, fracture toughness, and so on.

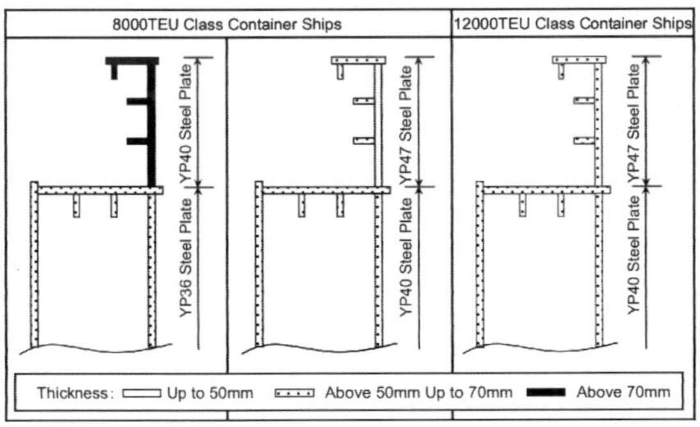

Figure 4: Example of Strength deck structure of ultra large containerships (application of YP47 steel plates)

3.2 FATIGUE STRENGTH

Fatigue strength is one of the most important issues to be resolved when higher tensile steel plates are applied to hull structures. ClassNK has already established adequately proven technical guidelines for fatigue strength assessment of hull structures (Guidelines for Fatigue Strength Assessment of Container Carrier Structures) when high tensile steel is used, based on the past experiences on fatigue damages and results of various related studies.

In the application of YP47 steel plates, appropriate fatigue tests are firstly required through the approval of the manufacturing process (i.e. the approval of steel material) so as to confirm that the fatigue strength of the intended YP47 steel plates is equivalent to or more than that of YP32 / YP36 / YP40 steel plates on mother plates and welded joints. Assessment of fatigue strength (cumulative fatigue damage) on the stress concentration parts such as hatch corners of hatch coaming top and upper deck etc. is to be carried out according to the Guidelines for Fatigue Strength Assessment mentioned above taking account of increasing hull girder stresses due to application of YP47 steel plates, which secures sufficient fatigue strength on the hull structures of YP47 steel plates.

Container ships have various kinds of outfitting such as hatch cover pads etc. on the strength deck structures especially on hatch coaming top. The guidelines require the fatigue strength assessment also on parts connecting the outfittings to the hull structure. Furthermore, the guidelines specify details of precautions to be taken related to ways of fitting the outfittings to the hull structures.

3.3 APPLICATION OF STEEL GRADES TO HULL STRUCTURAL MEMBERS

The steel grades applied to hull structural members need to be decided from the aspects of preventing brittle fracture occurrences. More specifically, it is important to set the application of steel grades to hull structural members for appropriate distribution of YP47 steel plate grades complying with the steel material requirements mentioned in Section 3.5 in this paper.

Figure 5 shows the conceptual sketch of assessment of fracture toughness based on the K concept, which aims to specify the steel grades applied to hull structural members. This assessment method uses the relationship between three fracture mechanics parameters, namely fracture stress, defect dimension and fracture toughness value. It assesses how best to select steel grade quality to be used (required fracture toughness value) to satisfy the required performance (fracture stress, defect dimension) of the structural member.

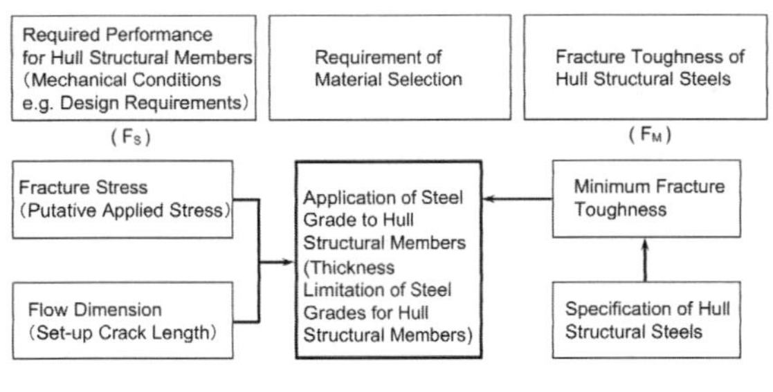

Figure 5: Assessment concept of fracture toughness based on fracture mechanics method (K concept)

Figure 6 shows the flow of the fracture toughness assessment method used for the establishment of application of steel grades to hull structural members specified in the Guidelines. In the assessment, the safety of the steel grade of YP47 steel plates applied to hatch side coaming can be verified from the view point that

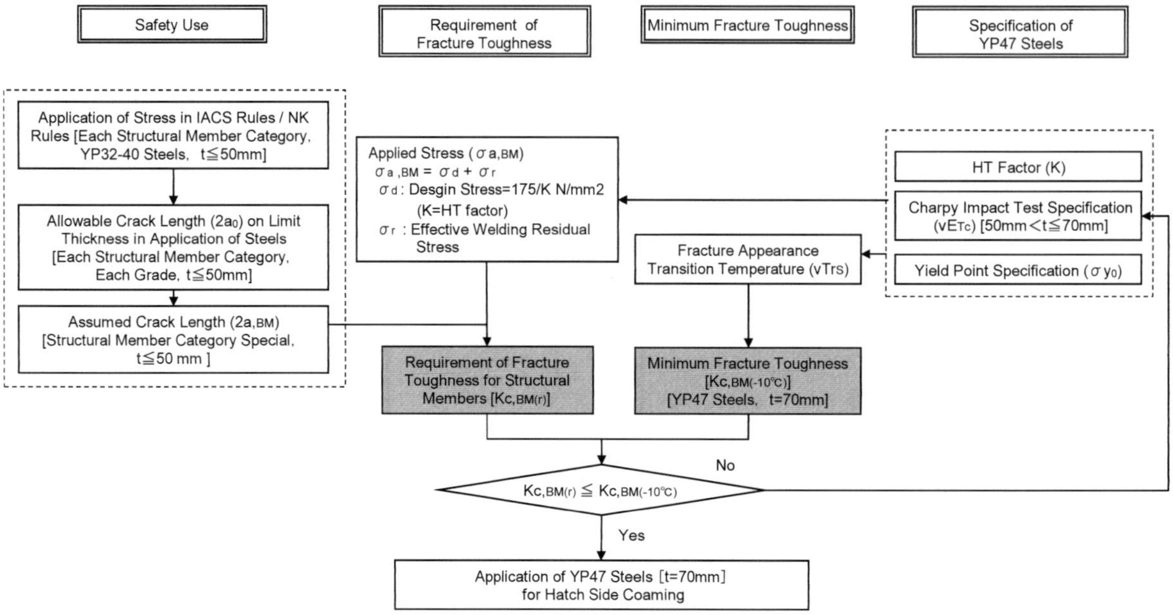

Figure 6: Flow for evaluation of application of YP47 steel plates

brittle fracture does not occur when the equation given below is satisfied. In the equation, $K_{C,BM(r)}$ is the required fracture toughness value set based on the safety usage records of application of steel grades specified in the IACS Rules and the ClassNK Rules. Also, $K_{C,BM(-10^{\circ}C)}$ is the minimum fracture toughness value at -10°C (minimum design temperature of ship operating in normal trading areas specified by IACS) of YP47 steel plates with thickness 70 mm that satisfies the minimum material requirements shown in 3.5 of Chap. 3.

$$K_{C,BM(r)} \leq K_{C,BM(-10^{\circ}C)}$$

To assess the fracture toughness, a new estimation equation for fracture toughness value (equation to estimate fracture toughness value from the results of V-notch Charpy impact tests) that can be used for YP47 steel plates was necessary. Thus, fracture toughness tests were carried out on several high strength, very thick steel plates including YP47 steel plates, and a fracture toughness estimation equation that can be applied to very thick YP47 steel plates has been derived.

Fracture toughness assessment has been performed on the base metal and welded joints of YP47 steel plates, and the application of steel grades for hull structural members has been established and toughness of welded joints has been ensured in the Guidelines so that the safety of YP47 steel plates can be ensured at a level equivalent to or greater than that in the present Rules.

3.4 BRITTLE CRACK ARREST DESIGN

The prevention of brittle fracture occurrences is of fundamental importance in ensuring safety of the ship. As described in Sections 3.3 and 3.6 in the paper, steel material requirements, application of steel grades to hull structural members, and welded joints requirements are established in the guidelines so that YP47 steel plates (base metal and welding joints) also possess the fracture toughness required to prevent brittle fracture occurrences comparable to the higher tensile steel plates presently being used (such as YP32, YP36, YP40).

In addition to preventing brittle fracture occurrences, the Guidelines require the provision of hull girder with brittle crack arrestability as a backup measure against brittle cracks that might unlikely occur considering that the acting stresses (hull girder stresses) will increase more when YP47 steel plates are used than when conventional higher tensile steel plates such as YP36, YP40 are used.

In recent years, various tests and studies are being carried out related to brittle crack propagation and its arrest on welded joints in thick steel plates. ClassNK has also been carrying out a research project related to brittle crack propagation and arrest to develop a brittle crack arrest design jointly with shipbuilders and steel mills.

The Guidelines give the following as concepts of the brittle crack arrest design, and details on the design will be specified based on the results of the above project:

- Arrest brittle crack by arrestability (brittle crack arrest toughness) of the steel material (base metal)
- Arrest brittle crack by using structural discontinuities
- Arrest brittle crack by a combination of the two approaches mentioned above.

Figure 7 shows an example of arrest design (concept of arresting brittle crack by arrestability of base metal) in which a brittle crack occurred in the hatch side coaming is arrested in the upper deck plates (mother plates).
The safety of large container ships to which YP47 steel plates are applied can be further enhanced by performing brittle crack arrest design.

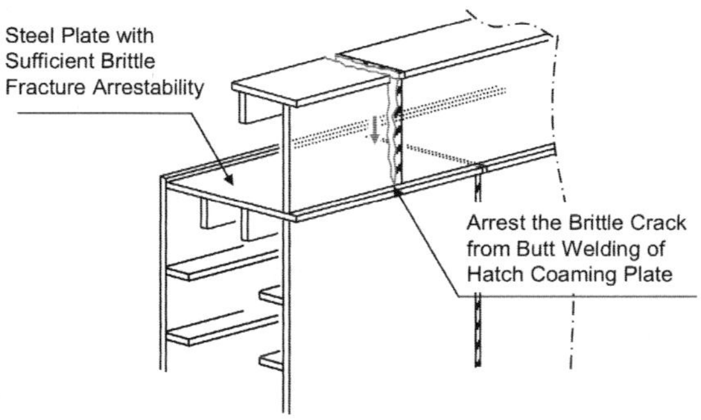

Figure 7: Brittle fracture arrest design (Example)

3.5 STEEL MATERIAL REQUIREMENTS

3.5.1 Heat Treatment

Weldability and workability that offer fracture toughness for high heat input welding and the ability to resist cold cracking for small heat input welding are necessary, in addition to the higher yield strength required to successfully use YP47 steel plates in hull structures. To simultaneously satisfy both these requirements (strength requirement and weldability / workability requirements), the heat treatment of YP47 steel plates is restricted only to Thermo Mechanical Controlled Process (TMCP) in the Guidelines so as to enable low carbon (C) and low carbon equivalent (Ceq) compositions to be attained.

3.5.2 Mechanical Properties

Table 1 show the required mechanical properties of YP47 steel plates specified in the Guidelines. The tensile strength and elongation are taken identical to those of the Quenched and Tempered high tensile steel plates with 47kgf/mm^2 yield point , which already has a proven track record in offshore structures.

Table 1: Specification for mechanical properties of YP47 steel plates (50mm < t ≤ 70mm)

Grade	Tensile test			V-notch Chary Impact Test	
	Yield Point (N/mm^2)	Tensile Strength (N/mm^2)	Elongation (5.65√S) (%)	Testing Temp. (°C)	Minimum Mean Absorbed Energy (J) L Direction
YP47	460 min.	570-720	17 min.	-40	53

Note : 1) Specification for less than 50mm or over 70mm in thickness is to be as deemed appropriate by the Society

As mentioned in Section 3.3, the requirements of V-notch Charpy impact test has been set in the Guidelines so that the fracture toughness of YP47 steel plates can be equal to or higher than those of mild steel plates and conventional higher tensile steel plates such as YP32, YP36, YP40 etc. while also considering the increase in acting stresses and the increase in plate thickness.

3.5.3 Approval of Manufacturing Process (Approval of Steel Materials)

In addition to the usual mechanical tests (tensile test, bending test, Charpy impact test, and so on) the Guidelines require brittle fracture tests (deep notch test or CTOD test, and NRL drop weight test), standard ESSO test, deep notch test or CTOD test for welded joints, fatigue tests (base metal and welded joint) etc. for approval of the manufacturing process (approval of steel materials).

This is to verify that the approved YP47 steel plates have adequate strength, fracture toughness, workability, weldability, and so on, as the steel plates for hull structure.

Small test specimens having an adequately long flat part are required to be used in tensile tests in the Guidelines so that the effect of softened part can be accurately verified, because the softening of heat affect zone in welded joints is a cause for concern in higher strength steel plates such as YP47 steel plates. The Guidelines also accept the use of large-width tensile test specimens which are close to actual hull structure welded joints when the results of small test specimens cannot satisfy the requirements.

3.6 APPROVAL OF WELDING WORK AND WELDING PROCEDURE

By using YP47 steel plates in strength deck structures, hull girder stresses rise much more and therefore more stringent quality assurance of welded joints becomes necessary.

Table 2 show the required mechanical properties of welded joints of YP47 steel plates specified in the Guidelines. Concerning the brittle fracture toughness, the requirement of V-notch Charpy impact test in the welding procedure approval tests has been established in the Guidelines by the assessment method described in Section 3.3 so that the welded joints have adequate brittle fracture toughness. The Guidelines also require an appropriate brittle fracture test to be performed in addition to the V-notch Charpy impact test to verify that the fracture toughness of the welded joint is adequately ensured in the welding methods applied in the shipyard.

Table 2: Specification for welded joints of YP47 steel plates (50mm < t ≤ 70mm)

Grade	Tensile test	V-notch Chary Impact Test	
	Tensile Strength (N/mm²)	Testing Temp. (°C)	Minimum Mean Absorbed Energy (J) L Direction
YP47	570min	-20	53

Note : 1) Specification for less than 50mm or over 70mm in thickness is to be as deemed appropriate by the Society

2) Deep notch test or othr fracture toughness tests which can assess effect of mismatching hardness of welded joints on fracture toughness are to be carried out

It has been reported in the test results of recent research related to fracture toughness of welded joints that the actual fracture toughness value might decrease to a value below the fracture toughness value estimated from the results of Charpy impact test because of the mis-matching hardness of welded joints.[2] The Guidelines require appropriate brittle fracture tests to assess the effects of mis-matching hardness on brittle fracture, if necessary, after giving consideration to mis-matching hardness of welded joints including adjacent base metal.

It is essential for welding works to implement the control standards of welding procedures applicable to YP47 steel plates concerning the weldability and workability, which has been set beforehand. The Guidelines have specified the control standards of welding procedures on the following based on the results of multiple performance assessment tests such as cold cracking tests, line heating tests (tensile tests, Charpy impact tests) and so on:

- Short beads
- Preheating in working
- Cold forming
- Line heating (Thermal fairing)

Furthermore, in order to detect and eliminate weld defects generated through the welding works during construction which is important for quality assurance of welded joints, the Guidelines have enhanced the non-destructive inspections on butt welded joints of YP47 steel plates as follows:

- Non-destructive inspection to be performed on all butt welded joints
- Structural members to be inspected are all longitudinal strength members in the strength deck structure (hatch side coaming including top plates, upper deck plates, shear strake, uppermost strake of longitudinal bulkheads) including longitudinal frames, beams, stiffeners with plate thickness exceeding 50 mm.

4. CONCLUSIONS

ClassNK has established the Guidelines for application of YP47 steel plates to the hull structures of large container ships based on the results of extensive research works and the experiences in the application of YP47 steel plates to actual large container ships, in response to the recent trend of using extremely thick steel plates to hull structures because of the increase in the size of container ships. The guidelines are anticipated to contribute to enhance the safety of ultra large container ships.

Various studies are now being carried out on thick steel plate structures of large container ships, and the findings of these studies will be incorporated in these guidelines from time to time.

5. REFERENCES

1. Y. Yamaguchi, T. Matsumoto, H. Yajima, 'Technical Requirements to ensure structural reliability for mega container ships - Application of New higher strength hull structure steels with heavy thickness', RINA International Conference of the Design and Operation of Container ships, November 2006

2. T. Ishikawa, T. Inoue, et al., 'Fracture toughness in welded joints of high strength shipbuilding steel plates with heavy-thickness', Proceeding of the 17th International Offshore and Polar Engineering Conference, July 2007

6. AUTHORS' BIOGRAPHIES

Yoshiya Yamaguchi, is currently the Manager of the materials and welding section of the Material and Equipment Department at Nippon Kaiji Kyokai (ClassNK)

Toshiyuki Shigemi, Ph.D. is currently the General Manager of the Development Department at Nippon Kaiji Kyokai (ClassNK)

Toshiyuki Matsumoto, is currently a Research Manager of the Research Institute at Nippon Kaiji Kyokai (ClassNK)

Akizumi Miura, is currently the Manager of the general cargo ships section of the Hull Department at Nippon Kaiji Kyokai (ClassNK)

SELECTION OF A DESIGN CONCEPT FOR THE CREATE3S PROJECT

P. Wright, **D. Broderick** and **M. Landamore,** Newcastle University, UK
J. Hopman, Technical University of Delft, Netherlands

SUMMARY

This paper reports on work undertaken as part of the current FP6 EU funded project 'CREATE3S', that seeks to identify new efficient short sea shipping solutions. Existing cargo handling techniques and hybrid solutions are summarised and new potential solution concepts to meet the objectives of the project are proposed. The rationale behind these concepts and the subsequent selection process are described. The work reported here brings together the contributions made by all partners in the project consortium and reported in [1].

1. INTRODUCTION

The 'CREATE3S Project' (Innovative Concepts Realised by Advanced design and production to improve Total Efficiency of new generation Short Sea Shipping) is funded by the EC 6th Framework. The objective of the project is to develop a ship concept that will allow European short sea shipping to compete with other transport methods, e.g. road, and logistics solutions established using CREATE3S vessels should be able to outpace current concepts in terms of logistic and economic performance, and possess improved enviromental performance.

It is also intended that by the use of novel structural design methods, the competitiveness of European shipbuilding can be increased through the production of such vessels. The expected number of vessels needed to replace older tonnage and to meet the increased demand for transport should exceed 350 ships per year for European short sea shipping alone. However the current production output for the short sea shipping freight sector is approximately 70 ships only. The application of an intelligent and rationalised production approach would allow for the potential mass production of such vessels, and hence reduction in first cost. This has been one of the key objectives for the CREATE3S project, in that more than one type of ship may be produced using the approach developed in the design of a fundamental CREATE3S ship platform.

This paper provides a background to some of the approaches proposed by the CREATE3S project to develop a new family of short sea vessels to address these objectives.

2. IMPROVEMENT OF LOGISTIC AND ECONOMIC PERFORMANCE

The scenario for door-to-door logistics needs to be considered as a complete system, the development of the ship alone without consideration of the whole logistic chain may simply lead to a shifting of the current bottleneck from the terminal-vessel interface to the terminal itself. Figure 1 shows the current logistic chain in door-to-door operations. The expected savings made from improved vessel operations and cargo handling must not be nullified by an increase in the complexity of terminal operations.

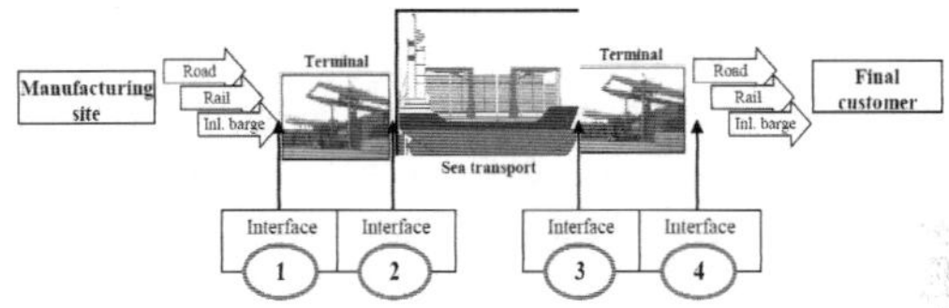

Figure1: Door-to-Door Multimodal Logistics Chain [1]

In the context of the CREATE3S project, intermodal transport is the freight of goods items packed in one load unit (e.g. container), using various transport modes and without handling the individual goods items during the transhipment from one mode to the other, namely they remain unitised during the entire transport chain. Intermodal transport implies that at least two distinct transport modes are utilised in the logistic chain.

Concepts developed here are intended to address the following sections of the logistic chain:

- Seaport to seaport, including terminal operations;
- Seaport to inland port, including terminal operations.

The basic operating principle of all the proposed concepts within the project is that the loading and unloading time for vessels will be reduced that in turn will result in less port congestion and shorter waiting times. Additionally, the concept also aims to make considerable reductions in both terminal handling time and cost. The approach proposed by the CREATE3S project to achieve this is, in essence, the simultaneous transfer of multiple cargo units comprising the complete cargo inventory. This it is hoped will facilitate the benefits sought and offer significant improvements over current systems, this could also be relevant to the inland waterways sectors, where such improvements could also provide significant benefits.

To increase the operational flexibility of the CREATE3S vessel and to allow for a standardized ship concept across different trades, the handling of cargoes in multiple units can be applied to existing unitised cargoes, such as ISO/ non-ISO containers and can be extended to other cargoes if they are unitized in an appropriate way.

For container operation, the handling of a number of units together has been termed a 'cargo package', with a number of packages forming the complete cargo envelope of a particular loading condition. These definitions are given in Figure 2 [2]. For unitising bulk or liquid cargo the new cargo unit for this purpose may be a package providing the ability to carry more than one cargo type (provided it is in a standard package) which would greatly increase the flexibility of the vessel concept. This could potentially lead to further reductions in door-to-door time, as the need for cargo transfer is removed.

	Cargo Unit: A fundamental unit of cargo, e.g. ISO/ non-ISO/ cube max/ reefer etc. container, tank container for liquid cargo or new 'cargo unit' to meet the needs of how a new cargo might be unitized.
	Cargo Package: A number of cargo units handled together to facilitate loading and unloading. Minimum size is a single 'cargo unit' and the maximum size is a 'cargo envelope'. Maximum package size (number of cargo units) is set by cargo handling and logistic constraints.
	Cargo envelope: A number of 'cargo packages' onboard the ship in a particular loaded condition or the nominal maximum capacity of the vessel assuming a specified weight per 'cargo unit' or 'cargo package'.

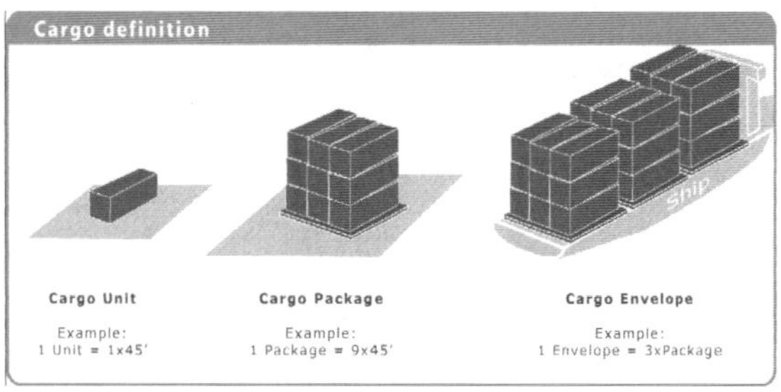

Figure 2: Cargo Package Definitions [2]

The total cost of any proposed concept needs to be considered in terms of first cost and through life operational cost, and cost benefits need to be achieved by consideration of both cost implications. In order to reduce operational costs this study has focused particularly on the reduction of handling cost, by adopting the philosophy described of reducing the number of handling operations. There is also the possibility that if turnaround time is reduced, the vessel can then reduce its service speed on the sea leg whilst maintaining the same frequency of service, with obvious benefits for fuel economy and maintenance costs.

Innovative approaches to ship structural design are required if reduction in first cost is to be achieved, by facilitating more cost effective production of weight efficient families of ship designs with reduced lead time, that also provide through life maintenance benefits. Such production benefits need to be achieved through structural configurations that allow for modularisation and standardisation of structural components (including piece-parts, sub-assemblies and blocks), a reduction in the inherent work content and facilitate ease of sub-system outfitting. Potential configurations could include advanced framing systems and steel and/or composite sandwich structures. Weight benefits achieved from such new structural configurations would allow for reduction in fixed weight and therefore more optimal 'payload weight/ cost performance' required for different vessel types as a function of cargo value and the required service speed.

Novel structural concepts could achieve this by the rationalisation of structural configurations and the elimination of complexity associated with such features as notch details and stiffener intersections to allow for greater use of automated fabrication processes. Facilitating smooth interior surfaces also greatly enhances the installation of piping and electrical systems, painting, inspection and repair. This may result in lower cost 'equivalent' structures or allow for weight optimised structures (where advantageous). A rational approach would allow the optimal weight-production cost balance to be achieved for different ship types. Such new configurations would also seek to achieve better crashworthiness due to collision or grounding by allowing deformation over a large area to dissipate load and ensuring effective load transfer.

3. CONVENTIONAL CARGO HANDLING SOLUTIONS

There are well established cargo handling operations that can be summarised under three main categories: Roll-on/ Roll-off; Float-on/ Float-off (Ballast-on/ Ballast-off); Load-on/ Load-off.

3.1 ROLL-ON /ROLL-OFF (RO-RO)

This principle has been very successfully applied in the short-sea sector for wheeled unitised cargo. Figure 2 shows the different stages of the Ro-Ro loading process. It can provide significant improvements in handling rate for container operation with container units on cassettes, perhaps in combination with Advanced Guided Vehicles (AGVs). Handling rates of up to 300 units per hour, become possible, which is an order of magnitude higher than conventional Lo-Lo operation. This can facilitate a fast turnaround time and is therefore a concept that could meet the objectives of the CREATE3S project. However, the disadvantage of this system is the large investment in port infrastructure needed to allow efficient loading of the vessel, and the relative cost and cargo carrying efficiency of the vessel.

Figure 3: The Ro-Ro Loading Sequence [1]

3.2 FLOAT- ON /FLOAT-OFF (FLO-FLO)

This principle utilises the ability to alteri the draft of the vessel using internal ballast tanks, to allow the free-floating cargo or tailored cargo barges to be floated on and off of the vessel. The solution does not need any investment in additional port infrastructure; it does however need sufficient water depth for ballast operation, and sheltered condition for handling of the individual barges or for forming barge trains. Flo-Flo systems are most prevalent in the heavy lift sector. There is a subset of Flo-Flo that uses ballast systems to lift a platform supported cargo load, referred to as Ballast-on/Ballast-off (Bo-Bo). This is more relevant approach to meeting the objectives of the CREATE3S project and Figure 3 shows the process.

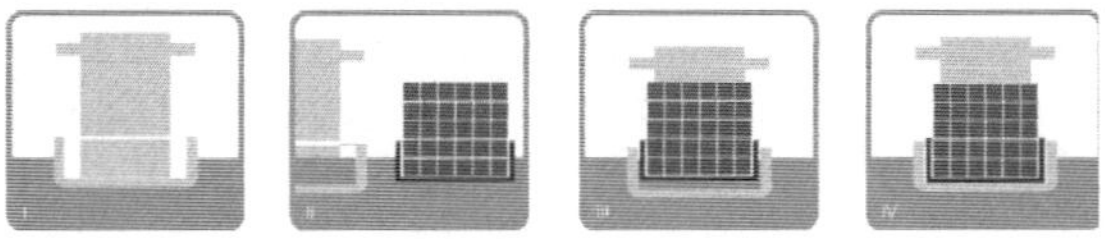

Figure 4: The Bo-Bo Loading Process [1]

3.3 LOAD-ON /LOAD-OFF (LO-LO)

This is the industry standard for (un)loading containerised, bulk and break bulk cargoes. The cargo is lifted in or off the vessel by a quay or ship mounted crane. Constant development in Lo-Lo equipment has led to increased unit handling speeds in the container sector where 'super-post panamax portainer' cranes provide the reach and capacity necessary for the latest ultra-large container carriers. Handling rates of around 30 units per hour per gantry crane are possible; this represents the current state of the art for container handling but has required significant investment in the cranes themselves, and an increasing amount of quay storage space for the associated stacking and movement of units (by straddle carriers etc.). To improve handling rates further the concept of a 'mega- container', combining several units, has been proposed, but exceeds current Lo-Lo capacity.

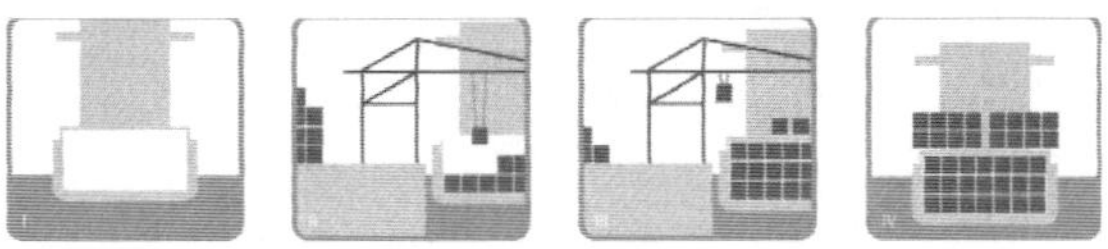

Figure 5: The Lo-Lo Loading Process [1]

4. EXISTING HYBRID SOLUTIONS

Based on combinations of these fundamental handling solutions, many hybrid vessels have been proposed and built, with varying degrees of success. Although not formally reviewed here, they can be considered under the general headings of; barge carriers, push barge systems, innovative Ro-Ro concepts and hybrid ships. Some examples of these are given below.

- Barge Carriers:
  - LASH (Lighter aboard ship);
  - SEABEE (Sea barge);
  - BACO (Barge/ Container liner);
  - BACAT (Barge aboard catamaran);
  - TSL (Trans sea lifter).
- Push Barge Systems:
  - ATB (Articulated Tug Barges).
- Innovative RO-RO based systems:
  - CTPB (Container Pallet Transfer System or 'mega pallet' concept);
  - Train Loader;
  - Roller Barge;
  - Schowple Barge;
  - IPSI
- Hybrid Lo-Lo/ Ro-Ro/ Container Carriers.

5. CONCEPT GENERATION

The CREATE3S project undertook an in-depth review of these concepts to develop an understanding of their relative strengths and weaknesses before proposing new concepts to take forward [1]. Using this background study and the aforementioned fundamental loading systems as a basis for brainstorming, four potential solutions were conjectured and initial design studies for each concept were progressed by separate working groups within the project consortium. These candidate solutions were: Ship 'Truck-Trailer' concept; 'Modular Ship' concept; 'Flo-Bo' concept; 'Mega-Pallet' concept.

5.1 CONCEPT 1: 'TRUCK-TRAILER'

The idea of this concept is a sea-going push barge, in essence the equivalent of a road truck-trailer combination (Figure 6). It consists of two individual parts, the propulsion unit (push or pull) and the cargo carrying platform (barge or ship). The cargo carrying platform, loaded with the required number of cargo units comprising the complete cargo envelope, would be loaded prior to the arrival of the propulsion unit so that it can proceed directly to sea. The loading of the cargo envelope could be by conventional Lo-Lo or Ro-Ro. As the cargo envelope is buoyant, it could also provide extra floating quay storage thereby improving current terminal handling operations, and is compatible with existing cargo handling infrastructure. One of the critical components is ensuring sufficient integrity of the link mechanism between the propulsion unit and the cargo carrying platform to resist the substantial forces and moments that would be experienced during short-sea operations.

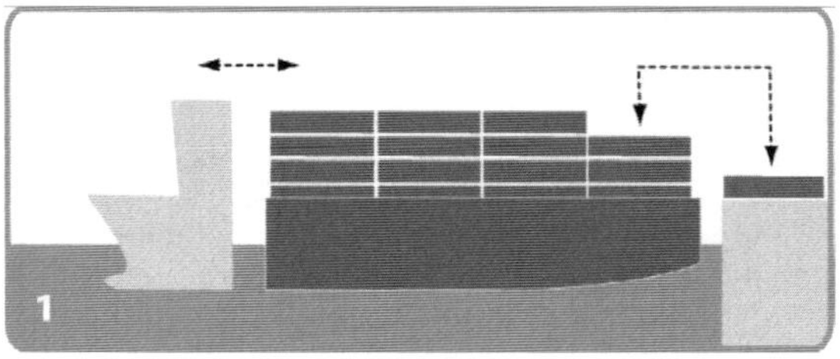

Figure 6: 'Trailer-Truck' Concept [1]

5.2 CONCEPT 2: 'MODULAR SHIP'

This is very similar to the 'truck-trailer' concept but it involves multiple cargo packages on dedicated barges, comprising the complete cargo envelope, that are carried onboard the ship and loaded by a Bo-Bo operation. Minimising ballast draft is a key feature of this solution. The ballast draft is reduced, as the packages are buoyant and also contribute to the ship's buoyancy when loaded. It was concluded that the ship must also be able to operate in a solely ballast condition to account for cargo flow imbalance between the head haul and the back haul. A potential advantage of this concept is that the floating package could also be used to increase the inland waterway penetration with river compatible package sizes (namely 11.4m or 22.8m) for level one or level two compatibility respectively.

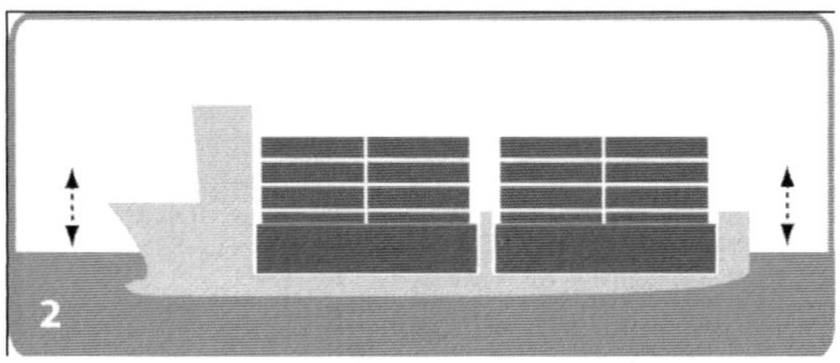

Figure 7: 'Modular Ship' Concept [1]

5.3 CONCEPT 3: 'FLO-BO'

The Float-on/ Ballast-off (Flo-Bo) concept seeks to maximise the use of inland waterways and provide additional floating quay storage [3]. Cargo units would be loaded conventionally through a Lo-Lo operation onto dedicated cargo package barges (level one or level two inland waterway compatible) which are then floated into an intermediary loading dock. The dock is de-ballasted to allow the cargo packages to be transferred from the barges to the dock and then subsequently ballasted back down to transfer the compete cargo envelope to the ship platform. The use of a floating dock means that the ship can be simplified and does not require a large ballast system onboard. However, in addition to the package barges, each port requires the dedicated dock infrastructure as well as sufficient depth of water to allow the transfer to take place. The cost and complexity associated with this facility has to be considered. Another possibility is to use other land-based height altering mechanisms, such as hydraulic elevator platforms, rather than a floating dock.

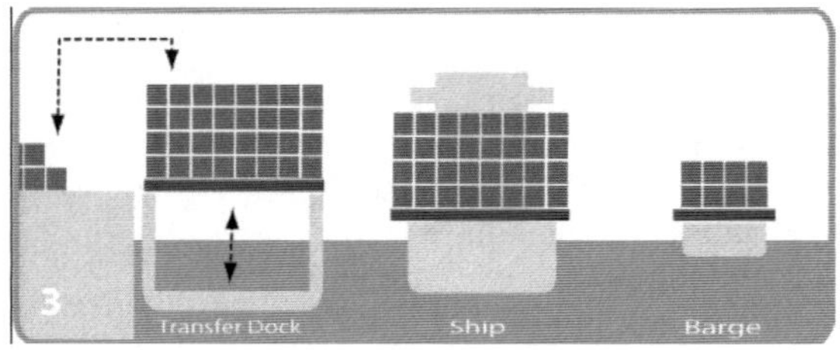

Figure 8: 'Flo-Bo' Concept [1]

5.4 CONCEPT 4: 'MEGA-PALLET'

The rationale behind this concept is to reduce the number of Lo-Lo operations needed to load a vessel by increasing the number cargo units transferred during one operation, namely handling cargo packages rather than individual units. By introducing a so-called 'mega pallet', ship loading and unloading time should be reduced proportionally. Such 'mega-pallets' could be made compatible with existing short sea lo-lo feeder designs. Cargo transfer could also be Ro-Ro based; additionally this concept could be used to transfer cargo packages from the quay to the associated floating platform for the previous three concepts. It again would require large investment in dedicated port facilities.

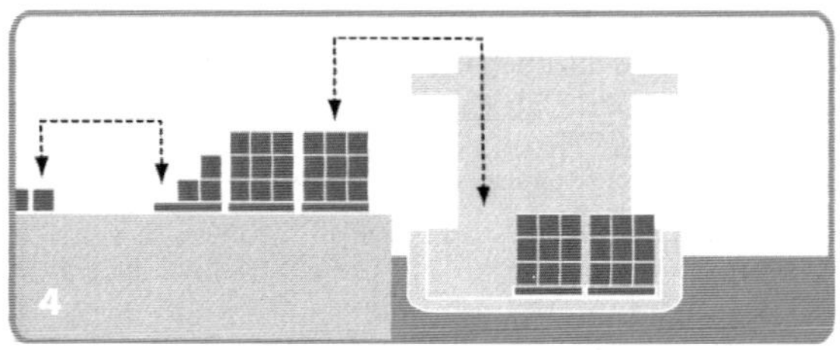

Figure 9: 'Mega Pallet' Concept [1]

6. CONCEPT EVALUATION

6.1 EVALUATION METHODOLOGY

The *MV Geeststroom* (a Damen 800 design) was selected to provide a benchmark to compare the performance of these candidate designs. This container feeder vessel design of 517 TEU (based on 14 tonnes/ TEU) operates around European routes. Figure 10 shows the vessel and its general arrangement. The general particulars are presented in Table 1.

To obtain comparable results it was necessary to apply techno-economic and operational evaluation criteria. The cargo size for each concept has been defined as: 515 TEU at 14 tonnes/ TEU to allow direct comparison with the *MV Geeststroom*.

Based on this weight per unit and the required capacity, a design deadweight of 7500 tonnes results. This was taken as the largest member of the family of vessels proposed by the project for two arbitrary A-B routes, such as Rotterdam-Tilbury and Rotterdam- Bilbao.

Although the initial focus and detailed evaluation has been on the container trade, all concepts were developed to allow for the possibility of carrying bulk or liquid bulk in appropriate units/ packages. The flexibility of each concept to be adapted to these cargoes was considered fundamental to the concept.

Figure 10: MV Geeststroom [1]

Particulars of MV *Geeststroom*	
Length between particulars (m)	130
Breadth (m)	21.8
Draught (m)	7.3
Depth (m)	9.5
Displacement (t)	14000
Light ship (t)	4850
Max. /Design speed (knots)	19/16
Main engine (kW)	8400
Payload dwt (tonnes)	7330
TEU capacity @ 14 ton	517
Max TEU capacity	812
45'cont capacity @ 22,28 ton	329

Table 1: General Particulars of MV Geestrom [1]

There are four main issues to consider when comparing the alternative concepts:

- The economic investment, and impact in terms of estimated operating and maintenance costs;
- The total logistic concept with respect to flexibility and growth potential;
- The technical feasibility of the concept;
- The environmental impact of the ship concept and any necessary related infrastructure to be developed or implemented.

The first of these means of comparison, namely an economic evaluation, can be done on a quantitative basis if appropriate estimation of associated costs can be made. There are a number of cost elements that can be attributed:

- Cargo unit costs per tonne km;
- Cargo unit lead-time on a door-to-door transit;
- Building costs of hydrodynamic ship platform;
- Building costs of Cargo Package(s);
- Building costs of Cargo Envelope;
- Time for loading/discharging units into Cargo Packages;
- Transit and loading/discharging time of the Cargo Package into the Cargo Envelope;
- Loading/discharging time of the Cargo Envelope and extra cargo to the hydrodynamic platform;
- Additional investment in infrastructure needed to load/discharge;
- Operational cost of equipment needed to handle cargo in terminals;
- Operational cost, including maintenance and port fees etc., of the hydrodynamic platform, holding envelope, cargo packages and cargo units.

The remaining logistic and technical issues selected for comparing the different concepts are more qualitative in nature, namely assessment of their relative:

- Flexibility in terms of suitability for liquid and dry bulk;
- Growth potential including geographic scope and increasing throughput;
- Scalability across the family of proposed designs (2500 and 5000 tonnes deadweight variants);
- Availability for service;
- Technical viability;
- Research implications to develop enabling technologies;
- Environmental sustainability.

The key results for one route for this study with respect to economic, logistic and technical viability are summarised in Table 2. The actual costs estimated are not presented due to their commercial nature but they are normalised with respect to the benchmark figure to provide some means of comparison. The estimation of these costs was considered in some detail but nevertheless the final values were considered as satisfactory for comparative purposes rather than as an accurate prediction of economic performance .The logistic and technical viability is simply ranked in relative order with first being best and fourth judged worst with respect to each criterion.

6.2 FURTHER APPRAISAL OF CONCEPTS

6.2 (a) Concept 1

This is a variation of the push barge system which is a relatively mature technology. It will provide a very fast turnaround time for the propulsion section of the vessel, however as the cargo envelope is similar in length to the *MV Geeststroom* it will not allow for greater utilisation of port facilities as the envelope is loaded using the conventional Lo-Lo method. Due to the size of the complete cargo envelope (Figure 11) this solution will not facilitate inland waterway penetration. Also, to allow for the greatest usage of the propulsion unit at least three cargo envelopes would be needed for each route, one for each port and one in transit. As the envelope will be a significant proportion of the overall build cost this has obvious cost implications.

Evaluation	Criteria	Concept 1	Concept 2	Concept 3	Concept 4
Economic performance	Unit cost (Euro per tonne km)	0.72	0.76	0.92	1.24
	Unit lead time (hours)	0.97	0.97	1.02	1.02
Logistic considerations	Infrastructure	Equal first	Equal first	Third	Fourth
	Cargo flexibility	Equal first	Equal third	Equal first	Equal third
	Scalability	Equal second	Equal second	First	Fourth
Technical viability	Critical technologies	Equal first	Equal first	Fourth	Equal first
	Production	Equal second	First	Fourth	Equal second
	Research need	Second	Equal third	First	Equal third
Suitability for CREATE3S		**Fourth**	**Equal first**	**Equal first**	**Third**

Table 2: Evaluation matrix

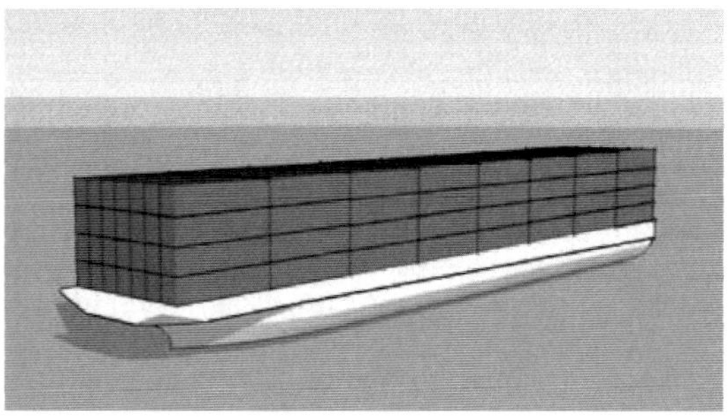

Figure 11: The Cargo Envelope of Concept 1 [1]

6.2 (b) Concept 2

This could provide a significant reduction in turnaround, providing the ballasting operations needed for the (un)loading of the vessel are fast enough. However, this concept raises issues of concurrent rather than sequential ballast exchange and so ensuring adequate trim and stability is an important operational consideration. As the vessel may need to be able to sail in the empty ballast condition, and relies on the cargo package barges for their buoyancy, there are concerns about technical viability. Two additional major concerns with this concept are compromise of the deadweight displacement ratio due to the weight of the package barges, and that the overall dimensions of the vessel will be greater than the basis vessel in order to accommodate the cargo envelope; this will have significant first cost implications. However the inland penetration will be far greater as the envelope is broken down into (up to) six packages compatible with both level 1 and level 2 inland waterways.

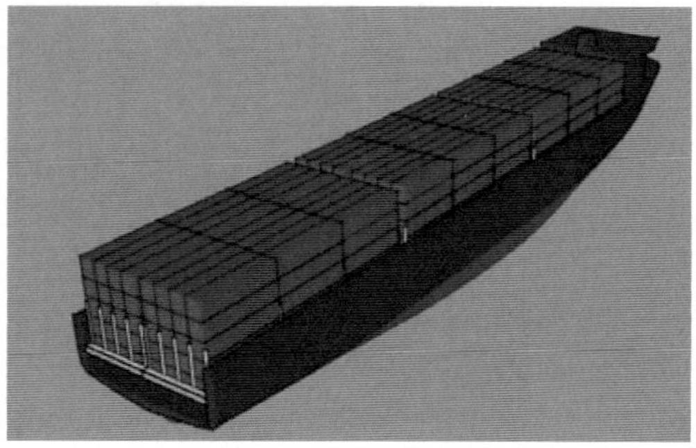

Figure 12: Vessel Loaded with the Cargo Envelope for Concept 2 [1]

Figure 13: Cargo Packages for Concept 2 Making up the Different Envelope Sizes Across the Family of Vessels [1]

6.2 (c) Concept 3

The ability to offload the cargo packages from the barges to avoid carrying them with the cargo envelope requires a 'pallet' like structure to support the packages as the barges are floated out from the dock and to allow the transfer of the cargo envelope from the barges to the vessel via the dock. As a result of this the complexity of the system will be greater than that of Concept 2; it will however have a better deadweight/displacement ratio and will allow a reduction in the overall dimensions of the vessel as the envelope is carried on a flat weather deck. As the cargo envelope does not have implications for internal hull volume or linear dimension requirements, the hull form can be optimised with respect to hydrodynamic performance. However, the concept will require additional investment in the sizeable floating dock, at least one floating dock per port or two if parallel loading and unloading operations are required to minimise the turnaround time.

Figure 14: Barges Being Floated into the Dock for Concept 3

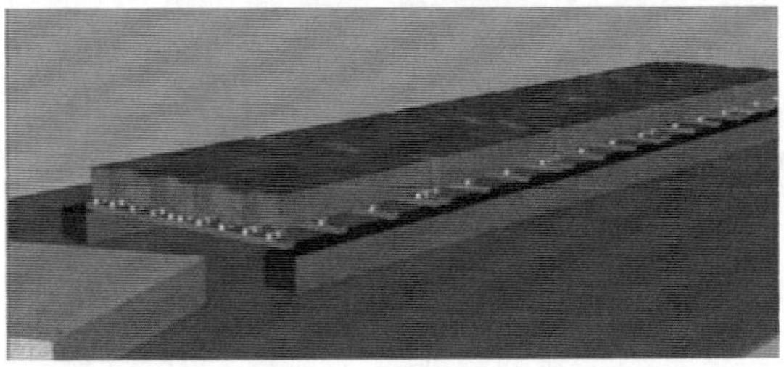

Figure 15: The Pallets Supporting the Packages for Concept 3

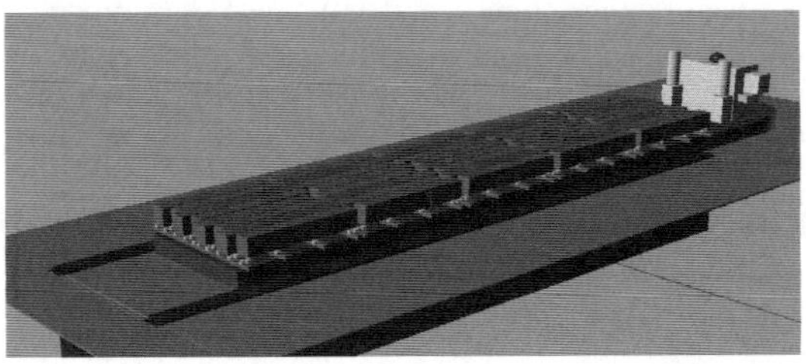

Figure 16: The Vessel Leaving with the Cargo Envelope for Concept 3

6.2 (d) Concept 4

This will require large investment in port infrastructure to optimise the (un)loading procedure as quay space is required on both sides of the vessel as shown in Figure 17. This would result in the need for new berths orthogonal to existing quay length. Additionally, the quay and capacity of the lifting gear (gantry crane, straddle carriers etc.) will need to be increased to handle the increased weight of the cargo packages being moved. The vessel itself could remain largely the same with only minor alterations to the holds and container bays, as well as local structure, to support the 'mega pallet' cargo packages. The trim and stability during the (un)loading process also needs to considered carefully due to the transfer of heavy cargo packages. To facilitate inland waterway penetration, the mega pallets will have to be loaded onto barges, thus increasing the number of handling operations.

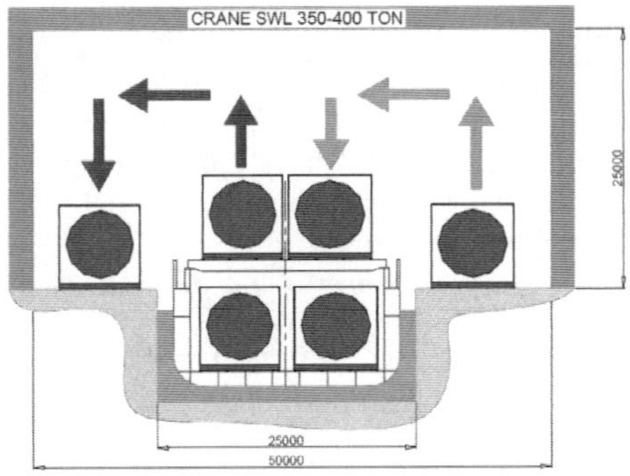

Figure 17: Concept 4 in the Loading Berth[1]

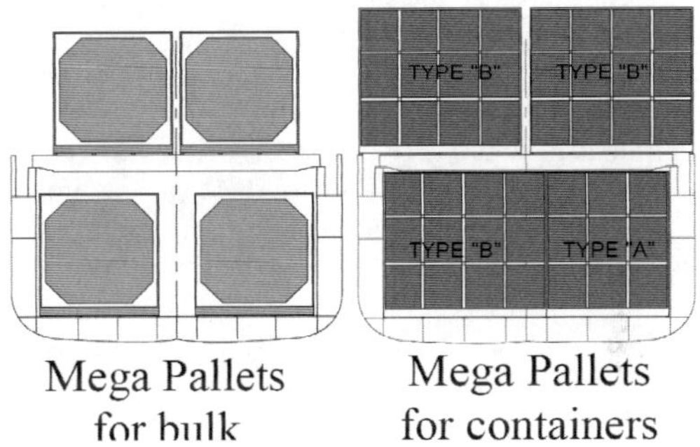

Figure 18: The Cargo Packages of Concept 4 [1]

7. SUMMARY AND FINAL CONCEPT SELECTION

As there are uncertainties in the estimation of first and operational cost, and the resultant ranking, it was judged prudent to combine these findings with the need to meet the following fundamental requirements:

- Minimal investment or changes to terminal/shore infrastructure;
- Inland waterway penetration capability;
- One type of (simple) cargo handling operation with maximum time saving.

From this a final summary of the relative merits of these concepts was considered:

- Concept 1 requires no additional terminal infrastructure, as it relies on conventional Lo-Lo cargo handling but is not inland waterway compatible. Once the propulsion unit has been disconnected from the cargo envelope it offers no logistic advantage over the conventional ship;
- Concept 2 provides good inland waterway penetration and handling rates but requires investment in cargo package barges.

- Concept 3 also requires significant investment in terminal infrastructure, especially the floating dock and the associated cargo pallets and barges. This concept, however, has the advantages of not having to carry the barges on the sea leg and providing both inland waterway potential and the opportunity to improve terminal utilisation.
- Concept 4 requires considerable investments in terminal infrastructure, dedicated berths and handling gear, and is not directly inland waterway compatible.

The final conclusion from this stage of the study was that rather than there being one clear 'best solution' there was the opportunity to combine elements of Concept 3, 'Flo-Bo', with the dock ship capability of Concept 2, 'Modular Ship'. Thus the need for the dedicated dock infrastructure required for Concept 3 can be removed, if the extra complexity of an onboard ballasting system is accepted. As a result of this Concept 2, in the form of a dock ship loaded with cargo package barges, was chosen as the candidate solution for further development. However two variations of this concept were also considered further before making the final selection:

- *Variant A*: A dock ship carrying the conventional cargo package barges onboard the vessel.
- *Variant B*: A dock ship using cargo package bases (pallets) to allow the pontoon cargo barges to be left in port.

The conventional cargo package barges of *Variant A* have a higher capacity compared to the simple flat top pontoon type barges used by *Variant B*, as they allow use of internal volume and provide better stack height before stability is compromised.

A conventional barge with a beam constrained to 11.4m for Level 1 waterway penetration is limited by stability considerations to a stack height of four, whereas a pontoon barge with the same beam is limited to a stack height of only two. Therefore a *Variant B* design will not allow the design requirement for the number of cargo units in the cargo envelope to be met.

Additionally, the combined pontoon barge and package base approach of *Variant B* has penalties associated with cost and complexity that are only justified if the weight saving benefits the payload displacement ratio when the maximum filling rate is reached. As this is a capacity limited design problem, this does not provide any advantage except that the weight saving could still facilitate a lower displacement and therefore reduced fuel consumption on the sea leg.

However this is, on balance, offset by the flexibility offered by carrying the barges associated with cargo packages onboard the dock ship. This maximises transfer rate and facilitates a reduced turnaround time, providing that the ballasting operation can be carried out quickly and efficiently while maintaining adequate trim and stability.

Therefore the final outcome of this study was to take forward *Variant A*; a dock ship concept carrying barged cargo packages compatible with both level 1 and 2 inland waterway penetration, Figure 19.

Figure 19: The Selected Dock Ship Concept

8. CONCLUSION

It is interesting that from a very broad consideration of cargo handling techniques, and an innovative approach to candidate design solution generation and selection for the CREATE3S project, that the final outcome, based on a methodical selection process, was a dock ship concept.

Further development of the design in the next phase of the project will hopefully confirm the concept's viability, and realisation of its objectives. It is of note that the recent 'Combi Dock' heavy lift ship, *Combi Dock I*, [4], although of larger deadweight is also intended to be flexible in operation, handling bulk, Ro-Ro and container cargoes. The augmentation of such a dock ship concept with cargo package barges should provide the further logistic benefits sought.

9. ACKNOWLEDGEMENTS

The work presented here was financially supported by the European Commission under the FP6 Sustainable Surface Transport Programme. The support was given under the scheme of STREP, Project No. 031488. The European Commission and the Authors shall not in any way be liable or responsible for any use that might be made of any knowledge, information or data contained in this paper, or the consequences thereof.

The authors would like to recognise the partners who contributed to this work and are part of the CREATE3S project consortium: Samskip Multimodal Container Logistics BV; PD Ports; Damen Shipyards Group; Estaleiros Navais de Viana de Castelo; Imtech Marine & Offshore; TTS Ships Equipment AB; LogIT AS; Center of Maritime Technologies; Norwegian Maritime

Technology Research Institute; Maritime Research Institute Netherlands; Delft University of Technology; Bureau Veritas; Centrum Techniki Okretowej SA; Newcastle University.

10. REFERENCES

1. Hopman, J., J., 'Generation and Basic Definition of Most Promising New Generation 3S Concept Based on Separation of Ship Hull and Cargo Containing Modules', Report D2.1/2.2, CREATE3S, Project Number 031488, February 2008, 35pp.

2. Wright, P. N. H., 'Clarification of Requirements and Definitions', Internal Communication, CREATE3S, Project Number 031488, 3pp.

3. Wright, P. N. H., and Broderick, D., R., 'The Flo-Bo Design Concept', Presentation to CREATE3S consortium, General Technical Meeting, Vianna, Portugal, 8th November 2007, 35pp.

4. The Naval Architect, 'Vuyk gets heavy', May 2008, p18-19.

11. AUTHORS' BIOGRAPHIES

Peter N H Wright is Senior Lecturer in Naval Architecture in the School of Marine Science and Technology at the University of Newcastle upon Tyne where he has been a member of the academic staff since 1995. He is responsible for the teaching of naval architecture and ship design and is involved in a number of international research programmes on aspects of warship and merchant ship design. He graduated in Naval Architecture and Shipbuilding from Sunderland Polytechnic in 1989 and from the University of Newcastle upon Tyne in 1991 with a BEng Naval Architecture and Shipbuilding with First Class Honours. He was sponsored by and worked for Vosper Thornycroft (UK) Ltd before returning to the University. He obtained his PhD in 2004. He has been involved with and led the Newcastle contribution to several European and national projects concerned with both naval and merchant vessel design; most notably EUCLID RTP3.21, Intermodeship and MARSTRUCT as well as the current CREATE3S project. He is a Chartered Engineer and corporate member of RINA.

Darren R Broderick is a researcher at Newcastle University. He is currently working on the application of longitudinal framing to short sea vessels as part of the CREATE3S project. He completed his under graduate degree at Newcastle University, after having completed a cadetship and four subsequent years of sea service, reaching the rank of second engineer, with P&O Nedlloyd (formally P&O Containers). He is a graduate member of RINA.

Melanie J Landamore is a qualified Naval Architect, specialising in Small Craft Systems and Environmental Sustainability of Marine Systems, and manages the CREATE3S project for Newcastle University. Her other current research fields include Stability of Anchor Handling Supply Vessels, Surface and Underwater Autonomous Craft for SAR and Survey applications, and the environmental assessment and regulation of SAR vessels.

J.J. Hopman is professor Ship Design at Delft University of Technology (DUT) and is as Work Package leader in the Creat3S project responsible for the development of the new ship concepts. J.J. Hopman completed his master degree in naval architecture in 1982. He started his professional career at the Directorate Materiel Royal Netherlands Navy where he worked for nearly 25 years in areas related to naval ship design and project management. In October 2006 he started his academic career at DUT.

ANALYSIS OF HIGH-SPEED TRANS-PACIFIC NUCLEAR CONTAINERSHIP SERVICE

G. Sawyer, J. Shirley, J. Stroud & E. Bartlett, General Management Partners, LLC, USA
C. McKesson, PE, MRINA, Centre for Commercial Deployment of Transportation Technology, California State University, Long Beach, USA

SUMMARY

The vast increase in the long distance U.S.-Far East Container Trade, the ever-increasing cargo values involved in this trade, and the increased size and speed of the vessels that service this trade are well-established phenomena. In recent years these trends have been accompanied by the higher fuel costs and ever more stringent governmental regulations. These issues combined have given rise to the question of whether nuclear powered ships could be both technically feasible and economically competitive in such service.

Accordingly, under the sponsorship of the Center For Commercial Deployment of Transportation Technologies (CCDoTT) a study was initiated about two years ago to address this issue by a team of professionals with requisite skills and experience in the major technical and economic issues involved. The study team developed a conceptual design of a very high speed (35 knots) large (9200 TEU) nuclear powered containership applicable to the Trans-Pacific container trade of Asia (HK or equivalent Asian mainland port) to U.S. (LA/LB or equivalent western port). This ship would be powered by a single Pressurized Water Reactor (PWR), supplemented with auxiliary and emergency diesel prime movers integrated in an all-electric power/propulsion system. Since there are currently no standards for design and deployment of commercial nuclear ships, a highly conservative approach was taken in the ship and plant design. Moreover, the design was based entirely on current state of the art commercial nuclear technology. Cost estimates were subsequently developed excluding non-recurring costs using an average of the 3rd to 10th vessels constructed and finally analyzed in a hypothetical three-ship express service making weekly calls between the two ports. This reference hypothetical service was compared with a four-ship 25-knot conventional service employing the same sized vessels employing current diesel technology.

The results have shown that under certain assumptions, the conceptual nuclear containership service would be economically viable with a crossover point compared to the diesel service at basic oil costs of about $89 per barrel. Should an extended service to the U.S. East Coast prove feasible, the economic feasibility of the concept would become even more persuasive. The authors opine that the study has demonstrated enough potential economic feasibility to motivate the execution of a preliminary design along with better definition for budgetary estimating and serious evaluation by cognizant regulatory agencies.

Figure 1 - Nuclear Ship SAVANNAH

NOMENCLATURE

CCDOTT	The Center for Commercial Deployment of Transportation Technology, of the California State University at Long Beach
MWt	MegaWatts - Thermal
MWe	MegaWatts – Electrical
NPV	Net Present Value
PWR	Pressurized Water Reactor

1. INTRODUCTION

1.1 PROLOGUE

It has been more than 36 years since the world's first nuclear merchant ship prototype, *NS Savannah*, was laid to rest after nearly 10 years of successful international operations and over 450,000 miles steamed with one partial core refueling (conducted after almost 2.5 years of equivalent full power output). *Savannah* employed a Pressurized Water Reactor with a conventional propulsion plant based on the just-proven technology of her era. Table 1 summarizes the principal characteristics of the vessel. As shown in Figure 1, the ship herself was a graceful design, but incorporated the modest cargo handling characteristics and other limitations of her era, including provisions for carriage of up to 60 passengers. Nevertheless, during her period of service in trade the vessel, sailing under charter to two US Flag fleet operators and manned by American merchant mariners, amply demonstrated the safety of nuclear ship operations through 77 port calls, including 45 visits to 29 countries, without significant incident. However, as could be expected of a lone prototype demonstration of an advanced propulsion technology in an era of relatively slow break-bulk service and plentiful alternate supplies of low cost conventional energy, she also proved to be uneconomic in service; and when her demonstration mission was completed, the ship was dutifully retired.

Since that time, little serious thought has been given to commercial nuclear ship applications in the United States other than a small number of conceptual studies of very high speed ship applications using unproven, hypothetical high temperature gas-cooled reactor (HTGR) technology. Because of the very high performance required and the weight and size limitations imposed by the particular ship designs employed in those studies, application of proven PWR technology was considered to be both technically and economically infeasible because of its lower specific performance characteristics (pounds and cubic feet per horsepower). Internationally, the same theme has evolved, with the sole exception of Russia where significant interest and activity has continued due to that nation's dependency on ice capable vessels employing PWR technology to maintain its vital northern sea routes in all seasons.

Since the era of *Savannah*, all aspects of commercial maritime commerce have undergone a considerable revolution, including vastly increased trade volumes accompanying the growth attendant to economic globalization. This growth for the U.S. in just the six years 2000 through 2005 is shown graphically in Figure 2 and the recent phenomenal growth trend in U.S./China trade statistics is depicted in Table 2.

This enormous growth inevitably will continue to increase in the years ahead. In meeting this challenge, the industry has responded with many impressive technological advances – particularly the high speed containership, advanced cargo handling methods facilitating rapid port turnarounds, reduced crew manning through automation, more powerful and efficient diesel powering, and more recently a pronounced increase in the size and speed of ships dedicated to international service.

Table 1 - NS Savannah Principal Characteristics

Length Overall	596.5 feet (180 m)
Beam	78 feet (23.8 m)
Draft Loaded	29.5 feet
Displacement	21,800 long tons
Mariner Complement	68, including 4 special nuclear technicians
Max. Speed	~23 knots
Power	Reactor 74 MWt; Propulsion 20,300 shp
Deadweight Capacity	~10,000 long tons
No. Holds	7
Shipbuilder	New York Shipbuilding
Reactor Type & Manufacturer	Pressurized Water; Babcock & Wilcox
Date Launched	July 21, 1959
Date Delivered	May 1, 1962
Date Deactivated	November 1971
Cost	$46.9 Million ($18.6 M ship $28.3 M nuclear plant)

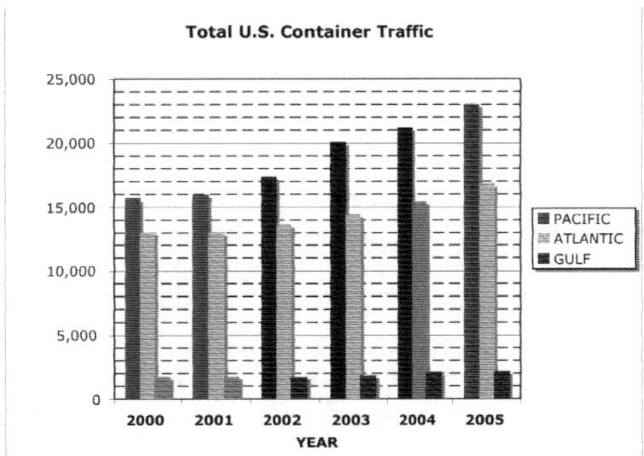

Figure 2 - Total US Container Traffic (thousands of container moves), showing a 45% increase in five years

Table 2 - Selected U.S.-China Trade Growth Statistics

Criterion	Statistic
30 Year Growth in China's Global Trade	Tenfold to the World's 3rd Largest Trading Economy
5 Year Average Annual Economic Growth of China	9.5%
5 Year Percent Growth by China of Total Global Trade	From <4% to >12%
5 Year Annual Waterborne Container Exports to U.S.A.	From 2.8 Million TEUs to & 7.4 Million/year.
5 Year Annual Waterborne Container Imports From USA	From 627 K TEUs to 1.65 Million/year.
5 Year Growth in U.S.- China Trade Deficit	From ~$80 Billion to $202 Billion
2000-2004 Growth in High Technology Exports to U.S.	From $10.7 Billion to $45.7 Billion

As a result of the enormous growth in the international container trades coupled with dramatic technology advances in prime movers and automation, the size and speed of individual ships engaged in this service have grown dramatically. This trend is well evidenced by the *Emma Maersk*, the first in a line of seven vessels being built by AP Moeller whose known characteristics are summarized in Table 3. The dimensions and power requirements of the *Emma Maersk* class leave little doubt even to a casual observer that a state-of-the-art PWR power plant might well be technically feasible in place of its advanced diesel propulsion system.

In the case of container services, the economic and technological trends have been extremely beneficial in improving operational efficiency and individual ship productivity, However, other rather recent developments have produced a counter effect on the cost of operations—namely, the alarming increase in the cost of basic marine fuel and the added regulatory constraints associated with environmental quality improvement. Of these two, the cost of energy has been of particular concern to containership operators since it tends to directly counterweigh the productivity enhancement associated with higher transit speed. Graphically illustrating this energy cost factor are Figure 3 illustrates the dramatic three year cost escalation for standard ship bunkers averaged at two major international ports of call on the Asia-U.S. trade.

The economic characteristics of nuclear power are different from those of fossil fuel power. The capital cost of a nuclear installation is high, but the fuel cost is low. In consequence, a nuclear facility should be worked hard, to maximize revenue for capital cost payback, with less concern for saving energy. It follows that the optimum economic speed may be higher for nuclear propulsion than for Diesel.

In addition to energy cost growth, the past two decades have seen a parallel increase in environmental awareness by all nations with evolving regulations impacting maritime operations worldwide. The result, particularly in the past five years, has been a marked and continually increased financial burden on maritime operations by furthering energy cost escalation and imposing operational restrictions. Table 4 summarizes the constraints that were to be imposed by the Board of Harbor Commissioners on ships calling at the Ports of Los Angeles and Long Beach. Although this imposition was only very recently over-ruled by Federal Court as usurping the constitutional powers of the U.S. Federal Government, it is likely that these or similar restrictions will eventually be imposed in California, and possibly, even extended to other jurisdictions in the United States through subsequent action by the U.S. Environmental Protection Administration.

The same environmental and energy impacts noted above have resurrected serious interest in using nuclear energy for land-based central-station power in a number of countries including the United States, where several serious projects have begun following over 30 years of relative inactivity. The proven safety record of the PWR technology for electric power generation in the United States and other developed nations together with the newly recognized phenomenon of Global Warming have added impetus to this recent activity.

Based on the above developments, a small study was initiated over two years ago to investigate the commercial feasibility of applying proven commercial PWR technology to a highly selective international maritime trade involving trans-pacific U.S.-Far East express service. This study was conducted by a small team of professionals with appropriate backgrounds in ship design, PWR propulsion technology, commercial maritime operations, and uranium fuel cycle economics under the sponsorship of CCDoTT – see the list of firms and key individuals shown in the Acknowledgements.

1.2 STUDY ASSUMPTIONS

The principal assumptions applied throughout the analysis are listed in Table 5. These assumptions have been validated by direct discussions with several Port Authorities and shipping company executives, and are appropriate for a hypothetical service beginning ten to twelve years from now.

Table 3 - Approximate EMMA Maersk Principal Characteristics

Length Overall	397 Meters (1,302 Feet)
Beam	56 Meters (184 Feet)
Projected Container Capacity	>11,000 TEU Stated (>13,000 Estimated)
Speed	25.5 Knots
Powering Type & Horsepower	Slow Speed Diesel 110,000hp

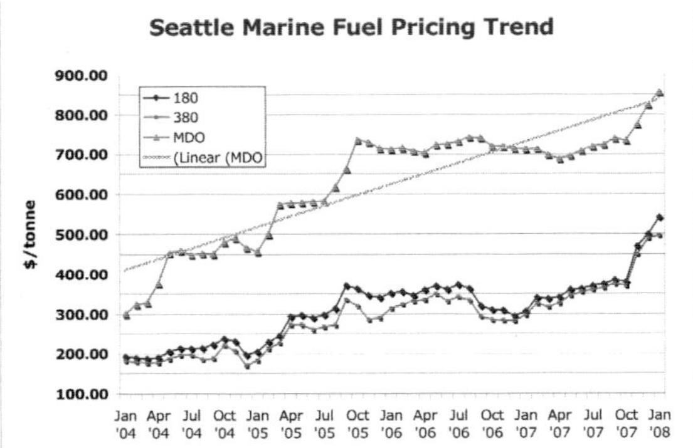

Figure 3 - US Pacific Coast Fuel Price Trend

Table 4 - Summary of Proposed Restrictions by California on Maritime Operations Recently Overturned By Federal Court Action

Restriction	Compliance Date
Vessel Speed Reduction to 12 Knots Currently 20 nautical miles—› 40 n.mi.	2008
Use of .2% Sulfur Marine Gas Oil Currently 20 nautical miles—› 40 n.mi.	2008
Use of Alternative Maritime Power for all Port of Los Angeles Terminals	2012
Use of Alternative Maritime Power for all Container and 1 Crude Oil Terminal at Port of Long Beach	2017
Use of Diesel Particulate & NOx Control Devices on all Newbuild Ships and Frequent Callers	2008

Table 5a - Principal Basic Assumptions Used in The Initial Study

Criterion	Assumption
Time Line For Initial Service	10 to 12 years
Reactor Technology Applied	State-of the Art Commercial PWR
Regulatory Agency Cognizance	USDOE- Nuclear Regulatory Commission (NRC), USDHS- US Coast Guard (USCG), USDOT-Maritime Administration (MARAD), International Maritime Organization (IMO)
Ship Type	State-of-the Art Large (≥ 9000 TEU) Containership
Ship Speed	35 Knots Sustained on Nuclear Power Only 15 Knots on near-shore diesel power
Hypothesized Service	Trans-Pacific, Express Shuttle, Two Ports
Security	Similar to LNG Service, Plus Added On Board Security Detail

Premium Service	LI-FO for 30% of U.S. Incoming Cargoes
Capital Costs	Average 3rd Thru 10th Vessel; Non Recurring Costs Excluded
Environmental Constraints	Fossil Fuel- Marpol Annex VI-Blend of <1% Sulfur Residual Fuel & MGO Nuclear Fuel- Current Market Costs of Low Enriched (~5%) UO2
Refueling Interval	Every 5 years; 35 Day Outage Period; Off-Load 'Cool" Spent Fuel From Special Storage Vessel On-Board; Surcharge Imposed to Maintain Service During Outage

Table 5b - Principal Alternatives Developed For Subsequent Investigation

Improved Security & Safety	Eliminate On-Board Spent Fuel Carriage & Refuel Directly From Core to Shore; Extend Refueling Outage to 40 Days
Refueling Interval	Extend to 7 Years Between Refuelings

2. SUMMARY TECHNICAL DESCRIPTION

2.1 THE SERVICE

The reference service selected for the initial study is a hypothetical non-stop Trans-Pacific express trade between a major port in China (HK) to California (LA/LB) using three 9000 plus TEU containerships calling at each port on a weekly service. The transits are made non-stop at 35 knots sustained speed over a 6,460 mile distance shuttling between the two ports. Thirty percent of the total container loads are loaded and discharged on a last-in first-out basis to provide minimal dead time in on and off-loading for which a $600 premium is paid above current per 40 foot box (FEU) rates. This weekly three ship high speed nuclear ship express service (10 days on-dock to on-dock transit time) equates to a four ship conventionally powered fleet of equivalent size and capacity transiting at 25 knots to the same ports (13.5 days on-dock to on-dock transit days). As noted earlier, this service is forecast to commence at least 10 to 12 years in the future. The reference operational protocol for the study assumes that port entry and egress will be conducted at low speed and probably under Coast Guard escort from pier to sea buoy. Maximum security of the ships and their cargoes is assumed, with a small security team riding each vessel during the transit. Under this very high utilization, the initial analysis concluded that refueling of the nuclear reactors must be conducted at slightly above five year intervals, with the refueling outage for each vessel consisting of 35 days at a nuclear capable shipyard employing the ship's on-board refueling system. The refueling process is similar to current nuclear central station capability and practice, where the evolution is routinely accomplished in under three weeks. A considerable economic penalty was included in the study analyses to account for both maintaining the continuity of service and the significant direct costs involved in these refueling outages.

Subsequent discussions with nuclear experts in the United States suggested that, for enhanced safety and security, the reference refueling interval and method should be amended to eliminate the long-term on-board storage of spent fuel and use direct refueling from the reactor itself to a spent fuel storage repository ashore, while at the same time increasing the design core lifetime from the conservative assumption used in the initial study from five years to a still feasible 7 years. These suggestions represent an alternative and slightly more aggressive approach to development of the power plant, and thus yield a net favorable impact to the economics derived in the study as well as allowing about a 23 foot reduction in the height of the reactor containment vessel itself. It may thus be considered that these two sets of assumptions effectively representing bounding values, and that final optima will lie somewhere within the range thus determined.

2.2 THE SHIP

The reference ship used in the study is a variant of an existing late generation containership, namely the *OOCL Shenzhen* shown in Figure 4. For the purposes of this study the basic lines of the *Shenzhen* were used, but have been modified to encompass two additional shafts (to a total of three) in order to accommodate the enormous increase in powering required to achieve the higher transit speed (35 knots). The ship was also lengthened 42 meters (172 feet) to 365m (1,198 feet) overall in order to better accommodate the increased powering required. This lengthening results in a 4 knot improvement in the speed of the vessel at its design horsepower and, because of the total weight saved by omitting approximately 8900 tons net of fuel, permits the load-out of an additional 1000+ 40 foot containers. Simplified inboard and outboard profile drawings of the as-modified vessel are depicted in Figure 5 below. Additional allowance for collision and inner-bottom structural protection is provided for the reactor plant consistent with *Savannah* practice and reasonable expectations of future Governmental regulatory requirements and Class rules. In addition, the entire nuclear plant is contained within a pressure containment vessel to absorb the maximum heat and pressure energy that could be released in a hypothetical massive loss of coolant accident. Special shielding of structure, lead, water, and neutron absorbing material is provided to ensure compliance with current nuclear regulatory body standards for personnel exposure outside the reactor plant to radiation under conditions of continual occupancy.

Excluding the five person security detail carried, the ship is assumed to be manned by a crew of 25 predominantly licensed mariners of whom 10 would possess special qualifications for nuclear operations and maintenance. Upward adjustments were made in the compensation of the nuclear qualified mariners and an economic penalty was assigned for their six months of individualized special training towards qualification.

Figure 4 - Containership OOCL Shenzhen

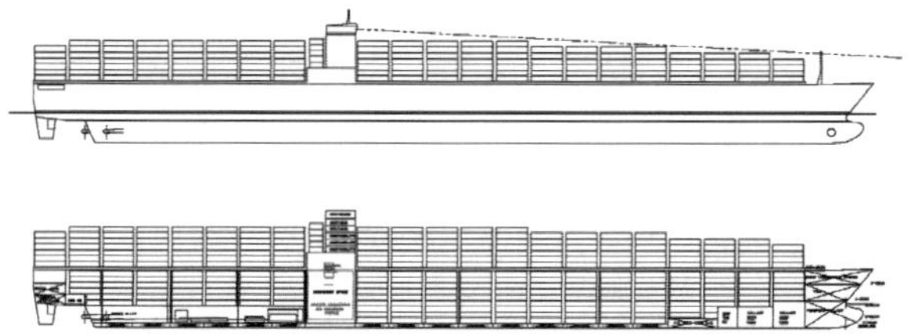

Figure 5 - Simplified Outline of Study Nuclear Containership Ship Concept

2.3 THE PROPULSION SYSTEM

2.3 (a) Reactor Plant

The ship is powered by an integrated nuclear and conventional propulsion and powering system consisting of a single state-of-the-art PWR utilizing Rolls-Royce provided *commercial* technology suitably modified for the ship motions, accelerations and transients expected of a high speed marine application. Figures 6, 7 and 8 depict the arrangement of the nuclear reactor system within its containment including the optional permanently carried refueling system similar to that employed on current land based nuclear systems. As noted above, although this arrangement was intended to simplify the refueling evolution itself, it has raised nuclear safety and security concerns and later design study will probably result in its elimination in favor of extending core lifetime and using a state-of-the-art direct refueling method. Selection of the final refueling method to be employed can await a full-scale preliminary design and review by cognizant regulatory authorities

As noted earlier the reactor core in the initial study was designed for a 5+ year life at the extremely high load factors developed for this specific service. Refueling is conducted on a 'batch' (all-out, all-in) basis. In the initial design concept spent fuel was to be removed from the reactor core under water and deposited in a special spent fuel repository located within the containment before the new virgin fuel was brought in and loaded from the refueling depot. With this approach, the radiation level of the spent fuel would have decayed to a level permitting its 'dry' transfer ashore before the then newly spent fuel from the reactor would be transferred under water to this repository.

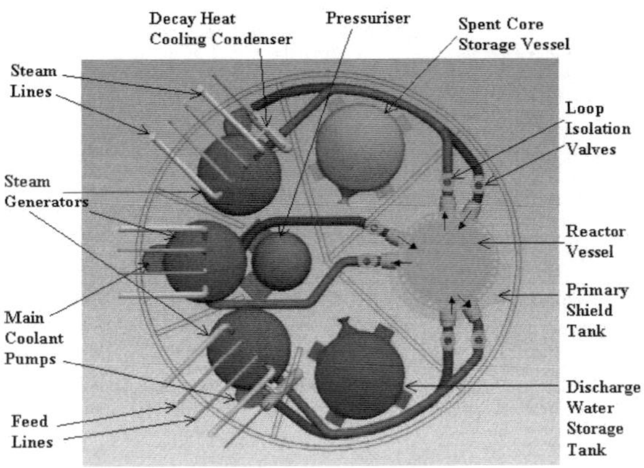

Figure 6 - Plan View of Reactor System Concept. The Spent Core Storage Vessel (top) is optional, and can be eliminated if spent fuel is not to be stored within the plant.

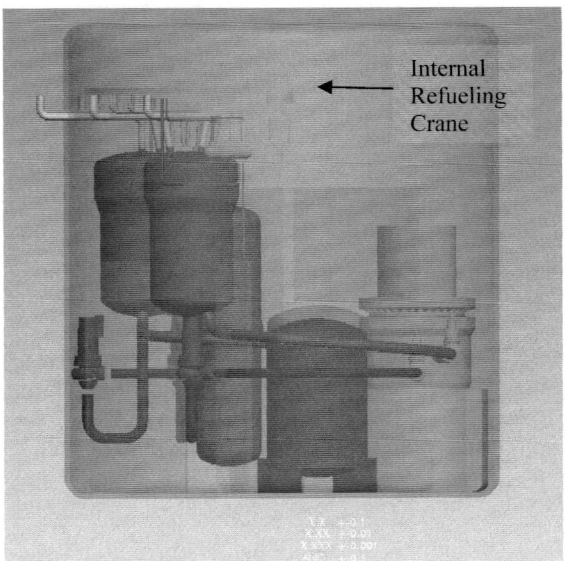

Figure 7 - Elevation View of Reactor System Concept. System is 22m tall if spent fuel is stored internally. If Direct Refueling is employed then the Internal Refueling Crane is eliminated and the system becomes 15m tall.

The reactor system also has an independent emergency decay heat removal system that minimizes the requirement for electrical power with the atmosphere providing the ultimate heat sink for the design base accident.

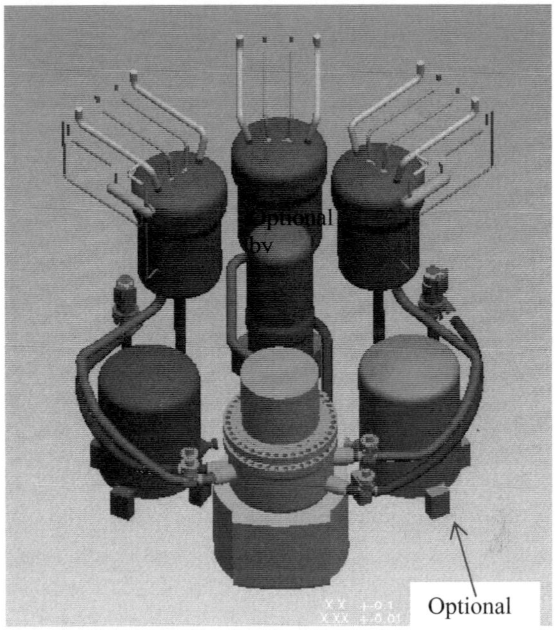

Figure 8 - Isometric View of Concept Reactor System Arrangement. The option Spent Core Storage Vessel is indicated.

2.3 (b) Integrated Propulsion & Powering Plant

As depicted in Figure 10, the propulsion-powering system used in the study assumes an all-electric system consisting of an integrated mix of primary nuclear, auxiliary diesel, and emergency diesel or battery powered generators all interconnected on a dual 4160 volt bus. The propulsion motor concept used in the study is the permanent magnet motor currently under development and full-scale demonstration by the U.S. Navy, but within the timeframe of the study, other types of high-energy power dense motor systems will likely be available. As the over-all one-line diagram of this system in Figure 9 depicts, the total system is highly flexible and facilitates both assurance of mobility under the most adverse conditions - the auxiliary diesels can drive the ship at 15 knots with the nuclear plant shut down - and enhancement of safety in the provision of multiple alternate and emergency sources of power to maintain vital loads during any emergency condition at sea.

The specific system selected is considered to be entirely feasible, particularly recognizing that the timeframe of reference is 10 years in the future. However, other propulsion and powering concepts could also be seriously considered - particularly the so-called 'hybrid' system that is already employed on a number of high-powered commercial and naval ships throughout the world. Final selection should be based on economic grounds, maintaining equivalency in safety and reliability, and left to the preliminary design stage and a better understanding of the precise rules that will be applied by cognizant regulatory agencies in the future.

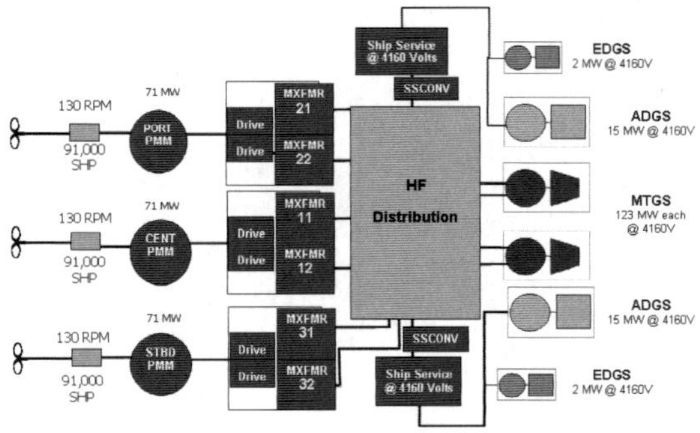

Figure 9 - Simplified Schematic of Integrated Power / Propulsion System

2.4 Ship and Propulsion Plant Summary

Table 6 provides a summary of the principal design features and characteristics of the ship and power plant concept used in the study.

The reader will be reminded that the major—in fact only—purpose in developing this complete conceptual design of the ship and its powering system was to create a credible basis for subsequent economic evaluation. It is the conviction of the authors that the accompanying economic analysis must be founded on professional principles and realistic extrapolation of current state-of-the-art nuclear technology available from a number of sources both here in the United States and elsewhere and within a relatively short 'look-ahead' of only ten to twelve years.

3 ECONOMIC RESULTS

The economic analysis consisted of three components: (1) a cost estimate, (2) the reduction of cost to a Net Present Value, and finally (3) a determination of the break-even price of petroleum for parity between diesel and nuclear options.

3.1 COST BREAKDOWN & COMPARISON

The cost breakdown is presented in Table 7. This cost estimate is necessarily based on one assumed value for diesel fuel price (US$455 / tonne blended average), and various assumed values for insurance, drydocking, and similar items. (It is interesting to note that a subsequent series of discussions suggests that the initial assumption for insurance costs is high, and that real rates would be slightly lower – but that is a subject for a later report.)

Note also that the cost analysis was only completed for the baseline option having a five year refueling cycle. The seven-year cycle option was not priced.

All economic figures are in 2007 US Dollars.

Table 6 - Summary of Principal Technical Characteristics

SHIP		
Length Over All	feet	1,198
Beam	feet	141
Draft	feet	43
Displacement	tonnes	121,100
Container Capacity	TEU	9,200
Reefer Plugs	No.	700
Speed Sustained Transit- Nuclear Only	Knots	35
Speed on Auxiliary Diesels Only	Knots	15
Port Turnaround Time	hours.	33
PROPULSION & POWER		
Propulsion Power	SHP	273,000
Number of Shafts	No.	3 (Equal P)
Total Installed Power	MWe	280
Nuclear Thermal Power	MWt	1,000
Auxiliary Power	MWe	30
Emergency Power	MWe	4
Refueling Interval	years	5.2
Refueling Outage Duration	Days	35 to 40
Conventional Fuel Carried	tonnes	1,000
Propulsion/Power Plant Weight	tonnes	3,724 to 4,501*

* Dependent on refueling method finally selected

3.2 NET PRESENT VALUE (NPV) COMPARISON

At current conventional marine fuel prices and assuming that large ships will be required to burn low sulphur marine diesel within 40 miles of shore, the Net Present Value @ 10%, for the conventional fleet is $259 million while the NPV of the base case nuclear fleet is $10 million. This gap of course depends upon the assumptions made in the cost estimate, including but not limited to the price of petroleum fuel. This gap is not too large to overcome and after analyzing some of the largest variables, the authors estimate that a long distance high-speed commercial nuclear service of the type presented will become viable within the ten to fifteen year planning horizon of this study if the trend in fossil fuel costs and availability continues.

Additionally, the reference fleet economic analysis assumed an express service between an Asian mainland port and a U.S. Pacific port. Subsequent discussions with major Port Authorities on both the East and West Coasts of the USA indicated significant interest and seemingly ready acceptance for a possible direct service from Asia to and from the East Coast via the enlarged Panama Canal whose improvements will be completed well before the 10 to 12 year time frame in commencement of this hypothetical service. Although intriguing as to its possibilities, extensive analysis has not been performed to date regarding the comparative economics of such a service, since it would require considerable further study involving both competing conventional containership and expanded rail 'land bridge' alternatives.

3.3 BREAK-EVEN CONVENTIONAL FUEL COST COMPARISON

As mentioned above, the economic comparison between diesel and nuclear fleets is strongly dependent upon diesel fuel price. If we assume that over time the cost of fossil fuels will continue to increase, the fuel surcharges already widely applied by ship owners worldwide will continue to increase top line revenue for both fleets of ships. However, since the nuclear ship will incur very little additional fuel related expense its bottom line will improve significantly and narrow the predicted NPV gap. Based upon the cost criteria developed above and modeled econometrically by Manalytics and GMP, it is estimated that on the basis of oil pricing alone the nuclear service could be economically equivalent to the conventional service with future bunker fuel prices of $455 per metric ton (or $89 per bbl) and MDO prices of $890 per metric ton, along with the requirement to burn MDO within 40 miles of shore. This results in a net blended fuel price of $585.50 as compared to the $455/tonne used initially. See Figure 10 below.

As was noted several times in previous sections, there are other forces at work beyond market supply and demand contributing to the price of marine fuels. California is already regulating large ships and the burning of residual fuels due to gas emission (NOx and SOx) and has enacted legislation to require ships to burn cleaner Marine Diesel Oil (MDO) low in sulphur content within 20 nautical miles of its coastline and by 2008 this requirement will be extended to 40 nautical miles. Should bunker fuels be prohibited entirely in the future, the result would have a dramatic effect on fuel costs. In this case of a complete ban on bunker fuels, fuel costs would double overnight making the nuclear containership service economically superior immediately with a NPV of $780 million compared to $259 million for the conventional service.

Further, current cost analyses do not make any provision for the cost of climate change. This is a new subject and is not uniformly implemented. The authors have seen discussions that suggest that the dollarized cost of carbon emissions lies in the neighborhood of US$100 / tonne-CO2. (We have seen figures ranging from US$50 to US$250/t.) Based on $100/t, this means that the cost of burning one barrel of petroleum should be increased by about US$40, to account for the cost of the carbon impact. Some jurisdictions are talking about handling this explicitly, in the form of a carbon tax. In other jurisdictions it is being ignored.

If this cost were indeed applied to the cost of petroleum, then it would mean that oil is today trading at about US$100+40 = $140/bbl – well above the break even point for the nuclear alternative.

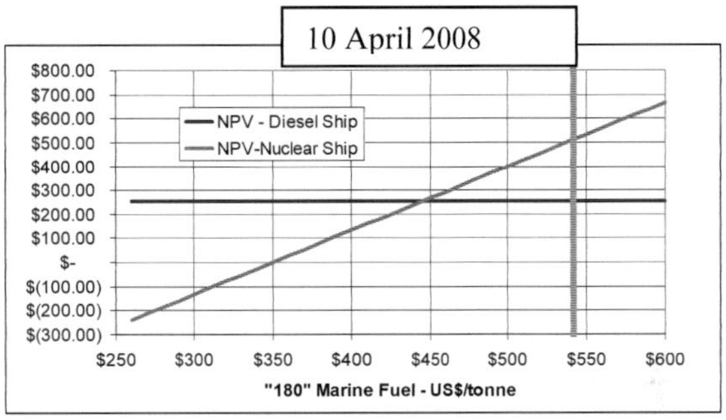

Figure 10 - Nuclear Containership 25 Year NPV Sensitivity to Marine Fuel Prices

Table 7 - General Economic Criteria and Assumptions

Revenue and Cost	Nuclear Vessel	Non-Nuclear Vessel
Ship Capital Cost	$722 million +$113M initial Core	$150 million
Ship Years of Life	25 years	25 years
# of Vessels Required	3 (with 13% contingency)	4 (with 12% contingency)
Operating Speed	35 knots	25 knots
Load /Discharge	300 moves/hour	300 moves/ hour
Maintenance & Repair	$1.5 million per year/vessel $2 million every 5 years for turbine overhaul & reactor maint./ vessel	$800,000 per year/vessel
Dry-docking /Class Inspections	$1.25 million every 2.5 yrs./vessel $1.75 million every 5 yrs./vessel	$1.25 M every 2.5 yrs./vessel $1.75 M every 5 yrs./ vessel

Table 8 - General Economic Criteria and Assumptions (continued)

Fossil Fuel @ blended rate of $455 / tonne	Burn rate: 17 tonnes / day at sea / vessel =$2.1 million per year /vessel =$6.3 million per year per fleet	Burn rate: 263 tonnes/day at sea / vessel =$33.6 million per year / vessel =$134 million per year / fleet
Oil	$.1 million year/vessel	$6.6 million year/vessel
Nuclear Refueling	$113 million to replace Reactor Core every 5 years *Note: Alternative of 7 year refueling interval not analyzed* $2 million for core removal & replacement *Note: Direct refueling alternative and additional spent fuel storage cost not analyzed*	N/A
Spent Fuel Disposal	$3 million beginning 3rd refueling	N/A
Manning	$16,000 per day/vessel (25 crew) including premium pay	$10,000 per day/vessel (19 crew)
Mgt/General/ Admin	$1,800 per nominal TEU capacity per year	$1,800 per nominal TEU capacity per year
Security	$3 million per year/Fleet	None
Insurance	1% of value of the vessel per year H&M+ $350K PI	$1 million per year/vessel
Scrap Value	$20 million steel and copper/vessel	$10 million steel only/vessel
Decommis-sioning	$20 million	None
Slot Charter Expense Re-fueling /Dry-docking	$71.1 million per 5 year period *Note: 7 year interval not analyzed*	$19.5 million per 5 year period

4 CONCLUSIONS

We believe that the above analysis is realistic and conservative. This analysis shows that the commercial nuclear containership is both technically and economically feasible.

As oil becomes more expensive, it seems inevitable that nuclear power will become competitive for commercial marine propulsion – initially for very large, fast ships as described here and subsequently for medium-size ships of moderate speed.

This conclusion does hinge upon the assumption that the major non-recurring costs and economic risks involved in re-starting a commercial nuclear ship program would be absorbed by the U.S. Government and that such an express service is deemed to have commercial merit by the industry.

Further, we believe that this application has potential national security benefits as well since any one of the vessels in this fleet could provide emergency electrical power to a significant fraction of a city's demand grid—and certainly to the industrial complex surrounding one of our major ports.

Further study of alternatives may well improve on the economic results depicted—for example:

- Extending core lifetimes and refueling intervals to 7 years vs. the 5 years of the reference design, as deemed appropriate by several nuclear experts who have read the full report from which this paper has been excerpted,
- Investigating the economic feasibility in enlarging the reference design ship to ≥12,000 TEU capacity vs. 9,200; (There is no inherent technical issue associated in such an enlargement).
- Examining the economic potential and comparative fleet sizing of a possible extended express service to the East Coast of the United States;
- Studying the potential economics involved should one of the carbon tax or trade concepts now being explored for CO2 abatement become a matter of law or regulation.

In summary, a large, fast commercial nuclear ship is technically feasible today using proven PWR technology that is both currently available and in wide service. The principal issue that now confronts the authors and sponsors of this study is not to engage in further paper studies per se, but to find an appropriate sponsor in industry and a capable, motivated agency within Government who together are willing to provide leadership and funding to get on with the hard work of preliminary design of such a vessel and develop the engineering details necessary to permit serious investigation by cognizant regulatory agencies within Government and Class. We estimate that such an effort would entail about two years of effort and cost approximately US$5 Million. It's scope would include the ship and propulsion systems design itself, appropriate review and analysis by Class, evaluation of the alternatives affecting principal characteristics of the vessel and service as described above, a budgetary estimate of both non-recurring and recurring costs, and submission of an engineered product in sufficient detail to permit a preliminary review by the Nuclear Regulatory Commission.

6. ACKNOWLEDGEMENTS

The authors thank CCDOTT for sponsoring this study, and the Institution for the opportunity to bring this to the industry, and the following individuals for their participation:

Nuclear Reactor Plant Design
Rolls Royce Nuclear
Patrick Marolda
Richard Swinburn
Stuart Holiday
Anthony Donaldson
Andrew Goldfinch
Michael Dolman
Robert Shooter
Samual Postans
Richard Partridge
Alan Copestake

Electric Propulsion Technology
DRS Technologies, Inc.
Edward Bartlett

Ship Design/Conversion
Alion/JJMA
Thomas Young
Peter Fleming
Steve Toby
George Knight (JJMA- Ret.)

Economics
Manalytics International
Douglas Coates
Carin Saunders

Nuclear Fueling/Disposal
United States Enrichment Corp.
John Welch
Russell Starkey
Mick Mastilovic

Ship's Manning & Voyage Information
McMunn Associates, Inc
Lawrence Cosgriff

8. AUTHORS' BIOGRAPHIES

Principal Author:

G. A. Sawyer is a founding partner of J.F. Lehman & Company and since January 2004 serves as Executive Advisor to J.F. Lehman & Company. In addition he serves as a Director of J.F.Lehman's portfolio companies. From 1993 to 1995 he served as President and Chief Executive Officer of Sperry Marine. Prior thereto, Mr. Sawyer held a number of prominent positions in private industry and the U.S. Government, serving as President of John J. McMullen Associates, President and Chief Operating Officer of TRE Corporation, Executive Vice President of General Dynamics Corporation, Vice President of International Operations for Bechtel Corporation and Assistant Secretary of the Navy for Shipbuilding and Logistics under Dr. Lehman. He graduated Phi Beta Kappa from Yale University and Completed graduate studies in nuclear engineering at the Knolls Atomic Power Laboratories. He is also the co-inventor of the Consolidated Nuclear Steam Generator II and served in the U.S. Navy for ten years as a nuclear submariner. Mr. Sawyer currently serves on the Board of Trustees of Alexandria INOVA Hospital and is a director of Atlantic Marine, Blacklight Power Inc., CHI Systems Inc., Hawaii Superferry and OAO Technology Solutions. Mr. Sawyer is also a managing member of Sawyer-Weems, LLC and General Management Partners, LLC.

Presenter: Chris B. McKesson PE, MRINA is an independent consulting naval architect who serves as a Subject Matter Expert for unconventional naval architecture for the Center for Commercial Deployment of Transportation Technology of California State University at Long Beach. He is also an Adjunct Instructor at the University of New Orleans School of Naval Architecture.

HYDRODYNAMIC ASPECTS OF CONTAINERSHIP PROPULSION - CHALLENGES AND SOLUTIONS

S. Whitworth, A. Boorsma, P. Fitzsimmons and **D. Radosavljevic**, Lloyd's Register – Technical Investigations (TID), UK

SUMMARY

Lloyd's Register's TID has been involved in a number of vibration and noise related habitability problems on feeder containerships. Propeller excitations play a major part in these vibration characteristics of container vessels and therefore it is important that radiated hull pressures are predicted accurately at the design stage.

Propeller observations on a feeder container ship clearly showed the cavitation phenomena that were present, namely sheet cavitation and tip vortex cavitation. Observations with and without vortex generators stressed the important effect of the propeller inflow on propeller cavitation. Furthermore, these observations demonstrated that in-service propeller cavitation can be examined, with little effort, through TID's borescope system.

Shortcomings in the traditional way of providing propeller inflows have been successfully addressed with Computational Fluid Dynamics (CFD) methods. Soon, these methods might render wake measurements at model scale obsolete and become part of routine computation procedures. Furthermore, CFD calculations can enhance our understanding of the flow and can highlight areas where caution is required when interpreting model test results.

1. INTRODUCTION

Recently, Lloyd's Register's TID has been involved in a number of vibration and noise related habitability problems on feeder containerships of 2000-3000 TEU capacity having all-aft deckhouses. The highest vibration levels measured on the bridge of one of these ships were of the order of 13 mm/s peak in vertical direction. These levels made it difficult to write down vibration readings and, possibly, for the crew to work. Indeed, they exceeded the maximum allowable level in the ISO6954-1984 standard, applicable at that time. Furthermore, they caused structural damage to the radar mast. Accompanying noise levels on the bridge were of the order of 70 dB(A) which is in excess of the IMO A.468 regulations for noise on board ships. Moreover, the comfort level in the crew and officers' accommodation was low. For the safe and sustainable operation of the vessel and the wellbeing of all on board, it would be beneficial to reduce vibration and noise levels to within acceptable limits.

The all-aft deckhouse on these ships was close to the main source of vibration excitation, i.e. the propeller, while the slender dimensions of the deckhouses reduced its resistance to vibration excitation. Spectral analysis of the vibration signal confirmed that propeller excitations, rather than main engine or other machinery excitations, were the source of the vibrations. For many vessels, a major contributor to vibration and noise problems is propeller excitation, originating from the propeller inflow and the resulting cavitation dynamics.

Lloyd's Register has been working actively on improving and validating methodologies for the prediction of propeller-radiated hull pressures. Validation material is provided by means of propeller observations through a borescope and hull pressure measurements at ship scale. Also, CFD methods are used to improve the prediction of propeller inflows, i.e. ship wakes.

These methodologies have been applied to an approximately 2000 TEU containership which suffered from high vibration and noise levels in the all-aft accommodation. Extensive pressure and vibration measurements were performed by Lloyd's Register-ODS on behalf of the builder, while the images of cavitation were obtained as part of ongoing research within the Cooperative Research Ship (CRS), a joint-industry research body. CFD calculations were performed at Lloyd's Register's initiative. To remedy the high vibration and noise levels, this ship was fitted with vortex generators to alter the inflow into the propeller causing a reduction in hull pressures radiated from the collapsing sheet and tip vortex cavitation on the propeller. As such, the focus of this paper is on ship scale cavitation-phenomena and the advances made in CFD methods to predict ship wakes. The influence of vortex generators is also addressed.

2. CAVITATION OBSERVATIONS

The exact levels of the vibration and noise in the all-aft deckhouse of the approximately 2000 TEU containership are not quantified here, however, they were sufficiently high to cause crew complaints and subsequent action from owner and builder. Hull pressure measurements, **Figure 1**, showed pressure energy was concentrated at the blade passing frequencies and its first three harmonics.

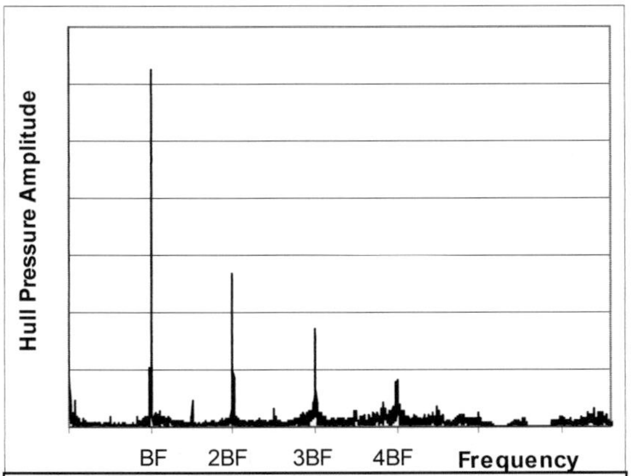

Figure 1, Measured Hull Pressure Spectrum at 85% MCR

Pressure pulses at the first and second blade passing frequency are generally ascribed to sheet cavitation and the ensuing observations showed that generous amounts of thick sheet cavitation grew and collapsed as each blade passed through the wake peak in the 12 o'clock position, **Figure 2**. The cavitation images were obtained using a low-light camera mounted on an industrial borescope.

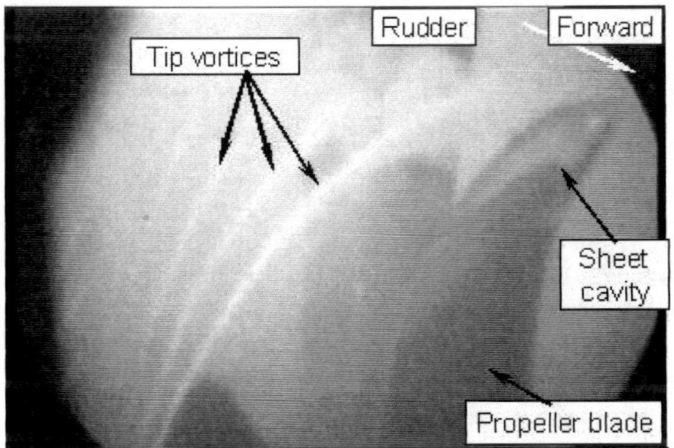

Figure 2, Cavitation without vortex generators, 85% MCR

This propeller operated in an inflow field that, at model scale, exhibited a strong and highly retarded axial wake peak between the 10 o'clock and 2 o'clock positions. Between the 10 o'clock and 12 o'clock position the propeller inflow angle increased, causing cavitation to grow. The subsequent decrease in inflow angle, between the 12 o'clock and 2 o'clock position, gave rise to a dynamic collapse of the cavitation volume. This process was possibly made more complicated and onerous due to the increasing tangential wake component in that region of the propeller disc. The radiated pressure field is directly related to the dynamics of the cavity volume variation and thus the strongest contribution to the hull pressure pulse is generated during the dynamic collapse phases of on-blade and off-blade cavitation volumes.

Figure 2 shows strong vortex cavitation emanating from the propeller tips. Tip vortex cavitation is, in general, associated with higher harmonics and broadband components of hull pressure spectra although the exact physics of the generation of these components by vortex cavitation are not yet fully understood. Broadband components are seen between the third and fourth harmonics in **Figure 1**. An in-plane view of the propeller was not obtained in the present case. However, interacting cavitating vortices have been observed, in the wake peak, on other container ships. **Figure 3**, from another vessel, shows interaction between adjacent tip vortices and, although no hull pressure measurements exist for this case, strong vibrations were experienced when this phenomena occurred. In other cases, the dynamics of the interacting vortices have produced strong cavitating vortices in the rotational direction of the tip vortex.

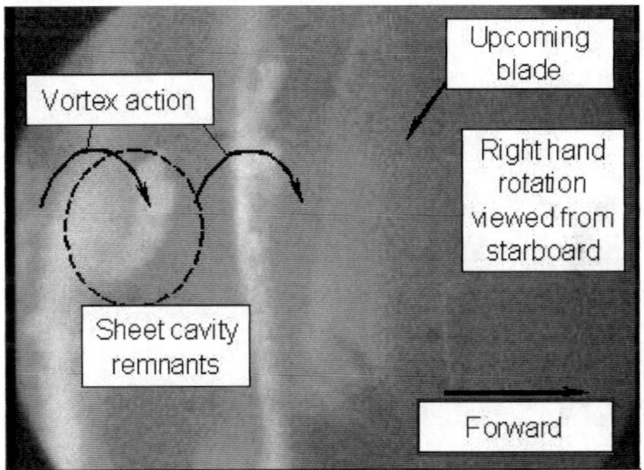

Figure 3, Interaction between sheet cavity and tip vortex

In the axially retarded flow in the wake peak, the cloudy remains of the collapsed sheet cavity are not transported aft fast enough, and are absorbed by the centrifugal flow of the vortex from the consecutive blade. As the clouds of cavitation from the first blade interact with the tip vortex from the second blade, their aggressive disintegration can give rise to large radiated pressures.

As a remedy against the large radiated pressure from the collapse of substantial sheet-cavitation volumes on the feeder ship, triangular fin-type vortex generators were fitted to the hull several propeller diameters upstream of the propeller plane. Such devices are placed sufficiently far ahead of any region of retarded hull flow to promote mixing of the low-energy boundary-layer flow with the more energetic free stream flow. This inhibits flow retardation in the streamwise direction thus reducing the axial velocity deficit in the wake peak. The present devices halved hull pressure amplitudes at blade passing frequencies as shown in **Table 1**.

P_{BF}	P_{2BF}	P_{3BF}	P_{4BF}
52	71	77	50

Table 1, % reduction of pressure with vortex generators

Subsequent video observations suggested cavitation volumes had decreased due to the fitting of vortex generators, **Figure 4**. However, video observations show only cavitation extents, and not the cavity volumes. Furthermore, the video observation were not synchronised to the pressure signals, thus making it difficult to relate observations to hull pressures. In subsequent studies, high speed borescope observations have been synchronised with time-series data.

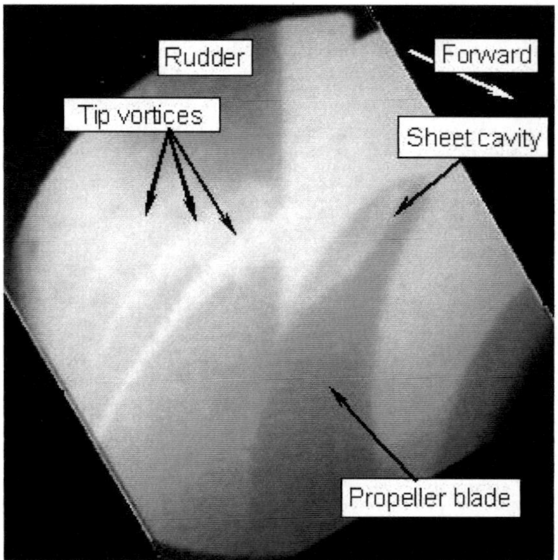

Figure 4, Sheet cavitation with vortex generators, 85% MCR

The above analysis shows borescope observations are a valuable tool in verifying propeller performance at ship scale and can provide essential information when diagnosing problems in service. Combined with the relatively small effort required, i.e. drilling and tapping of M20 holes in the shell, which can be performed afloat, this method becomes a viable tool to provide feedback for the propeller designer, in particular for first of class vessels.

3. PROPELLER INLFOW PREDICTION

As section 2 has shown, propeller performance is strongly dependent on the inflow provided. When propeller inflow is poor, i.e. there are large variations in propeller inflow angle, cavitation development can be rapid and could cause significant pressures. A good wake field, where propeller inflow angle variations are minimised, will reduce the dynamic growth and collapse of cavitation, with ensuing low hull pressure and vibration levels. Therefore, it is paramount that accurate estimation of propeller inflows is achieved at the design stage. For the past 15 years, TID has been actively engaged in studying wake fields with the use of CFD, resulting in some interesting insights to the developments of flows over the submerged afterbody.

3.1 CONVENTIONAL PREDICTION METHODS

The traditional way of generating propeller inflow wakes consists of measuring the wake at model scale, without an operating propeller. Although boundary layer flows are viscous dominated and should be scaled using the Reynolds number, this is not practicable; hence tests are performed based on Froude number-equality to ensure surface wave-patterns are accurately modelled. Measured wakes are then corrected for scale effects using empirical methods such as Tanaka's [**1**]. Furthermore, wakes should be corrected for the absence of the interaction between hull and propeller in the measurements, for instance with the force field method [**2**] developed by the CRS. **Figure 5** shows an example of this correction process.

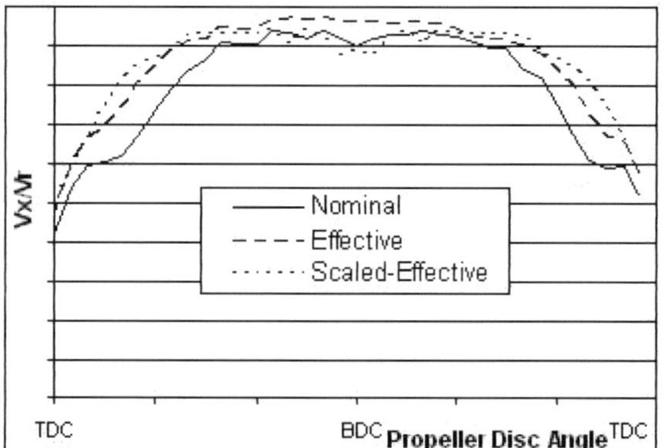

Figure 5, Nominal Wake Correction

Although such methods have, in the past, been used to provide hull pressure predictions, they have some shortcomings; the first being the scaling of the axial inflow with the use of empirical methods instead of first-principals based calculations. Secondly, only the axial wake is scaled; the transverse wake is not. Yet, tangential inflow components will influence the collapse of the cavitation on the down-going blade which can greatly influence hull pressure amplitudes. A third important component is the absence of the interaction between propeller and hull in these calculations which, as can be seen in **Figure 5**, is expected to significantly affect the propeller inflow. CFD methods can be used to overcome all three shortcomings, although only the first two will be addressed here.

3.1 CFD PREDICTION METHODS

CFD analysis was performed to predict the nominal wake of the feeder container ship travelling at 21.5kn with a draft of 11m. Calculations were performed at both ship scale and model scale. Comparisons are made between CFD results and model test data.

The CFD analyses were performed using STAR-CCM+, a commercial CFD solver based on the finite volume method. It was assumed that the flow around the ship was steady in time, and that the free surface was a flat, frictionless plane. Calculations have shown these to be reasonable assumptions for single screw ships that introduce negligible error into the results.

For all cases, the mesh was based on "trimmed-cell" topology with predominantly hexahedral cells and embedded refined mesh regions, mainly around the aft end. Its size varied between 1.2 million and 1.9 million cells depending on the particular case. The solution domain, with annotations indicating the boundaries, is shown in **Figure 6**.

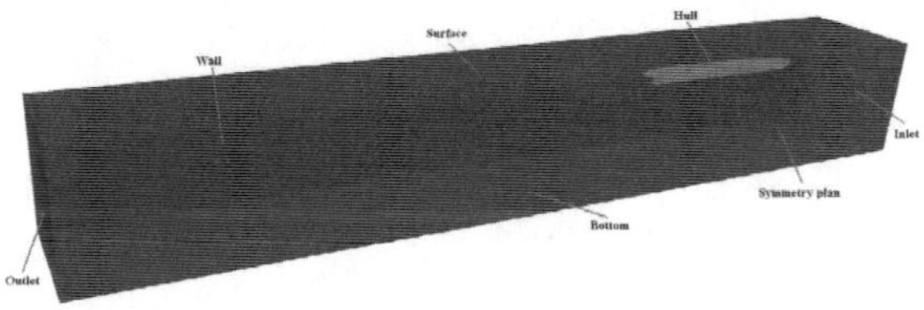

Figure 6. Geometry for CFD analysis

Several RANS turbulence models were employed and the results compared. The k-ω SST model is a 2-equation model, the use of which is well established in marine CFD. The Reynolds Stress model is a more complex, 6-equation model that is more sensitive to mesh quality and, for cases such as these, typically requires around 50% more computational time. The Menter model is also widely used and is a 1-equation model, which requires less computational effort than the other two.

3.1 (a) Axial Wake Results

Significant differences were observed between the results obtained using the k-ω SST model and those obtained using the Reynolds Stress model (RSM), as shown in the wake contour plots presented in **Figure 7**.

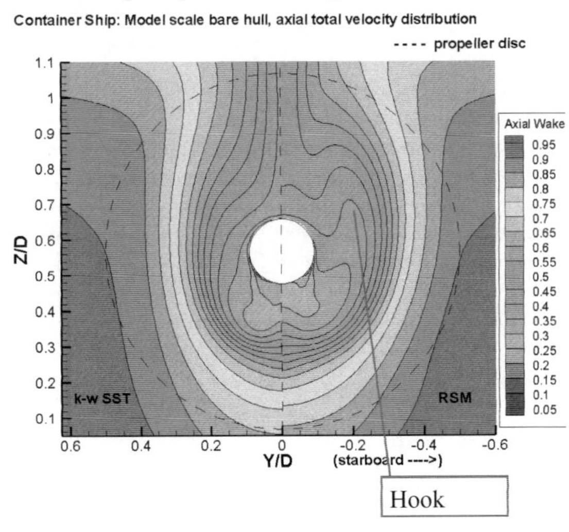

Figure 7. Axial wake profile at model scale comparing turbulence models

The result obtained using the k-ω SST model on the left of **Figure 7** fails to predict the 'hook' shape commonly observed in wake profiles. However, the result obtained using the Reynolds Stress model (RSM) is much closer to the model test results, **Figure 8**.

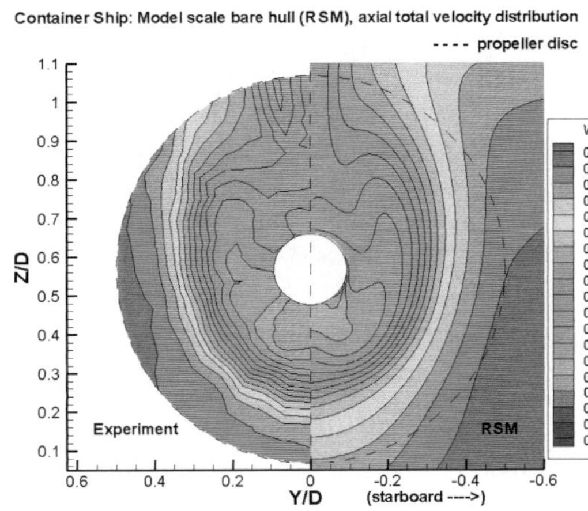

Figure 8. Axial wake profile at model scale comparing with experiment

Figure 9 shows a comparison between these results at 60% of the propeller radius. Also, experimental data and CFD results obtained using the 1-equation Menter model are shown.

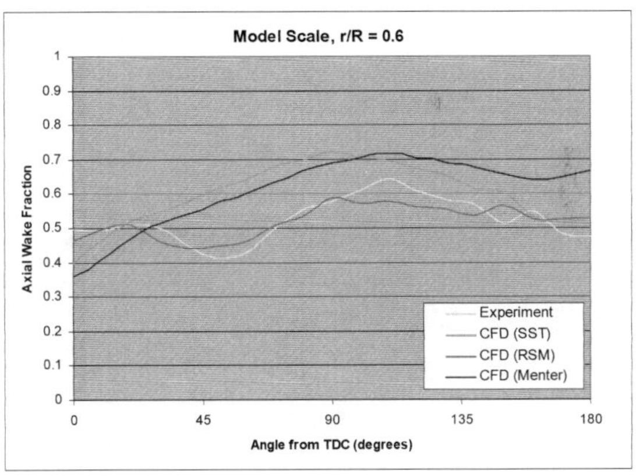

Figure 9. Axial wake at model scale at r/R = 0.6

Whilst all three CFD results show a reasonable correlation with the experimental data, only results obtained using the Reynolds Stress model (RSM) show a good match.

Ship scale calculations show less significant differences between results obtained with the two turbulence models than at model scale. However, the choice of turbulence model does appear to have a significant effect close to the propeller hub, as shown in **Figure 10**.

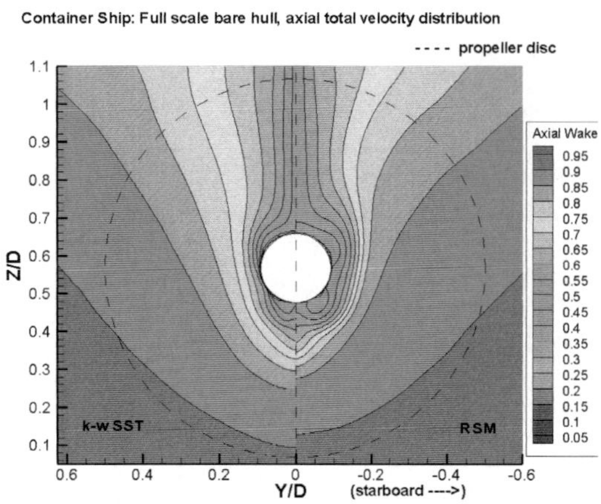

Figure 10. Axial wake profile at full scale comparing turbulence models

Measured ship scale wake data is rare, and none was available for this particular container ship. However, given that the techniques employed to predict wakes at ship scale are identical to those used at model scale, it is reasonable to assume that both sets of results will be of similar accuracy. Indeed, CFD wake analyses performed by TID for ships with available measured ship scale data have shown this to be the case. The large differences between full scale CFD results and those derived from the model test results using empirical scaling factors are shown in **Figure 11,** and these serve to emphasise the advantage of CFD over empirical methods. Such methods have been used extensively in the past but the increased accuracy of CFD wake predictions can now be utilised to provide better results.

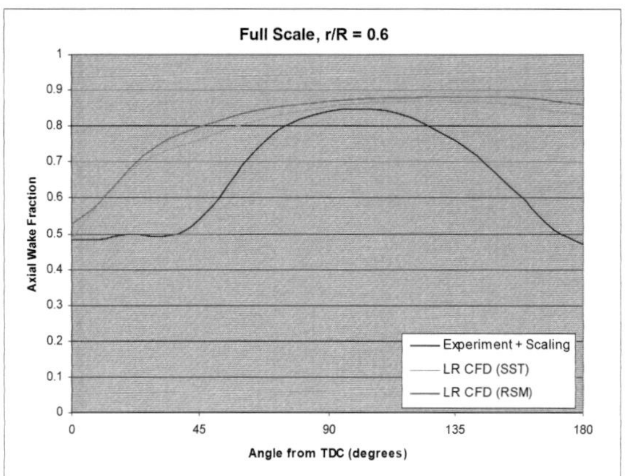

Figure 11. Axial wake at full scale at r/R = 0.6

3.1 (b) Tangential and Radial Wake Components

Accurate prediction of tangential and radial wake components is still difficult when using RANS type turbulence models. This is primarily due to the deficiencies of these turbulence models in predicting swirling flows. However, reasonable agreement with experiment is still attainable, as shown in **Figures 12 & 13**, and improvements to the calculation method are under continuous investigation.

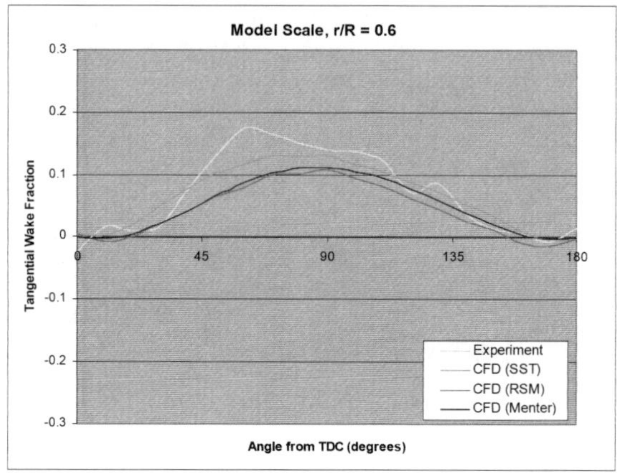

Figure 12. Tangential wake at model scale at r/R = 0.6

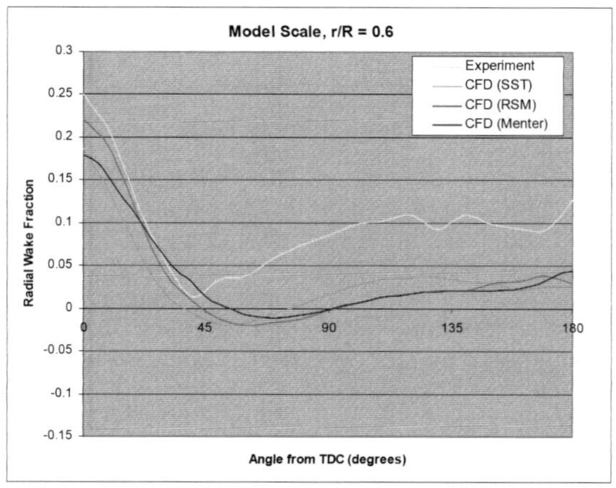

Figure 13. Radial wake at model scale at r/R = 0.6

Another important advantage of using CFD is that detailed flow features can be identified and immediately fed back into the design process. In **Figure 14**, for example, vectors of the tangential and radial velocity components in the propeller plane indicate the presence of vortices above and below the propeller hub. Generally, the grid of measurement locations used to obtain the model test data is too coarse to show such small-scale flow features.

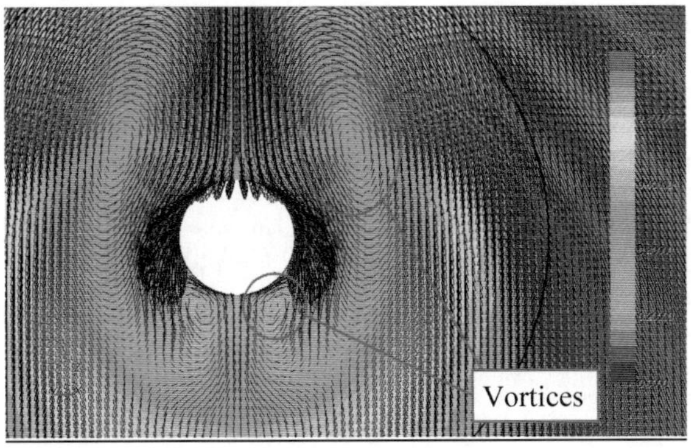

Figure 14. Vortices in wake indicated by tangential and radial wake component vectors

3.1 (c) Vortex Generator Analysis

In order to establish its effect, a vortex generator was added to the CFD model and the analyses were repeated. The vortex generators were intended to reduce the variation in inflow velocities to the propeller. **Figures 15 & 16** show that this aim was achieved at both model and ship scale.

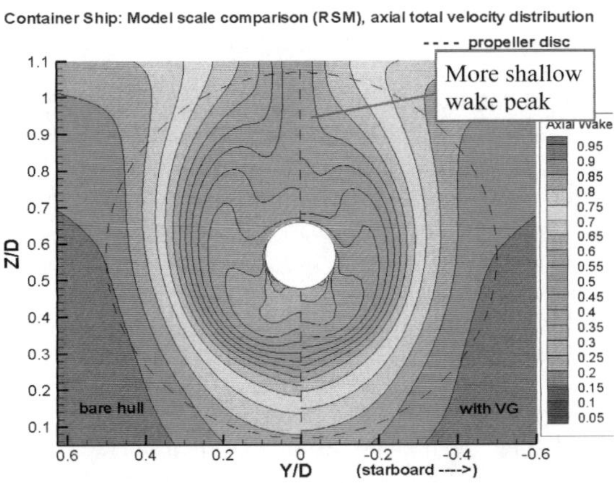

Figure 15. Axial wake profile at model scale with vortex generator

Ship scale calculations indicate that, as well as increasing the axial flow velocity in the region close to top dead centre, the vortex generator acts to reduce the velocity elsewhere, as shown in **Figure 16**.

The ability of the vortex generator to mix high velocity free stream flow into the boundary layer is reduced at ship scale due to the thinner boundary layer, which also causes the path of the vortex to be different, as shown in **Figure 17**. This phenomenon indicates that care should be taken when interpreting results from model scale tests with vortex generators.

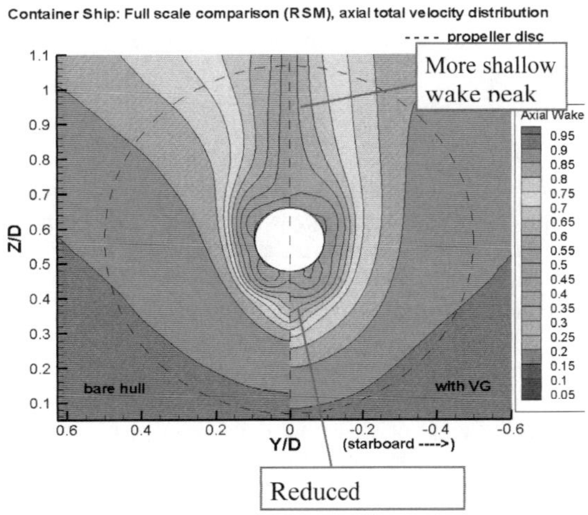

Figure 16. Axial wake profile at full scale with vortex generator

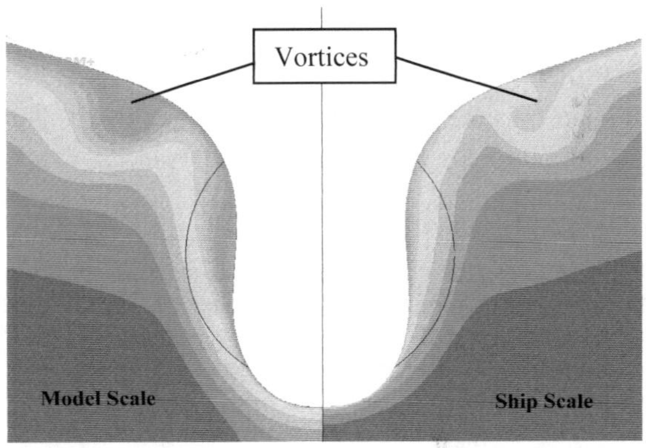

Figure 17 – Comparison of axial wake on a plane 11m upstream of the propeller plane

4. FUTURE DEVELOPMENTS

TID expects the trend for a growing use of CFD methods in the prediction of propeller inflows to continue. At present, we are working on validating ship scale propeller cavitation-characteristic predictions to study the accuracy of ship scale CFD calculated wakes. Once these studies have been concluded, CFD calculations are expected to become a standard method for determining propeller inflows.

Improvements to the prediction of tangential and radial propeller inflow components are currently being sought by improving the accuracy of our turbulence modelling through modifications to our RANS models, and also through the use of more advanced Detached Eddy Simulation (DES) models. DES models blend a RANS model in the boundary layer with a Large Eddy Simulation (LES) model in the free stream. DES models are thought to hold most promise for improving the transverse wake prediction, but require more computational time, since analyses have to be fully transient.

The combined CFD calculation of the flow around the ship's hull and propeller should further improve propeller inflow prediction methods, namely through the inclusion of the interaction between the hull and propeller. Research is under way within Technical Investigations into validating such methods using a wide range of appended geometries.

As observations have shown, not all hull pressures result from sheet cavitation, for which good calculation methods exist. Vortex cavitation plays a part in this too. Signal analysis has revealed the importance of vortex cavitation, and methods should be developed for better prediction of this phenomenon.

5. CONCLUSIONS

From observations of cavitation behaviour and calculation of ship wakes with CFD methods for a feeder containership, with and without vortex generators, it can be concluded that:

- Borescope observations of propeller cavitation can be routinely applied and can provide a significant insight into propeller cavitation phenomena. From this improved understanding of cavitation behaviour follows the possibility to target the problematic phenomena, and better correlate calculations and model tests to ship scale cavitation.
- CFD methodologies can accurately predict axial propeller inflows. The prediction of tangential and radial inflow-components requires improvement. This is attributed to shortcomings of RANS-type turbulence models.
- The choice of turbulence model, for instance the Reynolds Stress model, can significantly improve the accuracy of the prediction when compared to results obtained using eddy viscosity-based turbulence models.
- CFD methods predict a significant difference in flows from vortex generators at model and ship scale, suggesting care should be taken when interpreting results from model tests with vortex generators.
- Vortex generators might be applied in the design stage once calculation methods can accurately predict their effect on propeller inflows. Containership designers could use this to increase deadweight, while maintaining acceptable propeller-radiated hull pressures.

6. ACKNOWLEDGEMENTS

The photographs in **Figures 2 – 4** were obtained in conjunction with work part-sponsored by the CRS and Lloyd's Register.
The groundwork for the CFD analyses presented in this paper was performed by Peter Lindegard Nielsen, a student from the Danish Technical University who spent part of his MSc [**3**] working within Technical Investigations.

7. REFERENCES

1. I. Tanaka, 'Investigation of Scale Effects on Wake Distribution using Geosim Models', March 1984

2. W. van Gent & M. Hoekstra, 'Force Field Approach for Propeller-Wake Interaction', *Marin report no. 44303-7-SR*, 1985

3. Peter Lindegard Nielsen, 'Numerical Modelling of Ship Stern Flows', Danish Technical University, August 31st 2007.

8. AUTHORS' BIOGRAPHIES

Stewart Whitworth holds the current position of CFD Specialist at Lloyd's Register's Technical Investigations. He is responsible for executing and improving CFD methodologies in all marine industry fluid dynamics problems. His current focus is ship resistance and wake prediction.

Mr Anne Boorsma holds the current position of Senior Consultant at Lloyd's Register's Technical Investigations. He is responsible for performing hydrodynamic ship-propulsion investigations. In addition, he is part of a team developing cavitation-erosion detection methods using acoustic emissions.

Patrick Fitzsimmons holds the current position of Principal Consultant at Lloyd's Register's Technical Investigations. He is responsible for performing hydrodynamic ship-propulsion investigations. Furthermore he drives research and development into improvement of tools and methodologies and is responsible for developing the borescope observations system.

Dejan Radosavljevic holds the current position of head of the Fluid Dynamics group at Lloyd's Register's Technical Investigations. He is responsible for managing and improving CFD methodologies in all marine-related fluid dynamics problems.

MOORING LINE SOLUTIONS FOR STATE-OF-THE-ART CONTAINER VESSELS WITH DYNEEMA®

Steven Wardenier, Lankhorst Ropes, The Netherlands
Marc Eijssen and **Edwin Grootendorst**, DSM Dyneema B.V., The Netherlands

SUMMARY

The shipping industry is rapidly growing worldwide; new-build activity is high and the trend is towards larger vessels. At the same time, operational costs need to be trimmed. Harbour turnaround needs to be faster. On top of that, labour regulations are becoming tighter and safety awareness is growing.
To face these challenges, an increasing number of ship operators have chosen mooring lines with Dyneema®, a High Modulus PolyEthylene (HMPE) fiber. Mooring lines with Dyneema®, like the 12-strand braided Euroneema® rope manufactured by Lankhorst Ropes, have a proven track record, providing secure and safe mooring for LNG, VLCC and container ships. Ropes providing benefits to ship designers, operators and end users.

NOMENCLATURE

GDP	Gross Domestic Product
HMPE	High Modulus PolyEthylene
LNG	Liquefied Natural Gas
NEL	National Engineering Laboratory
PA	Polyamide
PES	Polyester
PP	Polypropylene
TEU	Twenty-foot Equivalent Unit
UV	Ultra Violet
VLCC	Very Large Crude Carrier

1. INTRODUCTION

In this paper we will demonstrate the benefits of these next generation mooring lines to ship designers and end users, illustrated by case studies.

Since the introduction of modern container shipping, this industry is rapidly growing worldwide and has become one of the most dynamic economic sectors of the last years. Trends on the demand as well as the supply side have triggered this tremendous expansion. Since 1990 the growth rate has averaged just under 10% p.a.. New-build activity is high and the trend is towards larger (container) vessels. Currently, container vessels with a capacity of 10,000 TEU and even larger plus are being designed and ordered. Initial designs have already been published for container vessels carrying over 20,000 TEU. At the same time, operational costs need to be trimmed. Harbour turnaround needs to be faster since some harbours are already almost chronically overloaded. On top of that, labour regulations are becoming tighter and safety awareness is growing.

To face these challenges, an increasing number of ship operators have chosen mooring lines with Dyneema®, a High Modulus PolyEthylene (HMPE) fiber. Mooring lines with Dyneema®, like the 12-strand braided Euroneema® rope manufactured by Lankhorst Ropes, have a proven track record, providing secure and safe mooring for LNG, VLCC and container carriers.

Nowadays container ships are mostly equipped with PA (polyamide, Atlas) and PP mooring lines. Comparing equal breaking strength, HMPE mooring lines will have 50 – 60% of the diameter and 30% of the weight of polyamide mooring lines. This could mean smaller winches can be designed.

The features of ropes with HMPE create valuable perspectives for the design and operation of today's and tomorrow's container shipping.

2. CONTAINERIZATION OF OCEAN SHIPPING

2.1 CONTAINER SHIPPING MARKET EXPANSION

Since the introduction in 1956, modern container shipping has brought about breath taking changes into the world economy. It transformed ocean shipping into a highly capital-intensive business and has become one of the most dynamic economic sectors of the last few years [1,5]. At the same time it has become the comber stone of global commercial integration [2].

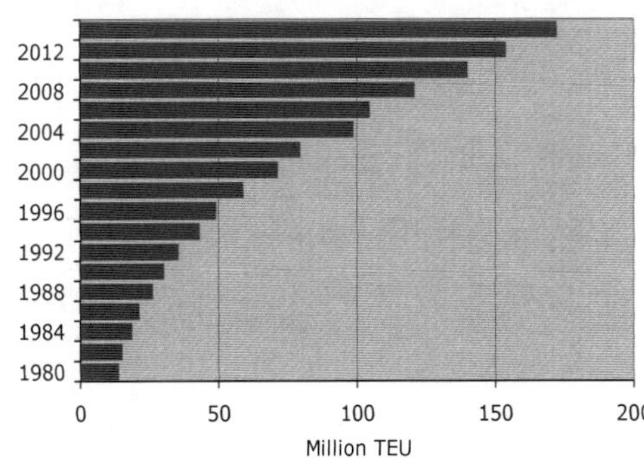

Figure 1. Global container vol. (1980-2015)

With an average increase of just under 10% 9% p.a. since 1990, container shipping has greatly surpassed seaborne trade overall. The annual growth is expected to continue for the next 10 years, with an only slightly reduced impetus (see figure 1) [1,5,6].

Major reasons on both the demand as well as the supply side have triggered the growth of modern container shipping.

The growth, on the demand side, has predominantly been caused by:

- High GDP growth in the last few years;
- The international division of labour in the course of liberalisation and the resulting trade movement;
- Decentralisation of production processes resulting in production at the component level occurring globally where it makes most sense to continue cost reductions;
- The shift from transport of raw materials to transport of (finished and part-finished) goods suited to transport by container as well as general cargo [2,4,5,6].

Trends on the supply side have as well significantly contributed, like:

- The considerable expansion of the container ship fleets (see figure 2);
- The development and building of larger (up to 10,000 TEU plus) and faster container ships;
- Increase flexibility through the combination of different (finished and part-finished) goods;
- Faster loading and unloading of container ships allowing shorter turnaround time [1,3,5,6].

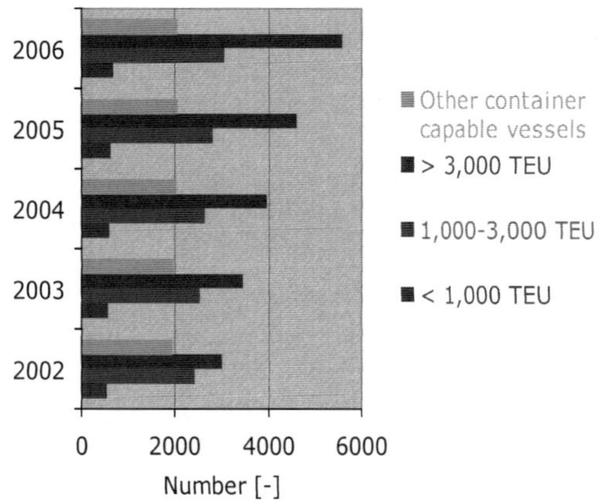

Figure 2. Increasing global container fleet

In this dynamic sector Asia is a key driving force, keeping into consideration China is dominating [3,5]. Of the 25 largest container ports in the world, 16 are in Asia and 7 of those are in China [5].

2.1 CHALLENGES OF AN EXPANDING CONTAINER SHIPPING MARKET

Despite the increasing demand for container transport freight rates are under considerable pressure. At the same time fuel rates are increasing. In view hereof improvements of productivity are of key importance. Shorter loading and unloading times, increasing size of the container fleet in terms of TEU (potentially by alliances/merges) as well as larger and faster ships contribute to the required improvement of operational excellence in which container ships and container handling form the capital-intensive parts [1,3,5,6].

The increasing demand for larger vessels induces infrastructural constraints. Congestions in and outside the ports are no rarity due to temporary capacity overloading, and canal sizes reduce flexibility and redundancy. This results in higher demands for the capacity and performance of infrastructure and superstructures to keep up with larger and faster ships [1,2,5,6,7,8]. Container terminals are moved to more exposed areas in order to provide easy access and new superstructures, like to the construction of new locks in the Panama Canal, are being build.

As a result stronger mooring and tugging ropes are required to minimize loading and unloading times and control and secure the container vessels.

Greater loads are placed on mooring lines in case terminals are exposed to currents and weather. Additional tugs may be needed for berthing container vessels at these terminals, and these tugs will also face greater marine exposure.

To provide the required strength to handle larger ships and harsher environments, traditional steel wire and conventional synthetic ropes must be made thicker in diameter – and this leads to a whole host of problems. First, these thicker ropes are heavier and more massive – and therefore more difficult to lift and manoeuvre. They put greater stress on handlers, mooring equipment and tugboats, and require more workers. The longer distances and increased exposures associated with offshore terminals, together with greater rope weight and mass, mean that the larger and heavier mooring lines can only be deployed one at a time. This slows down the mooring process and can add hours to turnaround times.

Safety is an ongoing concern for ships. More massive ropes have the potential to increase safety risks, such as injuries from backlash following a rope break, hand cuts from wire "fishhooks", and back strain from lifting and carrying the heavy ropes.

Similarly, the container shipping sector seeks to comply with environmental regulations. Not only by increasing fuel efficiency but also by eliminating the negative impact on water quality and sea life caused by grease applied to prevent corrosion of wire ropes.

As a result, container shipping companies seek mooring rope materials that can provide improved performance while directly addressing these challenges. Many are upgrading to specialized fibers from the conventional materials that have long dominated the shipping and towing industries.

3. NEW FIBER ROPE TECHNOLOGY SURPASSES TRADITIONAL ROPE MATERIALS

In the beginning, mooring lines were made of organic materials such as sisal and manila. Today, many vessels and tugs use lines made of common synthetics and steel wire. Especially on large vessels, the use of steel wire rope is the norm. Each of these materials offers advantages and disadvantages.

3.1 COMMON SYNTHETICS

PA, PES and PP have been used for many years, including for 11-meter stretching tail connected to steel wire mooring ropes. Although these are attractive for their lower cost, they have significant drawbacks. First, high elasticity means they will stretch under stress. This raises the safety risk of dangerous backlash if the rope breaks, and makes it difficult to position the ship at the quay. Failure to secure the vessel during rough weather can lead to shutdown of loading or offloading. For adequate mooring strength, synthetics require a large-diameter rope that takes up a large amount of space. In addition, these materials (except polypropylene) have a higher specific gravity than water, causing lines to sink and possibly become entangled in the propulsion screws of the tugs or the mooring boats. Nylon also absorbs water, lowering its breaking strength and making the ropes even heavier.

3.2 STEEL WIRE

Like traditional synthetics, steel wire ropes are also relatively inexpensive. An advantage over traditional synthetics is their low elongation, resulting in less ship movement when moored. Steel wire ropes have different disadvantages than synthetics. Steel corrodes in a marine environment, limiting the useful life of the rope and requiring lubrication with grease that can cause pollution both on the ship as well as in the environment. Compared to common synthetics of equal strength, steel wire is up to twice as heavy, posing lifting and carrying risks for workers. The rough surface of the rope can also damage workers' hands as well as fairleads, chocks and other ship components.

3.3 HIGH-PERFORMANCE FIBERS

To solve the performance issues of these materials, chemical companies have developed high-performance materials for rope fiber. Dyneema®, a leading HMPE material invented and manufactured by DSM, has been in production since the 1990s. Mooring ropes with Dyneema® made their entry into the shipping industry around 2002, and have been in use since 1994.

4. DESIRABLE PROPERTIES OF HMPE FIBER FOR MOORING AND TOWING ROPES

High modulus polyethylene offers many advantages over steel wire and conventional synthetics for mooring and towing ropes.

4.1 EXCEPTIONAL STRENGTH WITH MINIMAL WEIGHT

HMPE solves the problem of delivering enough strength to moor and tow larger container vessels without increasing the weight and mass of the ropes to a too high level. In fact, a mooring line made of HMPE fiber that is as strong as steel wire rope weighs only one-seventh as much as the wire. Compared to an HMPE fiber rope of equal strength, a PES and PA rope would have to be nearly twice as thick and would weigh four times as much.

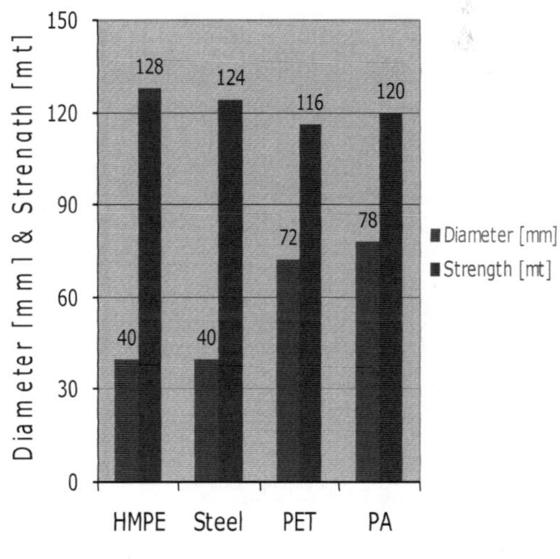

Figure 3. The diameter of ropes with equal breaking strength made of various materials

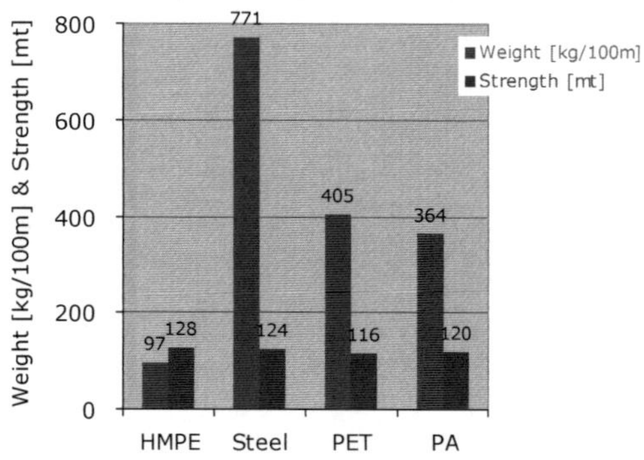

Figure 4. The weight of ropes with equal breaking strength made of various materials

Lankhorst Ropes, a manufacturer of Euroneema® ropes, with Dyneema® fiber, is seeing a major movement to replace steel wire rope, based on superior performance of the HMPE synthetic. "Low weight, high performance, and low elasticity are keys to producing rope for our markets," said Steven Wardenier, Commercial Director of Maritime Business with Lankhorst Ropes. "Dyneema® is the fiber to do this. We are now replacing both steel wire rope and chain. This is huge for us. Dyneema® allows us to do this at one-seventh (if not less) the weight of heavier materials. We partner with our end users (ship operators, tugboat companies, etc.) then tailor our product to meet their needs using Dyneema® fiber."

4.2 LOW ELONGATION

There can be a multitude of forces imposed on a container ship from wind and currents, speed and direction of waves, surge effects from passing vessels, tidal changes, and vessel draft changes due to loading or discharge of product. A mooring or towing rope, already subjected to stress from the weight of the ship alone, can build up dangerous energies under these additional forces. If the rope fails, the sudden release of energy - backlash - can lead to severe injury or death.

Dyneema® fiber provides a very low elongation of less than 2.5% at break. This low elongation results in a low energy storage level in the rope, so that there is hardly any backlash of the rope in case of failure. PA and PES both have higher elongation at break and are known to give a strong backlash when a rope breaks. Breaking steel wire ropes are known to have an unpredictable trajectory, leading to injuries, fatalities and material damage due to their high weight.

In actual mooring systems, steel wire ropes are often used in combination with an 11-meter mooring tail made from PET or PA. This is done to add some elasticity to the mooring system. A similar practice can be followed when using mooring ropes with Dyneema®.

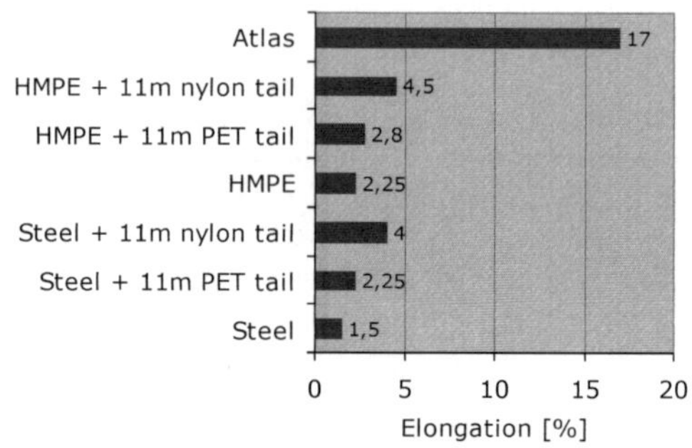

Figure 5. The elongation of mooring systems

Various components can be attached to the main mooring line to give the system some elasticity; this is done by attaching an 11-meter tail. The main mooring line is steel or HMPE. PET or PA is used as a tail to the main mooring line. The elongation of the mooring system at each load can be easily read from the graph.

4.3 DURABILITY

The work life of a rope can be affected by many environmental and usage factors, including degradation from exposure to sunlight and chemicals, corrosion from sea water, abrasion, bending fatigue and tension fatigue. Of course, for economic reasons, it is desirable to choose a rope that can offer the longest useful life.

Ropes made with HMPE exhibit excellent resistance to these wear factors. The material is resistant to most chemicals, salt and to UV light. It has superior tension fatigue performance, surpassing nylon and steel wire rope by several orders of magnitude. This superior tension fatigue performance plays an even more important role when mooring in offshore locations that are exposed to rough seas.

All these performance advantages add up to improved lifespan. It is for that reason that even though the initial investment in HMPE ropes is higher, the total cost over the lifespan of the product is lower due to exceptional durability and other factors.

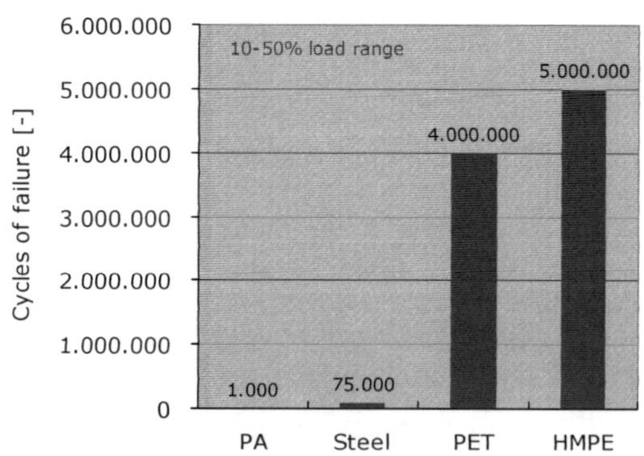

Figure 6. Tension-tension fatigue data

Each material is cycle loaded between 10% and 50% of minimum breaking strength of the ropes. Experiments have been performed at NEL in Scotland. As shown here, HMPE clearly outperforms steel wire with respect to tension-tension fatigue.

4.4 LOW MAINTANCE

As a synthetic with a smooth texture, HMPE fiber does not require lubrication to prevent corrosion or damage to ship surfaces. This saves the cost and labor of greasing steel wire ropes and repairing ship damage periodically. In fact, ropes made with HMPE fiber do not require any maintenance; however, some shippers wash them with fresh water after several moorings as a standard procedure.

5. BENEFITS OF UPGRADING TO HMPE MOORING AND TOWING ROPES

The safe, reliable and efficient mooring of container ships is more important today than ever. Ropes incorporating HMPE fiber can enhance the safety, reliability and speed of ship mooring and towing processes, while delivering excellent overall value.

5.1 IMPROVED SAFETY

Safety is not only related to the wellness of the people, it is also a matter of the image of the ship owner and operator; especially when sensitive cargo is carried. Together with environmental protection, safety is getting more and more important. Professional operations more and more want their partners to comply to environmental and safety policies.

There are a number of safety issues related to mooring a ship. For example, the risk of injury to crews is greatly increased during the mooring of a vessel when the first lines to be connected are under highest loads. This situation is exacerbated in bad weather. Workers in close proximity to mooring lines under tension face the risk of injury or even death from backlash if a rope fails.

On a less-dramatic scale, workers can be injured over time while trying to haul, lift and carry bulky, heavy ropes simply because of the sheer weight of the ropes. Broken wires in a steel wire rope can poke out and form "fishhooks" that can slice through workers' hands.

Unlike traditional rope materials, HMPE fiber addresses these safety concerns. The handling time of the ropes during mooring is shorter due to the lower weight involved, thus resulting in less exposure time to the workers. The backlash is minimal due to the low elongation of the HMPE ropes resulting in a minimal 'danger-zone'. Again, the workers are much less exposed to danger.

HMPE rope enables container ship operators to eliminate a number of safety risks associated with traditional mooring systems and to demonstrate their commitment to the highest safety standards.

5.2 SCALABILITY

One way of segmentation of container vessels is by the number of containers that can be loaded. For instance:

A) < 1,000 TEU;
B) 1,000 TEU – 10,000 TEU;
C) > 10,000 TEU;

The smallest segment refers to Coasters and River Ships. The size of the ropes is not so much the issue with these vessels, but the handling is. These ships have a tendency of very frequent mooring in combination with a minimal number of sailors on board. In numerous cases Captain/Owner are man and wife, with maybe one or two sailors on board. The HMPE ropes are easy to handle and thus quickly applied, they are comfortable in the hands as offer greater safety to all of them.

River Ships can be a combination of two or three vessels locked into each other the moment they set off on the river. Traditionally the connections were made with steel wire ropes. Lankhorst has developed a special Euroneema®, with Dyneema®, as an alternative for these applications. A six months trial period has been conducted with m/v Norma, sailing on the river Rhine and the results are excellent. Where several sailors were working with the steel wire ropes while making the connection, now only one sailor can make the connection in less time. Four different lines are put around the boulders in a specific sequence, on four individual winches these are pulled tight and the connection is made. These handlings can be done several times a day while operating in a port, the time saving in men hours is enormous.

In the middle segment of 1,000 up to 10,000 TEU, today's interest for HMPE is relatively small. By tradition these vessels sail with PA or PP ropes and the crews and line boats are familiar handling these. A few customers though did make the change to HMPE ropes and once accustomed to this, there is no turning back. A container ship carrying 8,000 TEU can be safely moored with a Euroneema® 40 mm in diameter, weighing 89 kg per 100 meters. A PP rope with the same breaking strength would measure 110 mm in diameter and weight 560 kg per 100 meter (figure 7 and 8).

The segment of super large vessels is still in the future but will have challenges of its own. Extrapolating the mooring lines of a container ship carrying 8,000 TEU to those carrying 22,000 TEU, each line with PP would measure 160 mm and weight 1160 kg per 100 meter. The same breaking strength in a Euroneema® mooring line with Dyneema®, would measure 60 mm and weight 202 kg per 100 meter (figure 7 and 8).

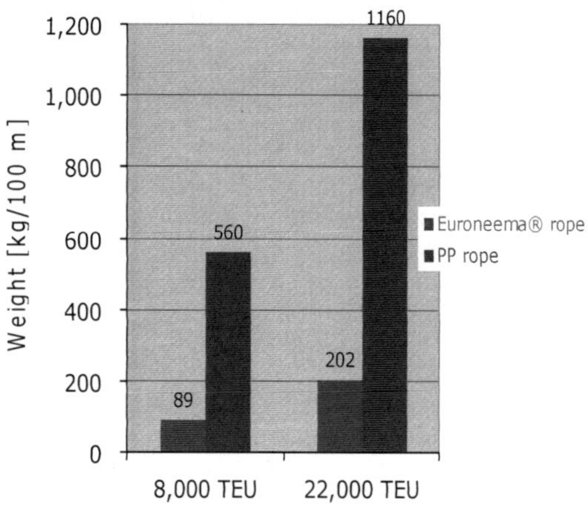

Figure 7. Weight of mooring systems with equal strength for 2 types of container ships

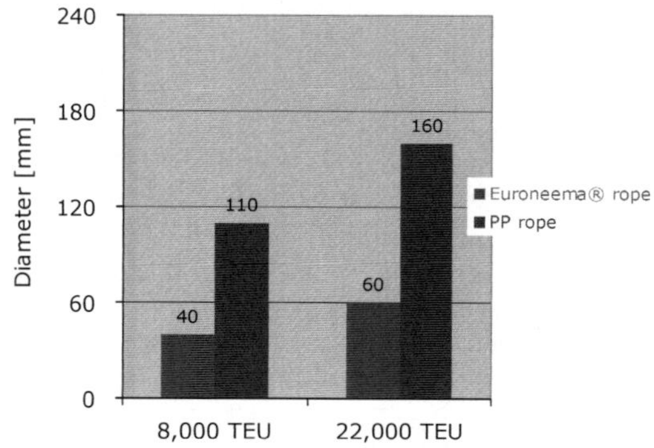

Figure 8. Diameter of mooring systems with equal strength for 2 types of container ships

In a new design, where the Naval Architects will calculate with the HMPE ropes, vast saving can be made on the sizes of the winches as well as the space that these winches require to operate. Less space means a higher payload. This may seem insignificant, but if due to smaller winches on a vessel 1 TEU could be added, over the life span of the vessel additional revenue up to one million US dollars could be achieved.

Tug boats that are handling all these container vessels when they come into port, more and more apply HMPE ropes in their towing system. The cargo vessels becoming bigger and bigger consequently have higher bows with the fairleads inside it. The connection has to be made by handing the rope from the tug boat up to the cargo vessel and the weight of the connecting line has to be transported across.

With the tug boat sailing very close to the vessel when making the connection, she cannot be seen by any of the people on the bridge, it is not a position the captain wants to stay long. Especially in case of high winds and high swell.

When the vessels are being disconnected, the HMPE ropes can simply be dropped from the main vessel and the rope will not damage the tug boat. Or in the event the lines drop in the water, they stay afloat and can easily be hauled inside.

Even though the operational costing of the HMPE ropes is higher than those of steel wire ropes, the ease of handling, the time involved and the safety of the people make the HMPE ropes a preferred proposition with tug boats. With less crew on board vessels, safety is becoming more and more important, let alone the down time of a vessel following an investigation after an accident would have occurred.

6. CONCLUSIONS

Naval Architects induce a quite rapidly changing maritime environment, implying newly designed vessels, challenging dimensions and new applications. An increasing number of ship operators have chosen mooring lines with Dyneema®, a High Modulus PolyEthylene (HMPE) fiber, to face these challenges. The 12-strand braided Euroneema® rope, with Dyneema® and manufactured by Lankhorst Ropes, has a proven track record, providing secure and safe mooring for LNG, VLCC and container ships.

Considering the container shipping market segmented into size of vessels and its applications, the stage of the product life cycle of HMPE ropes in these segments differ. Where handling is priority one, HMPE ropes has grown to the 'majority' stage. With container vessels carrying 1,000-10,000 TEU, the HMPE ropes are still at the stage of 'early adopters'.

Although the turning point to switch to HMPE ropes and the evolutions that will follow differs, sooner or later the strength-to-weight ratio of HMPE ropes will become an issue in each of the segments in order to face the challenges and create the benefits.

7. ACKNOWLEDGEMENTS

Dyneema® and Dyneema®, the world's strongest fiber™ are trademark(s) (applications) owned by Royal DSM N.V..

All information supplied by or on behalf of DSM Dyneema LLC and/or DSM Dyneema BV ("DSM") in relation to its products, whether in the nature of data, recommendations or otherwise, is supported by research and/or experience and believed reliable, but DSM gives no warranties of any kind, expressed or implied, including, but not limited to, those of correctness, completeness, merchantability or fitness for a particular purpose and DSM assumes no liability whatsoever in respect of application, processing, use of, or reliance on, the aforementioned information or products, or any consequence thereof, including but not limited to any infringement of the rights owned or controlled by a third party in intellectual, industrial or other property. Any information provided by DSM does not release the user from the obligation to verify such information and to perform its own testing and analysis to determine the suitability of the products for the intended process, use or specific application. The user accepts all liability in respect of or resulting from the application, processing, use of, or reliance on, the aforementioned information or products or any consequences thereof.

8. REFERENCES

1. MARC LEVINSON, New York Economist, 'Building With Boxes – The Containership Market Explosion', The Journal of OCEAN TECHNOLOGY Vol. 2, No. 1, Pg. 1-5, ISSN 1718-3200, Winter 2007;

2 CAPT. RAUL BROSTELLA, Autoridad del Canal de Panama, 'The Re-invention of the Panama Canal', The Journal of OCEAN TECHNOLOGY Vol. 2, No. 1, Pg. 6-15, ISSN 1718-3200, Winter 2007;

3. YUANMING ZHAO, Dalian Shipbuilding Industry (Group) Co. Ltd., 'Containership Building and Market Development in China', The Journal of OCEAN TECHNOLOGY Vol. 2, No. 1, Pg. 17-20, ISSN 1718-3200, Winter 2007;

4. STEPHEN M. CARMEL, Senior Vice President of Maersk Line Limited, 'Pollution Regulations and Unintended Consequences in the Container Shipping Business', The Journal of OCEAN TECHNOLOGY Vol. 2, No. 1, Pg. 21-28, ISSN 1718-3200, Winter 2007;

5. ERIC HAUMANN, Deutsche Bank Research, 'Container shipping – Overcapacity inevitable despite increasing demand', International topics – Current Issues, ISSN Print: 1612-314X / ISSN Internet and e-mail: 1612-3158, April 25, 2006;

6. E. POLMAN, SirLittle BV, 'Matching Port Positioning with Trends in Shipping', April 2007;

7. Executive Boars, Germanischer Lloyd AG, 'Annual Report 2007', Pg. 33-37, April 2008;

8. STX Shipbuilding, 'STX 22,000 TEU Ultra Mega Full Container Ship – The World's First Ultra Mega Container Ship', May 2008;

9. AUTHORS' BIOGRAPHIES

Steven Wardenier holds the current of position Commercial Director for Maritime Business at Lankhorst Ropes in The Netherlands. He is responsible for the world wide sales and marketing of the Lankhorst branded products.

Marc Eijssen, PhD, holds the current position of Application Manager EMEA within the Commercial Marine Group of DSM Dyneema B.V. in The Netherlands. He is responsible for application development and technical service of Dyneema® fiber for mooring and towing ropes as well as other industrial and offshore applications.

Edwin Grootendorst, PhD, holds the current position of Market Segment Manager Heavy Marine within the Commercial Marine Group of DSM Dyneema B.V. in The Netherlands. He is responsible for global marketing of Dyneema® fiber in mooring and towing ropes.

EXPERIENCE OF REPLACING STANDARD MECHANICAL SEAL PUMPS WITH HERMETICALLY SEALED MAGNETIC COUPLED BOOSTER PUMPS ON CONTAINER VESSELS

T Flauger, KRAL AG, A

SUMMARY

Residual heavy fuel oils are the common fuel on board of Container vessels. These residuals are a product of the refining process of crude oil, which extracts the components of higher volatility in a fractioned distillation process. Thus, the composition of the residual oils clearly depends on the grade and quality of the crude oil.

The presentation will describe the experience of KRAL AG with booster pumps and with damages on mechanical sealed pumps. Most of the failures lead to a heavy HFO leakage, which always has the risk of engine room fire and in case of a dirty engine room of State port detentions.

A complete elimination of the mechanical shaft seal is the most reasonable way to prevent engine room fire at the booster pumps (low pressure fuel system). The concept and the acceptance of a hermetically sealed, magnetic coupled pump by container ship operators will be presented.

NOMENCLATURE

HFO = Heavy Fuel Oil.
IMO = International Maritime Organization.
CIMAC = Conseil International des Machines a Combustion = International Council of Combustion Engines.
FY = Fiscal year.

1. INTRODUCTION

1.1 QUALITY OF HEAVY FUEL OIL

[1] Marine fuels are divided into heavy fuel oil (HFO) and distillates. Crude oil will always remain residuals after being processed. HFO is a product made from residual oil. High quality crude oil has a low amount of residuals. The prize of HFO is lower than the prize of crude oil. Therefore the refinery process is optimized for distillates. The higher the demand for (expensive) distillates, the lower is the quality of (cheap) HFO.

The refineries ability of cracking crude oil into distillates increases since 1980. Since then the amount and the quality of HFO decreases continuously. In addition globally the HFO quality depends on the local quality of crude oil and the refinery technology. To harmonize the HFO quality within the competing interests of quality and prize, the British Standard Institute and CIMAC published fuel oil specifications in 1982. The IMO followed in 1987. Today IMO's ISO 8217 is accepted as the international standard. ISO 8217 classifies HFO by viscosity and defines limits for the density and ingredients like water-, ash- and sulphur content.

1.2 CONTAINER VESSELS RUN ON HFO

Container vessels mainly operate on HFO. Therefore Container vessels are affected by the HFO properties and quality.

HFO as bunkered can't be injected by the diesel engines injection pumps without a prior treatment. A conditioning is absolutely necessary. In a first step a separator separates water and solid particles. In the following step a booster-module as shown in figure 1 reduces the viscosity and pressurizes the HFO. HFO is a high viscous fluid. RMH 380 has 380 mm^2/s at 50 °C and RMH 700 has 700 mm^2/s at 50 °C. A reduction of the viscosity is necessary to achieve an injection viscosity of app 12 mm^2/s. The low injection viscosity is reached at temperatures up to 150 °C. At this high temperature deairiation can cause gas bubbles which can affect a proper operation of the injection pumps. The HFO is pressurized to app 12 to prevent deairiation. Commonly the pressurization is done by two screw pumps, so called circulation or booster pumps. Two pumps are used for redundancy reasons. Defect booster pumps have operational risks on board a container vessel.

Figure 1: Booster Module to reduce HFO viscosity and for pressurization.

1.3 MINIMIZING OPERATIONAL RISKS WITH HFO

[1] Several measures are taken to reduce the risk of operational risks with fuel

- The HFO quality shall be fixed in an early phase during negotiating the charter contract.
- Before bunkering the available amount and the quality of the fuel will be checked with the supplier. Fuel samples should be taken on board, if possible by classification societies like GL, DNV, etc. Barge-samples from the supplier should be rejected. The samples should be sealed up and inscribed.
- A complete recording on board about fuel-order, bunkering, consumption, fuel-analysis, etc. is highly recommended.

The recommended and published list of recommendations to minimize breakdowns caused by fuels is much longer than mentioned here. That indicates the importance of a safe operation during a long passage.

2 PUMP FAILURES

C. Schneider (KRAL AG) and others reported a systematic investigation about pump failures done in 2002 [2]. 9 container vessels and 36 pumps have been considered. Defects have been claimed for 18 pumps, a considerable high rate of 50%. The claimed defects were restricted only to 3 damages:

- damaged couplings
- damaged bearings
- extended leakage of the mechanical shaft seal

Figure 2 shows the locations of the claimed parts in a typical booster pump.

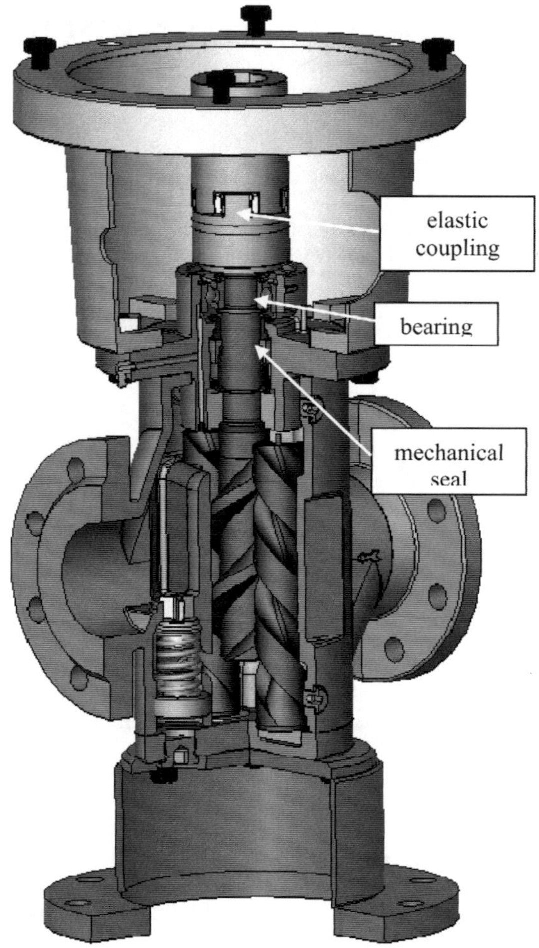

Figure 2: The 3 locations of the claimed parts in a screw pump.

2 (a) Failures of the elastic coupling

A flexible geared elastomer separates the coupling shoes of the coupling between the electrical motor and the shaft of the main spindle of the screw spindle pump. Figure 3 shows the damaged coupling insert.

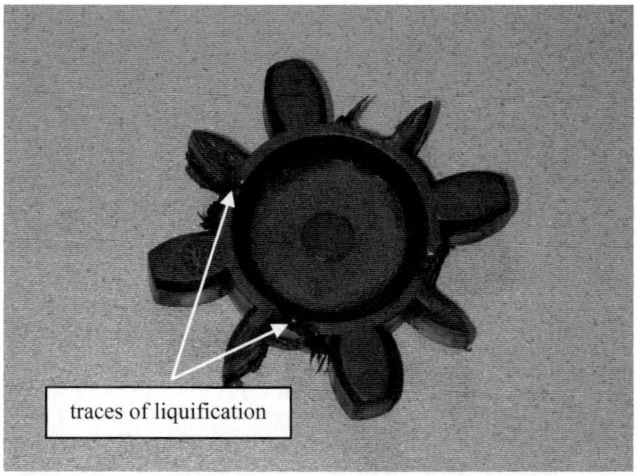

Figure 3: The elastomer coupling shows traces of liquification.

Because the torque is unidirectional it was expected that only every second gear is damaged. Remarkable is that only one side is destroyed and shows traces of liquification. The polymer was damaged by overheating. The applied elastomer is heat resistant up to 150 °C, normal operating temperature at this location is 90 °C. The damaged bearing where found to be the reason for this extraordinary operating condition.

2 (b) Failures of the bearing

Figure 4 shows typical damaged bearing after only app 2.000 hours of operation.

Figure 4: Damaged bearings.

Balls and rings are heavily worn out and covered with coke and carbon residues. Because of the increased friction such a damaged bearing runs hot within a short time. Figure 5 shows the temperature increase.

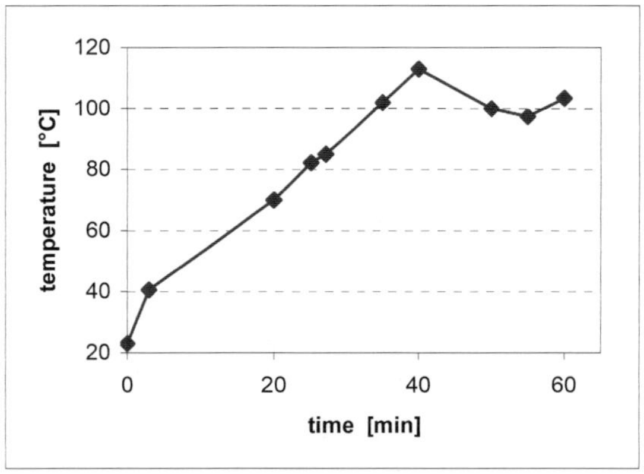

Figure 5: Temperature increase caused by higher friction of a damaged bearing.

The damaged bearing runs hot up to 110°C after only 40min. Again, the common temperature at the elastomer coupling is 90 °C. Such a damaged bearing with such a significant temperature increase can easily produce a temperature higher than the plastification or even liquification temperature of the coupling elastomer. HFO leakage was found as the reason for the damaged bearings.

2 (c) Extended leakage

To rotate the spindles of a screw pump, the main spindle is driven by an electrical motor. Where the shaft of the spindle comes out of the pump-casing to be coupled to the motor, a sealing is required. Mechanical seals are standard in booster pumps. Every mechanical seal manufacturer specifies a certain design leakage to maintain the lubrication film on the sliding surfaces of about a few cm³/h. This amount will not become evident due to evaporation. Figure 6 shows a chart to calculate the specified leakage.

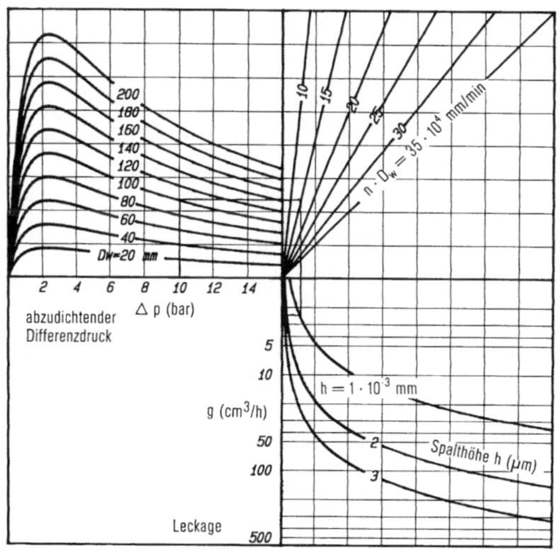

Figure 5: Example chart of a mechanical seal manufacturer to calculate the required leakage.

Figure 6 shows an accumulation of leakage residues in the lantern of a booster pump.

Figure 6: Accumulation of leakage residues.

It is obvious that this amount of residues can damage a bearing. The damaged coupling and the damaged bearings are secondary damages followed by a mechanical seal failure.

The stationary and rotating ring of a mechanical seal are fitted onto the shaft with O-rings. Reasons for mechanical seal failures where found in a damage of this O-rings. Most O-rings lost their elasticity, some show traces of liquification. Some of the mechanical seals show excessive wear on the sliding surfaces.

3 ADVANCED PUMP DESIGN

In [2] a new design of booster pump with mechanical sealing is comprehensively explained. Improvements are:

- improved elastormers for the O-rings
- advanced SiC qualities of the mechanical seal
- separation of the bearing and the leakage drain
- improved venting and flushing .

This so called K-pump is launched since 2003 and the most sold pump of this manufacturer.

3.1 THE SHAFT SEAL REMAINS CRITICAL

Despite of all improvements, the mechanical sealing will always remain the critical element of a screw pump operating with heavy fuel oil. Apart from SiC as material of mechanical seals there is no further option to reduce wear due to abrasive solids in the fuel. Specifications like ISO 8217 are very important, but they don't answer questions regarding:

- the abrasivity of HFO
- gluing of HFO residuals in contact with the atmosphere
- forming of residuals in contact with the atmosphere
- impact to elastomers.

And not to forget, there will always be "operational" leakage to lubricate the surfaces of the sliding rings as shown in figure 5.

3.2 PREVENTION OF A MECHANICAL SEAL BY MAGNETIC COUPLING

The only way to avoid the restricted life of a mechanical seal is to avoid this type of sealing. A successfully proven technology is the magnetic coupled screw pump.

The sealing element of a magnatic coupled pump is the containment can as shown in figure 7.

Figure 7: The containment can covers the shaft of the main spindle to seal the magnetic coupled pump.

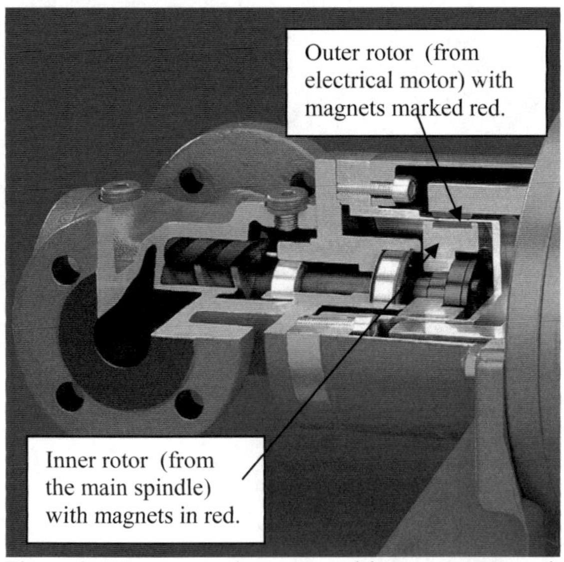

Figure 8: Magnets on the outer and inner rotor transmit the torque from the electrical motor to the pump.

It easy to see, that the pump is hermetically sealed when the pipes are connected. The torque required for the transmission of power to the pump shaft is transmitted through the walls of the containment can via magnets. The outer rotor from the electrical motor and the inner rotor of the pump-shaft are equipped with a set of extremely strong permanent magnets. Figure 8 shows the arrangement.

4 DECISSION AND SERVICE ON MAGNETIC COUPLING

Hermetically sealed pumps with magnetic power transmission are successfully operating in booster modules of container vessels for approximately ten years. Magnetic coupled pumps show a significant less failure rate than pumps with mechanical seals. Apart from the reduced risk of contamination with HFO leakage they offer the big advantage of distinctly reduced break down time. Increased investment costs will be reduced by operating costs for maintenance and spare parts. Magnetic coupled pumps will end up with lower lifecycle costs. The following sections show that a significant amount of appr 35% decisions are made for the magnetic coupled booster pump for container vessels. during the last fiscal year (FY) (1. May 2007 to 30. April 2008).

4 (a) Delivered pumps by sealing type

The K-series is the pumps type from a much larger program of screw spindle pumps which are mainly delivered to fuel supply applications. Diagram 1 shows the amount of K-pumps delivered in FY 07/08. 1.223 pumps had a magnetic coupling, 4.227 pumps a mechanical seal.

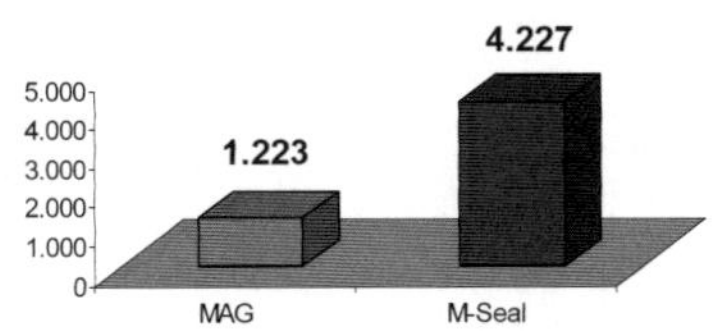

Diagram 1: Magnetic coupled and mechanical sealed pumps delivered during FY 07/08.

4 (b) Delivered pumps into the marine industry

Diagram 2 shows the amount of K-series pumps which are delivered to the Marine. Compared with the total amount of delivery (Diagram 1) 68% of the magnetic coupled and 69% of the mechanical sealed K-pumps are used in marine applications.

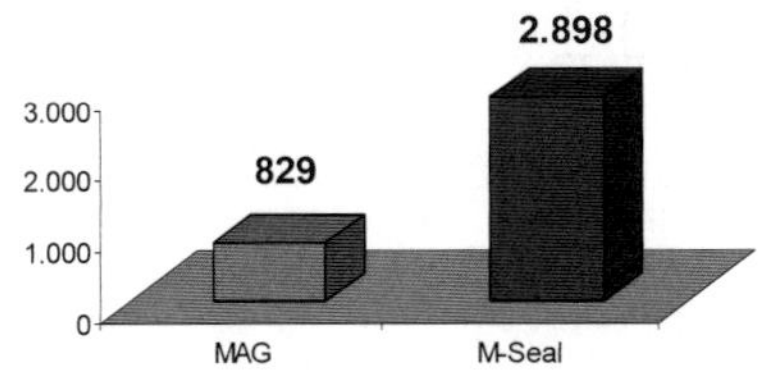

Diagram 2: Magnetic coupled and mechanical sealed pumps delivered to the marine during FY 07/08.

4 (c) Delivered pumps into container vessels

Diagram 3 shows the amount of K-series pumps which are delivered to container vessels. 86% of all magnetic coupled marine K-pumps and 80% of all mechanical sealed marine K-pumps are used on board container vessels.

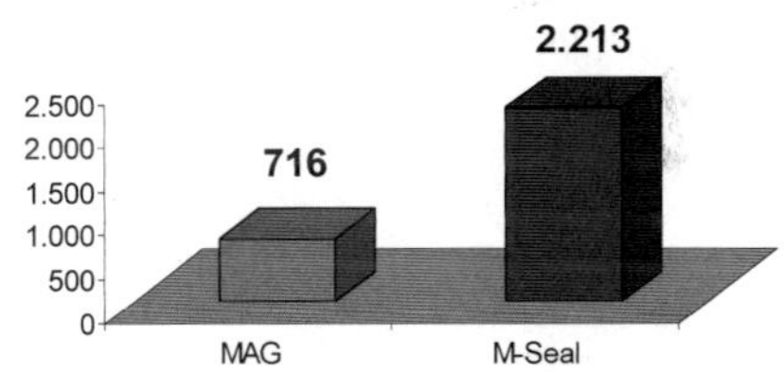

Diagram 3: Magnetic coupled and mechanical sealed pumps delivered to container vessels during FY 07/08.

Diagrams 3 show that today app 35% of orders for booster pumps for container vessels are magnetic coupled pumps.

4 (d) On site customer services

Interesting is the distribution of an on site service by project type shown in table 1. The "Pump Upgrade" project is a special incentive about to replace mechanical sealed pumps by magnetic coupled pumps. "Claims" means a service based on a damage.

Only 3 times there was a request to support the replacement of a mechanical sealed pump on container vessels. Obviously most crews can do the work by themselves.

3 services are caused by claimed damages. Compared with to total delivery of 716 magnetic coupled pumps, this is a low rate.

Container Vessel	Cruise Vessel	Power Plant	Project Type
3	1		Pump Upgrade
3		1	Claims

Table 1: On site service for support (Pump Upgrade) and damages (Claims).

5 MINIMIZING OPERATIONAL RISKS

Section 1.3 gives recommendations to minimize operational risks with HFO. In this section some recommendations for a safe run of booster pumps with magnetic coupling should be shared:

- When upgrading a mechanical sealed pump by a magnetic coupled pump, questions regarding pump size, operating data and installation requirement arise. It is highly recommended to check these requirements before quoting.
- The magnetic field is not fixed as the mechanical seal. There is some kind of dynamic. Extreme operating conditions like stand-by start and starting with cold HFO require a "smooth" start. A ramp of app 2 seconds is enough to prevent a de-synchronisation of the magnets.

6. CONCLUSIONS

Because of the HFO quality, failures of mechanical sealed booster pumps have reached a concerning frequency. Almost all damages are based on coupling failures, bearing damages and leaking mechanical shaft seals. Coupling and bearing damages are second order failures based on HFO leakage. State of the art mechanical sealed pumps are on a high technology level. However, the possibilities of a mechanical seal are limited with respect to materials for the wear of the sliding faces. A magnetic coupled pump replaces a mechanical seal completely. App 35% of all marine pump (K-pump) deliveries for container vessels are magnetic coupled pumps today. On site service for support or claim handling is on a very low level. But there are also special recommendations for a safe operation of a magnetic coupled pump as checking the operation data and prevent the magnets to de-synchronize by designing a ramp for the start up phase.

7. REFERENCES

[1] Hansheinrich Meier-Peter, Frank Bernhard. Handbuch der Schiffsbetriebstechnik. ISBN 3-87743-816-4. 2006.

[2]. Dr. Christoph Schneider, KRAL AG – Hans J. Naucke, Peene-Werft GmbH – Jürgen M. Wolf, Kupke+Wolf GmbH, Advanced screw pumps for heavy fuel oil conditioning systems. ICES 2003-636. 2003.

8. AUTHORS' BIOGRAPHIES

Thomas Flauger is marketing manager of the KRAL AG, Austria. He graduated as an engineer in physics at the University Wuppertal (Germany). He is member of CIMAC.

The Royal Institution of Naval Architects
8-9 Northumberland Street
London, WC2N 5DA
United Kingdom

ISBN 978-1-5108-8557-8

CURRAN ASSOCIATES INC.
proceedings.com